Quantitative Sociodynamics

Second Edition

Dirk Helbing

Quantitative Sociodynamics

Stochastic Methods and Models of Social Interaction Processes

Second Edition

Springer

Prof. Dr. Dirk Helbing
ETH Zurich
Swiss Federal Institute of Technology
Chair of Sociology, in particular of Modeling
and Simulation
CLU E 1
Clausiusstrasse 50
8092 Zurich
Switzerland
dhelbing@ethz.ch

ISBN 978-3-642-11545-5 e-ISBN 978-3-642-11546-2
DOI 10.1007/978-3-642-11546-2
Springer Heidelberg Dordrecht London New York

Library of Congress Control Number: 2010936007

Cover design: Integra Software Services Pvt. Ltd., Pondicherry

Printed on acid-free paper

Springer is part of Springer Science+Business Media (www.springer.com)

Preface

When I wrote the book *Quantitative Sociodynamics*, it was an early attempt to make methods from statistical physics and complex systems theory fruitful for the modeling and understanding of social phenomena. Unfortunately, the first edition appeared at a quite prohibitive price. This was one reason to make these chapters available again by a new edition. The other reason is that, in the meantime, many of the methods discussed in this book are more and more used in a variety of different fields. Among the ideas worked out in this book are:

- a statistical theory of binary social interactions,[1]
- a mathematical formulation of social field theory, which is the basis of social force models,[2]
- a microscopic foundation of evolutionary game theory, based on what is known today as 'proportional imitation rule', a stochastic treatment of interactions in evolutionary game theory, and a model for the self-organization of behavioral conventions in a coordination game.[3]

It, therefore, appeared reasonable to make this book available again, but at a more affordable price. To keep its original character, the translation of this book, which

---

[1] D. Helbing, Interrelations between stochastic equations for systems with pair interactions. *Physica A* **181**, 29–52 (1992); D. Helbing, Boltzmann-like and Boltzmann-Fokker-Planck equations as a foundation of behavioral models. *Physica A* **196**, 546–573 (1993).

[2] D. Helbing, Boltzmann-like and Boltzmann-Fokker-Planck equations as a foundation of behavioral models. *Physica A* **196**, 546–573 (1993); D. Helbing, A mathematical model for the behavior of individuals in a social field. *Journal of Mathematical Sociology* **19**(3), 189–219 (1994); D. Helbing and P. Molnár, Social force model for pedestrian dynamics. *Physical Review E* **51**, 4282–4286 (1995).

[3] D. Helbing, A mathematical model for behavioral changes by pair interactions. Pages 330–348 in: G. Haag, U. Mueller, and K. G. Troitzsch (eds.) *Economic Evolution and Demographic Change. Formal Models in Social Sciences* (Springer, Berlin, 1992); D. Helbing (1996) A stochastic behavioral model and a 'microscopic' foundation of evolutionary game theory. *Theory and Decision* **40**, 149–179; D. Helbing (1998) Microscopic foundation of stochastic game dynamical equations. Pages 211–224 in: W. Leinfellner and E. Köhler (eds.) *Game Theory, Experience, Rationality* (Kluwer Academic, Dordrecht); D. Helbing (1991) A mathematical model for the behavior of pedestrians. *Behavioral Science* **36**, 298–310 (1991).

is based on my PhD thesis, has not been changed. For readers whose excitement for mathematical details is limited, I would suggest to start reading the 'Introduction and Summary'. Chapter 2 offers a synopsis of the technical parts. The further chapters elaborate mathematical methods that are important to capture typical features of socio-economics systems such as time-dependence, non-linear interaction, and randomness. These are essential prerequisites to model social and economic self-organization phenomena.

Finally, I would like to thank Peter Felten and Pratik Mukerji for supporting the preparation of this book.

Zurich, September 2010 Dirk Helbing

Preface of the First Edition

This book presents various new and powerful methods for a *quantitative* and *dynamic* description of *social* phenomena based on models for behavioural changes due to individual interaction processes. Because of the enormous complexity of social systems, models of this kind seemed to be impossible for a long time. However, during the recent years very general strategies have been developed for the modelling of complex systems that consist of many coupled subsystems. These mainly stem from statistical physics (which delineates fluctuation affected processes by stochastic methods), from synergetics (which describes phase transitions, i.e. self-organization phenomena of non-linearly interacting elements), and from chaos theory (which allows an understanding of the unpredictability and sensitivity of many non-linear systems of equations).

Originally, these fascinating concepts have been developed in physics and mathematics. Nevertheless, they have often proved their explanatory power in various other fields like chemistry, biology, and economics. Meanwhile they also receive an increasing attention in the social sciences. However, most of the related work in the social sciences is still carried out on a qualitative level though in complex systems some phenomena can only be adequately understood by means of a mathematical description.

Up to now, there is not very much literature available that deals with a quantitative theory of social processes. Apart from the approaches of game theory, pioneering models have been developed by WEIDLICH and HAAG concerning spontaneous behavioural changes due to indirect interactions [285, 289, 292]. The work presented in this book was stimulated by both approaches which first seemed to be incompatible. However, the main focus was put on the concept of direct pair interactions. This was first applied to the description of dynamic pedestrian behaviour [123, 125] but it can be extended to the delineation of a variety of other social processes, e.g. of opinion formation.

It turned out that the concepts of indirect interactions and pair interactions can be unified in one general model. Surprisingly, by incorporating decision theoretical approaches a very fundamental model resulted which seems to open new perspectives in the social sciences. It includes as special cases many established models, e.g. the logistic equation for limited growth processes, the gravity model for exchange processes, some diffusion models for the spreading of information, the game dynamical equations for cooperation and competition processes, and LEWIN's social field theory that describes behavioural changes by dynamic force fields. Therefore, this model is well-founded and could be viewed as some kind of metatheory. In addition, it implies many new results. Consequently, in the course of time a large amount of material was accumulated concerning the foundation, generalization, and unification of behavioural models most of which is included in this book. Except for the omitted chapter on pedestrian behaviour, this book is essentially the translation of an expanded version of my PhD thesis [132] which has won two research prizes. It includes new results and many fruits of discussions I had in connection with presentations at a number of international conferences (e.g. the STATPHYS 18 in Berlin, 1992, and the XIIIth World Congress of Sociology in Bielefeld, 1994).

'Quantitative Sociodynamics' is directed to the following readership: First, to natural scientists, especially physicists and mathematicians who are interested in a comprehensive overview of stochastic methods and/or their application to interdisciplinary topics. Second, to social scientists who are engaged in mathematical sociology, decision theory, rational choice approaches, or general system theory. Moreover, this book supplies a number of relevant methods and models for economists, regional scientists, demographers, opinion pollsters, and market researchers.

It is assumed that the reader possesses basic mathematical knowledge in calculus and linear algebra. Readers who are not interested in mathematical details may continue with Part II after having read the different introductions preceding it. Moreover, those who are interested in the basic ideas only may skip the formulas during reading. The underlying terms, principles, or equations can always be looked up in Part I if necessary. An additional help is provided by the detailled 'Index' and the 'List of Symbols' which contains supplementary information about mathematical definitions and relations. Readers interested in an introduction to calculus, linear algebra, and complex analysis can, for example, consult some of the References [20, 140, 164, 165, 246, 266]. A more detailled discussion of stochastic methods, synergetics, and chaos theory can be found in [84, 106, 107, 135, 151, 183, 254].

Here I want to thank W. WEIDLICH for teaching me his fascinating ideas, for his encouragement, his numerous useful comments, and the freedom he gave me concerning my work. Moreover, I am grateful to A. DIEKMANN, W. EBELING, N. EMPACHER, M. EIGEN, G. HAAG, H. HAKEN, I. PRIGOGINE, R. REINER, P. SCHUSTER, F. SCHWEITZER, K. TROITZSCH, and many others for stimulating discussions and/or the inspiration by their work. I am very much indebted to R. ČALEK for his tireless engagement in connection with the translation of the German manuscript and to A. DRALLE as well as R. Helbing for their many useful advices concerning the final corrections. Moreover, I would like to thank M. SCHANZ for his help regarding the numerical calculation of the LIAPUNOV exponents in Sect. 10.3.4. Finally, I am pleased about the support, motivation, and understanding of my friends, relatives, and collegues.

Stuttgart, 1995 Dirk Helbing

Contents

List of Symbols

Notation and Conventions

x $= (x_1, \ldots, x_i, \ldots, x_n)^{\text{tr}} \equiv \begin{pmatrix} x_1 \\ \vdots \\ x_i \\ \vdots \\ x_n \end{pmatrix}$: *Vector; as subscript or superscript not printed in boldface!* Vectors x_i for which the equation $\sum_i c_i x_i = \mathbf{0}$ is only solvable for $c_i \equiv 0$ are called *linearly independent,* otherwise *linearly dependent.* A set $\{x_i\}$ of linearly independent vectors x_i build a *basis* of a set $\mathcal{M}$ of vectors if each vectors $x \in \mathcal{M}$ is representable as *linear combination* $x = \sum_i c_i x_i$ with suitable numbers c_i. Given that the vectors x_i of a basis are normalized and pairwise orthogonal to each other, the basis is called *orthonormal.* Cf. [20], pp. 55ff., 139, 292ff.

x_i $= (x)_i$: Component i of the vector x.

$(x_1, \ldots, x_N)^{\text{tr}}$ $= \begin{pmatrix} x_1 \\ \vdots \\ x_N \end{pmatrix}$.

$x \cdot y$ $= \sum_i x_i y_i$: *Scalar product (inner product), dot product)* of the vectors x and y. If we have $x \cdot y = 0$, the vectors x and y are *orthogonal,* i.e. perpendicular to each other. Cf. [20], pp. 130ff., 290ff.

$\underline{M}$ $\equiv \left(M_{ij}\right) := \begin{pmatrix} M_{11} & \cdots & M_{1j} & \cdots & M_{1n} \\ \vdots & \ddots & \vdots & & \vdots \\ M_{i1} & \cdots & M_{ij} & \cdots & M_{in} \\ \vdots & & \vdots & \ddots & \vdots \\ M_{n1} & \cdots & M_{nj} & \cdots & M_{nn} \end{pmatrix}$: (Quadratic) *matrix*.

M_{ij} $= (\underline{M})_{ij}$: Matrix elements (components) of the matrix $\underline{M}$.

$\underline{M}^{-1}$ $\equiv \left(M_{ij}^{-1}\right)$: *Inverse* of the matrix $\underline{M}$. Important relations: $\underline{M}\underline{M}^{-1} = \underline{1} = \underline{M}^{-1}\underline{M}$ and $[\underline{L}\underline{M}]^{-1} = \underline{M}^{-1}\underline{L}^{-1}$.

$\underline{M}^{\mathrm{tr}}$ $\equiv \left(M_{ji}\right)$: *Transpose* of the matrix $\underline{M} \equiv \left(M_{ij}\right)$. Important relation: $[\underline{L}\underline{M}]^{\mathrm{tr}} = \underline{M}^{\mathrm{tr}}\underline{L}^{\mathrm{tr}}$.

$\underline{L} + \underline{M}$ $= \left(L_{ij} + M_{ij}\right)$: Sum of two matrices.

$\underline{L}\underline{M}$ $= \left(\sum_k L_{ik} M_{kj}\right)$: Product of two matrices; normally we have $\underline{L}\underline{M} \neq \underline{M}\underline{L}$.

$\underline{M}x$ $= \left(\sum_j M_{1j}x_j, \ldots, \sum_j M_{ij}x_j, \ldots, \sum_j M_{nj}x_j\right)^{\mathrm{tr}}$: Multiplication of the matrix $\underline{M}$ with a vector x (which has the meaning of a transformation of vector x). The vectors x_i which satisfy $\underline{M}x_i = \lambda_i x_i$ are called *eigenvectors* to the (possibly complex) *eigenvalues* λ_i. Provided that all eigenvalues λ_i are greater than zero, the matrix $\underline{M}$ is called *'positive definite'*.

Mostly the n eigenvalues of the matrix $\underline{M}$ are different from each other. In this case the representation $\underline{M} = \underline{S}^{-1}\underline{D}\underline{S}$ is possible where $\underline{D} = \left(\lambda_i \delta_{ij}\right)$ is a *diagonal matrix* and $\underline{S}$ a transformation matrix. The *columns* of $\underline{S}^{-1}$ are eigenvectors of $\underline{M}$. If the matrix $\underline{M}$ is *symmetrical* (i.e. $\underline{M} = \underline{M}^{\mathrm{tr}}$), all its eigenvalues λ_i are *real* and $\underline{S}$ can be chosen as *rotation matrix*. In comparison with *orthogonal matrices* which satisfy the relation $\underline{S}^{-1} = \underline{S}^{\mathrm{tr}}$, rotation matrices additionally fulfil the relation $|\underline{S}| = 1$. Cf. [20], pp. 147ff., 212ff., 226ff.

$f(..)$ Function.

$f^{(k)}(x)$ kth approximation of the function $f(x)$. For a convergent sequence of functions we have $\lim_{k\to\infty} f^{(k)}(x) = f(x)$.

$f_*(x)$, $f^*(x)$ Estimate for $f(x)$.

$f_{\mathrm{e}}(x)$, $f^{\mathrm{e}}(x)$ Empirical value of the model quantity $f(x)$.

$p^{\cdot\cdot}_{\cdot\cdot}(.\vert..)$	*Transition probability.* Important relations: $p^{\cdot\cdot}_{\cdot\cdot}(.\vert..) \geq 0$ and $\sum\limits_{.} p^{\cdot\cdot}_{\cdot\cdot}(.\vert..) = 1$.
$P^{\cdot\cdot}_{\cdot\cdot}(.)$	*Probability.* Important relations: $P^{\cdot\cdot}_{\cdot\cdot}(.) \geq 0$ and $\sum\limits_{.} P^{\cdot\cdot}_{\cdot\cdot}(.) = 1$.
$P(..\vert.)$	$= P(..,.)/P(.)$: *Conditional probability.* Important relations: $P(..\vert.) \geq 0$ and $\sum\limits_{..} P(..\vert.) = 1$. In the special case $P(..,.) = P(..) \cdot P(.)$ or $P(..\vert.) = P(..)$ the variables . and .. are called *'statistically independent'*.
$w^{\cdot\cdot}_{\cdot\cdot}(..\vert..)$	*Transition rate.* Important relation: $w^{\cdot\cdot}_{\cdot\cdot}(..\vert..) \geq 0$.
$\nu^{\cdot\cdot}_{\cdot\cdot}(.)$	Rate. Important relation: $\nu^{\cdot\cdot}_{\cdot\cdot}(.) \geq 0$.
$\mathcal{A}$	*Operator* of a quantum mechanical *observable.*
A_{XY}	$= (\mathcal{A})_{XY} = \langle X\vert\mathcal{A}\vert Y\rangle$: Components of the operator $\mathcal{A}$ in *matrix representation.*
A_X	Eigenvalue of the operator $\mathcal{A}$ to the eigenvector $\vert X\rangle$ (i.e. we have $\mathcal{A}\vert X\rangle = A_X\vert X\rangle$).
$\underline{\mathcal{B}}$	*Tetradic operator (superoperator).*
$B_{XYX'Y'}$	$= (\underline{\mathcal{B}})_{XYX'Y'}$: Components of the tetradic operator $\underline{\mathcal{B}}$.

Frequently Occuring Symbols

$\mathbf{0}$	m-dimensional zero vector (null vector). Important relations: $x + \mathbf{0} = x$ and $c\mathbf{0} = \mathbf{0}$.
$\underline{0}$	Zero matrix (null matrix). Important relations: $\underline{M} + \underline{0} = \underline{M}$, $\underline{0M} = \underline{0} = \underline{M0}$, and $\underline{0}x = \mathbf{0}$.
$\mathit{0}$	Null operator. Important relations: $\mathcal{A} + \mathit{0} = \mathcal{A}$ and $\mathit{0}\mathcal{A} = \mathit{0} = \mathcal{A}\mathit{0}$.
$\underline{\mathit{0}}$	Tetradic null operator. Important relations: $\underline{\mathcal{B}} + \underline{\mathit{0}} = \underline{\mathcal{B}}$, $\underline{\mathit{0}\mathcal{B}} = \underline{\mathit{0}} = \underline{\mathcal{B}\mathit{0}}$, and $\underline{\mathit{0}}\mathcal{A} = \mathit{0}$.
$\underline{1}$	Identity matrix. Important relations: $\underline{1M} = \underline{M} = \underline{M1}$ and $\underline{1}x = x$.
$\mathit{1}$	Identity operator (unity). Important relations: $\mathit{1}\mathcal{A} = \mathcal{A} = \mathcal{A}\mathit{1}$ and $\mathit{1}\vert X\rangle = \vert X\rangle$.
$\underline{\mathit{1}}$	Tetradic identity operator (unity). Important relations: $\underline{\mathit{1}\mathcal{B}} = \underline{\mathcal{B}} = \underline{\mathcal{B}\mathit{1}}$ and $\underline{\mathit{1}}\mathcal{A} = \mathcal{A}$. Cf. (3.58).
a, a, b, c	Types of subsystems, behavioural types, subpopulations.

A	Number of distinguished types a of subsystems.
$A_{ab}(i, j; t)$	$= r_{ab}(t)E_{ab}(i, j; t)$: *Payoff* of strategy i for an individual of subpopulation a when confronted with strategy j in interactions with individuals of subpopulation b. Cf. (12.2b).
$C_m(t)$	Relative central moments. Cf. (12.70).
$\dim \Gamma$	Dimension of the Γ-space.
$D_a(t)$	Unit of distance. Cf. (9.38b).
$D_a(x', x; t)$	$= D'_a(x', x; t)/D_a(t)$: Scaled distance (dissimilarity) of two behaviours x and x' for spontaneous behavioural changes from the perspective of the individuals of subpopulation a. Important relation: Cf. (9.37b).
$D'_a(x', x; t)$	$= \mathrm{e}^{S'_a(x',x;t)} = D'_a(x, x'; t) \geq 0$: Distance of two behaviours x and x' for spontaneous behavioural changes from the viewpoint of the individuals of subpopulation a, measure for the incompatibility of x and x'.
$D^a(x', x; t)$	$= D^a(x, x'; t) \geq 0$: Distance (dissimilarity) of two behaviours x and x' in pair interactions from the perspective of the individuals of subpopulation a. Important relation: Cf. (9.61b).
e	$= 2.71828\ldots$: EULER *constant*.
$\mathrm{e}^x = \exp x$	$= \sum_{k=0}^{\infty} \frac{x^k}{k!}$: *Exponential function.*
$E_a(i, t)$	Expected success of strategy i for an individual of subpopulation a. Cf. (12.2a).
$E_a(i, n; t)$	Configuration-dependent expected success of strategy i for an individual of subpopulation a.
$E_{ab}(i, j; t)$	Success of strategy i for an individual of subpopulation a in interactions with individuals of subpopulation b who pursue strategy j.
$\widetilde{f}(\omega)$	$= \int_{-\infty}^{+\infty} dt\, \mathrm{e}^{\mathrm{i}\omega t}\, y(t_0 + t)$: FOURIER *transform* of function $y(t_0 + t)$. Important relations: $y(t_0 + t) = \frac{1}{2\pi} \int_{-\infty}^{+\infty} d\omega\, \mathrm{e}^{-\mathrm{i}\omega t}\, \widetilde{f}(\omega)$ *(inverse* FOURIER *transformation)* and $\left(\int_{-\infty}^{+\infty} dt\, \mathrm{e}^{\mathrm{i}\omega t}\, y_1(t) \right) \left(\int_{-\infty}^{+\infty} dt\, \mathrm{e}^{\mathrm{i}\omega t}\, y_2(t) \right) = \int_{-\infty}^{+\infty} dt\, \mathrm{e}^{\mathrm{i}\omega t} \int_0^t dt'\, y_1(t')\, y_2(t - t')$ *(convolution theorem)*. Cf. [218].

$f(\omega)$	$= \lvert\widetilde{f}(\omega)\rvert^2$: Power spectrum of the function $y(t)$.
$f_{ab}^k(t)$	Interaction frequency of kind k for an individual of subpopulation a in pair interactions with individuals of subpopulation b. Cf. (10.28).
$G(u)$	$= \int\limits_0^\infty dt\, \mathrm{e}^{-ut} P(t_0+t)$: LAPLACE *transform* of $P(t_0+t)$. Important relation: $P(t_0+t) = \frac{1}{2\pi \mathrm{i}} \int\limits_{c-\mathrm{i}\infty}^{c+\mathrm{i}\infty} du\, \mathrm{e}^{ut} G(u)$ *(inverse* LAPLACE *transformation)*. Cf. [218].
$\mathcal{G}(u)$	LAPLACE transform of the statistical operator $\rho(t)$. Important relation: $\left(\int\limits_0^\infty dt\, \mathrm{e}^{-ut} \rho_1(t)\right)\left(\int\limits_0^\infty dt\, \mathrm{e}^{-ut} \rho_2(t)\right) = \int\limits_0^\infty dt\, \mathrm{e}^{-ut} \int\limits_0^t dt'\, \rho_1(t')\, \rho_2(t-t')$ *(convolution theorem)*. Cf. (3.59) and [218].
$\hbar$	$= \mathrm{h}/(2\pi)$ where $\mathrm{h} = 6.6262 \cdot 10^{-34}$Js is PLANCK'S *quantum of action*.
$\mathcal{H}$	HAMILTON operator of a system.
i	$= \sqrt{-1}$: *Imaginary unit*.
i, j	(Behavioural) strategies, (expressed) opinions.
I, J, K, L	(Vectorial) subscripts each of which combines various subscripts/superscripts, e.g. $I = (\alpha, i)^{\mathrm{tr}}$.
$\mathrm{Im}(x)$	*Imaginary part* of a *complex number* $x = \mathrm{Re}(x) + \mathrm{i}\,\mathrm{Im}(x)$.
k	$= 1.38062 \cdot 10^{-23}\mathrm{JK}^{-1}$: BOLTZMANN *constant*.
$k_I(Z,t)$	Scaled first jump moments (drift coefficients). Cf. (6.14b).
$K(t)$	LIAPUNOV function. Cf. (3.109), (6.25).
$K_a(x,t)$	$= (K_{a1}(x,t), \dots, K_{ai}(x,t), \dots, K_{am}(x,t))^{\mathrm{tr}}$: Vector of the effective drift coefficients K_{ai}; social force that acts on an individual of subpopulation a who shows behaviour x.
$K_{ai}(x,t)$	Effective drift coefficients in γ_a-space. Cf. (6.64b).
$\lim\limits_{x\to y} f(x)$	$= f(y)$: Limit of $f(x)$ as x approaches y.
$\ln x$	$= \log_{\mathrm{e}} x$: Natural *logarithm* of x. Important relations: $\ln(xy) = \ln x + \ln y$, $\ln x^a = a \ln x$, and $\frac{d}{dx} \ln x = \frac{1}{x}$.

$\log x$	$= \log_{10} x$: Decadic logarithm of x.
$\log_y x$	$= \dfrac{\ln x}{\ln y}$: Logarithm of x to the base y.
$\underline{L}$	LIOUVILLE matrix. Cf. (3.100).
m	$= \dim \gamma_a$: Dimension of the γ_a-space.
$m_{l_1 \ldots l_M}(X, t)$	lth jump moments in Γ-space ($l = l_1 + \cdots + l_M$). Cf. (6.2a).
$m_x^a(n, t)$	First jump moments (drift coefficients) in Γ'-space. Cf. (5.15).
$m_{xx'}^{ab}(n, t)$	Second jump moments (diffusion coefficients) in Γ'-space. Cf. (5.24).
$m_{x_{i_1} \ldots x_{i_m}}^{a_{i_1} \ldots a_{i_m}}(n, t)$	mth jump moments in Γ'-space. Cf. (5.38).
$m_I(X, t), m_{(1)}(X, t)$	First jump moments (drift coefficients) in Γ-space.
$m_{IJ}(X, t), m_{(2)}(X, t)$	Second jump moments (diffusion coefficients) in Γ-space.
$\max(x, y)$	$= \begin{cases} x \text{ if } x \geq y \\ y \text{ if } y > x \end{cases}$: *Maximum* of the numbers x and y.
$a \bmod b$	*a modulo b*. Important relation: $x \equiv a \bmod b$ if and only if $x = a + zb$ with an integer z.
M	$= \dim \Gamma$: Dimension of the state space Γ.
$M_{l_1 \ldots l_M}(Z, t)$	Scaled lth jump moments ($l = l_1 + \cdots + l_M$). Cf. (6.12b).
n	$= (n^1, \ldots, n^A)^{\text{tr}}$: Configuration of a system that consists of many elements.
n^a	$= (n_{x_1}^a, \ldots, n_{x_S}^a)^{\text{tr}}$.
n_x	$= \sum_a n_x^a$: Number of subsystems (of arbitrary type) that are in state x.
n_x^a	Occupation number, i.e. number of subsystems of type a that are in state x.
$n_{x_1' \ldots x_k' x_1 \ldots x_k}^{a_1' \ldots a_k' a_1 \ldots a_k}$	$= (\ldots, (n_{x_1'}^{a_1'} + 1), \ldots, (n_{x_1}^{a_1} - 1), \ldots, (n_{x_k'}^{a_k'} + 1), \ldots, (n_{x_k}^{a_k} - 1), \ldots)^{\text{tr}}$.
N	$= \sum_{a,x} n_x^a$: Number of subsystems of a considered system.

N_a	$= \sum_x n_x^a$: Number of subsystems of type a.
$O(x^n)$	represents terms $f(x)$ the absolute value $\|f(x)\|$ of which converges, for $x \to 0$, to zero like the function $\|x\|^n$, i.e. there exists a constant C so that $\|f(x)\| \leq C\|x\|^n$.
$p_a(x'\|x;t)$	$\overset{x' \neq x}{=} w_a(x'\|x;t)\Delta t$: Probability for a subsystem of type a to change from state x to state x' during a time period Δt.
$p_{ab}(x', y'\|x, y; t)$	$= w_{ab}(x', y'\|x, y; t)/\nu_{ab}(t)$: Probability of an individual of subpopulation a to exchange the behaviour x for x' in an interaction with an individual of subpopulation b who changes the behaviour from y to y'. Cf. (9.50).
$p_{ab}^k(x'\|x;t)$	$\overset{x' \neq x}{=} f_{ab}^k(t) R^a(x'\|x;t)$: Probability with which an individual of subpopulation a exchanges the behaviour x for x' in an interaction of kind k with an individual of subpopulation b.
$P(n, t)$	Probability of configuration n to occur at the time t.
$P(t)$	Vector with the components $P(X, t)$.
$P(x, t)$	$= \sum_a \frac{N_a}{N} P_a(x, t)$: Probability of a subsystem to be in state x; proportion of subsystems that are in state x.
$P(X, t)$	$= \rho_{XX}(t)$: Probability of state X to occur at the time t. In the case of a continuous state space: Probability density. Cf. (3.52).
$P(X, t\|X_0, t_0)$	Transition probability from state X_0 to state X during the time period $(t - t_0)$.
$P(X_n, t_n; X_{n-1}, t_{n-1}; \dots; X_0, t_0)$	Joint probability to find the system in states $X_0, \dots, X_n$ at the times $t_0, \dots, t_n$.
$P(X_n, t_n\|X_{n-1}, t_{n-1}; \dots; X_0, t_0)$	Conditional probability to find the considered system in state X_n at the time t_n given that it was in states $X_0, \dots, X_{n-1}$ at the times $t_0, \dots, t_{n-1}$.
$P(a, x; t)$	$= \frac{N_a}{N} P_a(x, t)$: Probability to find a subsystem of type a in state x. In the case of a continuous state space: Probability density.
$P^\lambda(X)$	Eigenfunction to eigenvalue λ.

$P^0(X)$	Stationary probability distribution.
$P_a(x,t)$	$\stackrel{N_a \gg 1}{=} \dfrac{\langle n_x^a \rangle}{N_a}$: Probability of a subsystem of type a to be in state x; (relative) proportion of subsystems of type a that are in state x.
$\text{Prob}(\mathcal{M})$	Probability with which a value *(event)* x is element of the set $\mathcal{M}$.
$q_{IJ}(Z,t)$	Scaled second jump moments (diffusion coefficients). Cf. (6.14c).
$Q_{aij}(x,t)$	Effective diffusion coefficients in γ_a-space. Cf. (6.64c).
r	Location.
$r_{ab}(t)$	Relative frequency of interactions of an individuals of subpopulation a with individuals of subpopulation b. Cf. (12.1).
$R_a(x'\|x;t)$	$\stackrel{x' \neq x}{=} w_a(x'\|x;t)/\nu_a(t)$: Readiness of individuals of subpopulation a to spontaneously change their behaviour from x to x'. Cf. (9.1b).
$R^a(x'\|x;t)$	Readiness of an individual of subpopulation a to exchange the behaviour x for x' due to a pair interaction with another individual. Cf. (9.62a).
$R^a(x'\|x;n;t)$	Configuration-dependent readiness R^a.
$\text{Re}(x)$	*Real part* of a complex number $x = \text{Re}(x) + \text{i}\,\text{Im}(x)$.
$\sin x$	Sine of x.
$S, \widehat{S}_a$	Number of states x belonging to the γ_a-space.
$S'_a(x',x;t)$	$= S'_a(x,x';t)$: Symmetrical part of the scaled conditional utility $U'_a(x'\|x;t)$; subjective effort of a behavioural change from x to x' for an individual of subpopulation a. Cf. (9.30b).
$S_a(x',x;t)$	$= \ln D_a(x',x;t) = S_a(x,x';t) = G_a(x',\ x;t) + G_a(x,x';t)$: (Scaled) effort of a behavioural change from x to x' (or vice versa) for individuals of subpopulation a.
t, t', t''	Times (points in time).
t_0	Initial time.
$T, \widehat{T}$	Time periods.
T', T'_a	Absolute temperature in K = temperature in °C + 273.15 K.

$\mathrm{T}\exp\Big(\underline{L}, t, t_0\Big)$ $= \underline{1} + \int\limits_{t_0}^{t} dt' \underline{L}(t') + \int\limits_{t_0}^{t} dt' \int\limits_{t_0}^{t'} dt'' \, \underline{L}(t')\underline{L}(t'') + \dots$: *Time-ordered exponential function.* (T indicates DYSON's *chronological operator,* here.) Important relations: $\frac{d}{dt}\mathrm{T}\exp\Big(\underline{L}, t, t_0\Big) = \underline{L}(t)\mathrm{T}\exp\Big(\underline{L}, t, t_0\Big)$. For the *time-independent* case $\underline{L}(t) \equiv \underline{L}$, by carrying out the time integrations, one obtains $\mathrm{T}\exp\Big(\underline{L}, t, t_0\Big) = \sum\limits_{k=0}^{\infty} \frac{\underline{L}^k (t - t_0)^k}{k!} = \exp\Big(\underline{L}(t - t_0)\Big)$.

$\mathrm{Tr}(\mathcal{A})$ $= \sum\limits_{Y} \langle Y|\mathcal{A}|Y\rangle$: *Trace* of the operator $\mathcal{A}$.

u Argument (independent variable) of a LAPLACE transform.

$U_a(x,t)$ (Scaled) utility of behaviour x for spontaneous behavioural changes. For reasons of uniqueness one usually demands $\sum\limits_{x} U_a(x,t) \stackrel{!}{=} 0$.

$U_a(x,n;t)$ Configuration-dependent utility U_a of behaviour x.

$U^a(x,t)$ Utility of behaviour x for individuals of subpopulation a in pair interactions. Normally one demands $\sum\limits_{x} U^a(x,t) \stackrel{!}{=} 0$.

$U_a^*(x'|x;t)$ $= \widehat{U}_a(x'|x;t) + \varepsilon_a(x'|x;t)$: Estimated conditional utility of a spontaneous behavioural change to behaviour x' for an individual of subpopulation a who pursues behaviour x.

$\widehat{U}_a(x'|x;t)$ Known part of the conditional utility U_a^* $(x'|x;t)$.

$U_a'(x'|x;t)$ Scaled known conditional utility. Cf. (9.13).

v, w Velocities.

V Volume.

$V_a(x,t)$ Social field for individuals of subpopulation a with behaviour x. Cf. (11.16a).

$V_a(x'|x;t)$ Objective advantage of a behavioural change from x to x' for an individual of subpopulation a. Important relation: $V_a(x'|x;t) \geq 0$. Cf. (9.27).

$w(n'|n;t)$ Configurational transition rate with which a transition from configuration n to configuration n' occurs. Cf. (5.34), (5.50b).

$w(X'|X;t)$ $\stackrel{X' \neq X}{=} \lim\limits_{\Delta t \to 0} \frac{P(X', t + \Delta t|X,t)}{\Delta t}$: Transition rate from state X to state X'. Cf. (3.22b), (3.91).

$w_a(x'\|x;t) = \widehat{w}_a(x'\|x;t)$	Individual transition rate for spontaneous transitions from state x to state x' for systems of type a.
$w_a(x'\|x;n;t)$	Effective configuration-dependent individual transition rate. Cf. (5.15b).
$\widehat{w}_a(x'\|x;n;t)$	Configuration-dependent individual transition rate.
$w^a(x'\|x;t)$	Effective transition rate from state x to state x' by spontaneous transitions and interactions for a subsystem of type a. Cf. (4.22b).
$w_{ab}(x',y'\|x,y;t)$	$\stackrel{N_a \gg 1}{=} N_b \widehat{w}_{ab}(x',y'\|x,y;t)$: Pair interaction rate for a subsystem of type a with subsystems of type b due to which the element of type a changes from state x to x' and the element of type b from state y to y'. Cf. (4.20b).
$\widehat{w}_{ab}(x',y'\|x,y;t)$	Individual pair interaction rate (transition rate for pair interactions) of two concrete subsystems of types a and b due to which the element of type a exchanges the state x for x' and the element of type b exchanges the state y for y'.
$\widehat{w}_{ab}(x',y'\|x,y;n;t)$	Configuration-dependent pair interaction rate $\widehat{w}_{ab}$.
$w_{t-t'}(X'\|X;t')$	Transition rate of the *generalized* master equation. Cf. (3.87).
x, x', y, z	States of elements.
x_α	State of element α.
X	$= (x_1, \ldots, x_N)^{\text{tr}}$: System state in Γ-space.
$\widehat{X}(t)$	Most probable state of a system at time t.
X_0	Initial state at time t_0.
X_I	Component I of the state vector X.
Z	$= \varepsilon X$ with $\varepsilon = 1/\Omega$: Scaled state vector X.

Greek Symbols

α, β, γ	Subsystems; individuals.
γ_a	γ_a-space: Set of all possible states x for subsystems of type a.
Γ	State space (LIOUVILLE *space*), set of all possible states X of the considered system.
Γ'	Configuration space. Cf. (5.2).

$\delta(x-y)$ DIRAC *delta function.* Important relation:
$$\int_{y_0}^{y_1} dy\, f(y)\delta(x-y) = f(x)\cdot \begin{cases} 1 \text{ if } y_0 \le x \le y_1 \\ 0 \text{ otherwise} \end{cases}.$$

δ_{xy} $= \begin{cases} 1 \text{ if } x = y \\ 0 \text{ otherwise} \end{cases}$: KRONECKER *symbol.*

Δt Small time period.

ΔV Volume element.

$\Delta x'$ $= x' - x$.

$\Delta_{ji} f$ $= f(j) - f(i)$.

$\Delta_x f(x)$ $= \sum_i \left(\frac{\partial}{\partial x_i}\right)^2 f(x)$: LAPLACE *operator* applied to the function $f(x)$.

ε Small number. Normally one assumes $\varepsilon \ll 1$.

$\varepsilon_a(x'|x;t)$ Unknown part of the conditional utility $U_a^*(x'|x;t)$ that can only be estimated.

κ Control parameter.

λ_i, λ Eigenvalues (which are possibly complex numbers).

$\lambda_i(\kappa)$ LIAPUNOV exponent. Cf. (7.43).

$\nu_a(t)$ $= \dfrac{1}{D_a(t)\Delta t}$: Rate which is a measure for the frequency of spontaneous behavioural changes of individuals of subpopulation a.

$\nu^a(t)$ $= \nu_{aa}(t)$: Meeting rate of an individual of subpopulation a with other individuals of the same subpopulation.

$\nu^a(x'|x;t)$ Rates that are a measure for the overall frequency of interactions of an individual of subpopulation a with other individuals. Cf. (9.62c).

$\nu_{ab}(t)$ Meeting rate of an individual of subpopulation a with individuals of subpopulation b.

$\widehat{\nu}_{ab}(t)$ $= \nu_{ab}(t)/N_b$.

$\nu_{ab}^k(t)$ $= \nu_{ab}(t) f_{ab}^k(t)$: Interaction rate of kind k between an individual of subpopulation a and individuals of subpopulation b.

$\xi(t)$ Fluctuation at the time t.

π $= 3.1415926\ldots$: Pi.

$\rho(a,x;t)$ $= NP(a,x;t) = N_a P_a(x,t)$: Density of subsystems which are of type a and in state x.

$\rho(t)$ Statistical operator. Cf. Sect. 3.2.4.1.

$\sigma_{IJ}(t)$ $= \langle X_I X_J \rangle - \langle X_I \rangle \langle X_J \rangle$: Covariances in Γ-space. The *correlation* of two randomly distributed quantities X_I and X_J is measured with the *correlation coefficient* $r_{IJ} := \frac{\sigma_{IJ}}{\sqrt{\sigma_{II}\sigma_{JJ}}}$. Important relation: $-1 \leq r_{IJ} \leq 1$. For $r_{IJ} = 0$ the quantities X_I and X_J are called *uncorrelated*.

$\sigma^{ab}_{x\,x'}$ $= \langle n^a_x n^b_{x'} \rangle - \langle n^a_x \rangle \langle n^b_{x'} \rangle$: Covariances of the occupation numbers n^a_x.

τ $\equiv \tau(t)$: Transformed time.

ω $= 2\pi/T$: (Angular) frequency of oscillation.

ω_0 (Angular) frequency of the fundamental mode of oscillation.

Ω System size. Cf. (6.9).

Operators and Special Symbols

$x := y,\ y =: x$ x is defined by y.

$x \widehat{=} y$ x corresponds to y.

$x \stackrel{!}{=} y$ x must be equal to y.

$x \stackrel{a=b}{=} y$ x ist equal to y on condition $a = b$.

$x \equiv y$ x is identical with y.

$x \ll y$ x is much less than y.

$x \gg y$ x is much greater than y.

$x \sim y$ x is similar to y, i.e. x and y have the same order of magnitude.

$x \longrightarrow y,\ y \longleftarrow x$ Transition from x to y.

$x \leftrightarrow y$ Interchange of x and y.

$n!$ $= 1 \cdot 2 \cdot \dots \cdot (n-1) \cdot n$: n *factorial*.

x^y x to the *power* of y. Important relations: $x^{y+z} = x^y \cdot x^z$, $x^y = x \cdot x^{y-1}$, $x^1 = x$, and $x^0 = 1$.

x^{-1} $= 1/x$: *Reciprocal* of x.

$f(x)\Big|_{x=y}$ $= f(y)$: Value of function f for $x = y$.

$|x|$ $= \sqrt{[\mathrm{Re}(x)]^2 + [\mathrm{Im}(x)]^2}$: *Absolute value* of a (possibly complex) number x. Important relation: $|x+y| \leq |x|+|y|$ *(triangle inequality)*.

$\vert\underline{M}\vert$	$= \vert\underline{M}^{\mathrm{tr}}\vert$: *Determinant* of the matrix $\underline{M}$. Cf. [20], pp. 154ff., 193ff.
$\Vert x\Vert$	$= \sqrt{\sum_i (x_i)^2}$: *Magnitude* (length) of the vector x. Vectors x with $\Vert x\Vert = 1$ (i.e. with magnitude 1) are called *normalized.* Vectors of arbitrary magnitude fulfil the *triangle inequality* $\Vert x + y\Vert \leq \Vert x\Vert + \Vert y\Vert$ $\left(\text{or } \left\Vert \sum_i x_i \right\Vert \leq \sum_i \Vert x_i\Vert\right)$ and the CAUCHY-SCHWARZ *inequality* $\sum_i c_i\, \vert x_i y_i\vert \leq \sqrt{\sum_i c_i\, (x_i)^2} \cdot \sqrt{\sum_i c_i\, (y_i)^2}$ $(c_i \geq 0)$, in particular, $\vert x \cdot y\vert \leq \Vert x\Vert \cdot \Vert y\Vert$. Cf. [20], pp. 133f., 288f.
$\vert X\rangle$	State vector of a quantum mechanical system.
$\vert X, t\rangle$	State vector of a quantum mechanical system at time t.
$\langle X\vert$	Adjoint of the state vector $\vert X\rangle$.
$\langle X\vert Y\rangle$	*Scalar product* of the state vectors $\vert X\rangle$ and $\vert Y\rangle$ in HILBERT space.
$\langle f\rangle$, $\langle f(x,t)\rangle_t$	$= \sum_x f(x,t)P(x,t)$: *Mean value (expected value, average)* of the function $f(x,t)$.
$\cdot\cdot, \cdot$	represent arguments (independent variables) or subscripts/superscripts of a function.
$\ldots$	And so on; up to.
$\times$	Indicates the continuation of a formula.
$\{x_1, \ldots, x_S\}$	*Set* consisting of the *elements* $x_1, \ldots, x_S$.
$[x, y]$	$= \{z \text{ with } x \leq z \leq y\}$: *Closed interval* reaching from x to y (set of all values z with $x \leq z \leq y$).
$x \in \mathcal{M}$	x is element of the set $\mathcal{M}$.
$\forall x$	For all admissible values of x.
∞	Infinity.
$\prod_{x=1}^{n} f(x)$	$= f(1) \cdot \ldots \cdot f(n)$.

$\prod_{x}$	*Product* over all admissible values of x.
$\sum_{x=1}^{n} f(x)$	$= f(1) + \cdots + f(n)$.
$\sum_{x}$	*Sum* over all admissible values of x.
$\sum_{x(\neq x')}$	Sum over all admissible values of x except for x'.
$\sum_{x \neq x'}$	$= \sum_{x} \sum_{x'(\neq x)}$.
$\sum_{x_1,\ldots,x_n}$	$= \sum_{x_1} \cdots \sum_{x_n}$: Sum over all *n-tuples* $(x_1, \ldots, x_n)$.
$\sum_{\substack{x_1,\ldots,x_n \\ (f(x_1,\ldots,x_n)=g(x_1,\ldots,x_n))}}$	Sum over all n-tuples $(x_1, \ldots, x_n)$ which satisfy the condition $f(x_1, \ldots, x_n) = g(x_1, \ldots, x_n)$.
$d^n x$	n-dimensional differential volume element (for $n = 2$: Surface element, $n = 1$: Line element).
$\int_{x_0}^{x} dx' f(x')$	$= \lim_{n\to\infty} \sum_{i=1}^{n} \Delta x\, f(x_i)$ with $\Delta x = \frac{x - x_0}{n}$ and $x_i = x_0 + i\, \Delta x$: *Integral* of the function $f(x)$ from x_0 to x.
$\int d^n x\, f(x)$	$= \int dx_1 \ldots \int dx_n\, f(x)$: n-dimensional *volume integral* over $f(x)$.
$\int dx \cdot f(x)$	*Line integral* of the *vector field* $f(x)$. Cf. [246], pp. 1027ff.
$\frac{d}{dx} f(x)$	$= \lim_{\Delta x\to 0} \frac{f(x+\Delta x) - f(x)}{\Delta x}$: *Derivative* of the function $f(x)$ with respect to x. Important relations: $\frac{d}{dx}\Big[f(x)g(x)\Big] = \left(\frac{d}{dx} f(x)\right) g(x) + f(x) \left(\frac{d}{dx} g(x)\right)$ *(product rule)*, $\frac{d}{dx}\left(\frac{f(x)}{g(x)}\right) = \frac{\left(\frac{d}{dx} f(x)\right) g(x) - f(x) \left(\frac{d}{dx} g(x)\right)}{\Big(g(x)\Big)^2}$ *(quotient rule)*, $\frac{d}{dx} f\Big(g(x)\Big) = \frac{d}{dy} f(y)\Big\vert_{y=g(x)} \cdot \frac{d}{dx} g(x)$ *(chain rule)*, and

	$\frac{d}{dx}\int\limits_{x_0}^{x} dx'\, f(x') = f(x)$ *(second fundamental theorem of calculus).*
∂V	Surface of the volume V.
$\frac{\partial}{\partial x} f(x, y)$	$= \lim\limits_{\Delta x \to 0} \frac{f(x+\Delta x, y) - f(x, y)}{\Delta x}$: *Partial derivative* of $f(x, y)$ with respect to x.
$\frac{d}{dx} f(x, y)$	$= \frac{\partial}{\partial x} f(x, y) + \frac{\partial y}{\partial x}\frac{\partial}{\partial y} f(x, y)$: *(Total, substantial) derivative* of the function $f(x, y)$ with respect to x.
$\frac{\delta}{\delta f} F\big(f(x)\big)$	*Functional derivative (variational derivative)* of F with respect to the function $f(x)$.
∇	$= \left(\frac{\partial}{\partial x_1}, \ldots, \frac{\partial}{\partial x_n}\right)^{\mathrm{tr}}$: *Nabla operator.*
$\nabla_x f(x, y)$	$= \left(\frac{\partial f(x, y)}{\partial x_1}, \ldots, \frac{\partial f(x, y)}{\partial x_n}\right)^{\mathrm{tr}}$: *Gradient* of the function $f(x, y)$ with respect to $x = (x_1, \ldots, x_n)^{\mathrm{tr}}$.
$\nabla \cdot f(x)$	$= \sum\limits_{i=1}^{n} \frac{\partial f_i(x)}{\partial x_i}$: *Divergence* of the *vector field* $f(x)$.

Chapter 1
Introduction and Summary

The field of *quantitative sociodynamics* is still a rather young and very thrilling interdisciplinary research area which deals with the mathematical modelling of the temporal evolution of social systems. In view of the growing complexity of social, economic, and political developments quantitative models are becoming more and more important—also as an aid to decision-making. From a scientific point of view as well, a *mathematical formulation* of social interrelations has long been overdue. Compared with purely qualitative considerations it allows clearer definitions of the terms used, a more concentrated reduction to the interesting interrelations, more precise and more compact descriptions of structures and relations, more reliable conclusions, better forecasts and, thus, statements which are easier to verify [102, 231]. Apart from this it turned out that many social phenomena cannot even *approximately* be understood with static concepts. Some dynamic social processes are not comprehensible as a sequence of time-dependent equilibrium structures. The description of self-organization and structure formation processes requires dynamic mathematical concepts which also can describe *non-equilibrium phenomena,* i.e. the temporal evolution of systems which become unstable.

Because of the complexity of social systems a mathematical formulation of social processes was, for a long time, considered to be almost impossible. Nevertheless, again and again corresponding formulations have been developed by social as well as natural scientists. However, the proposed concepts often were too simple and did not sufficiently do justice to those phenomena which they should describe.

Only in the recent years, powerful methods have been developed which make it possible to describe complex systems consisting of many elements (subsystems) that interact with each other. Here pioneering were particularly the concepts of *statistical physics* for the description of chance affected processes and discoveries in the field of *non-linear dynamics* which was shaped by *synergetics* [106, 107], *chaos theory* [16, 135, 183, 254], *catastrophe theory* [273, 302], as well as by the theory of *phase transitions* and *critical phenomena* [145, 177, 202, 263]. These again and again proved their interdisciplinary explanatory power and allow the development of models for social processes at an actually much sounder level than it was feasible only a short time ago. In the meantime a remarkable list of physicists exists who already dealt with topics from the field of mathematical sociology: WEIDLICH

D. Helbing, *Quantitative Sociodynamics*, 2nd ed.,
DOI 10.1007/978-3-642-11546-2_1, © Springer-Verlag Berlin Heidelberg 2010

[285, 286, 288–290], HAAG [103], HAKEN [107, 300], EBELING [56, 57, 59], MALCHOW [179], SCHWEITZER [257], PRIGOGINE [225, 226], ALLEN [2–6], MONTROLL [91, 185–187], MOSEKILDE [191, 192], HUBERMAN [89, 90, 147], WALLS [284], and LEWENSTEIN [168].

During the next years, in the field of the mathematical description of social phenomena we can surely expect a rapid scientific development and fundamentally new results from the various disciplines involved.

1.1 Quantitative Models in the Social Sciences

Despite the difficulties with the mathematical modelling of social processes already for some time a number of formal models has been developed by social scientists. These include OSGOOD and TANNENBAUM's *principle of congruity* [213], HEIDER's *balance theory* [119, 120], and FESTINGER's *dissonance theory* [74] which deal with the stability of attitude structures, to mention a few. Very interesting is also the model suggested by LEWIN which assumes that the behaviour of an individual is determined by a *'social field'* [169]. Further mathematical models were developed for *learning processes* (cf. [231], Chap. 14) and *decision processes* [48, 53, 174, 209, 297]. The first *stochastic models* for social processes go back to COLEMAN [43] and BARTHOLOMEW [18]. In addition, numerous *diffusion models* [50, 98, 110, 157, 178, 247] are worth mentioning. Of big importance is also the description of the competition and cooperation of individuals within the framework of *game theory* which was founded by VON NEUMANN and MORGENSTERN [196]. Last but not least the *event analysis* or *survival analysis* received considerable attention [51, 275]. That is, meanwhile a broad and substantial literature exists in the field of *mathematical sociology* (cf. also [44, 187, 231]). In the following those models which play a particular role in the course of this book will be discussed in more detail.

1.1.1 The Logistic Model

For *growth processes* an exponential temporal development was assumed for a long time. In many cases, however, this is not empirically confirmed. Rather it turned out that growth processes with *limited resources* can well be characterized by the *logistic equation* of PEARL and VERHULST [217, 281]. According to the logistic equation growth processes start in a roughly exponential way and then turn into a saturation phase in which a certain system dependent maximum value is gradually approached (cf. Sect. 4.5.2). Exponential growth occurs for unlimited resources only.

The applications of the logistic equation reach from the description of chemical reaction rates, the growth of animal and plant populations in the absence of enemies, and the growth of towns up to the description of the spreading of information or innovations [18, 43, 91, 110, 130, 187, 206].

1.1.2 Diffusion Models

The topic *'distribution processes'* is generally the subject of diffusion models. In detail these deal with the spreading of such different 'things' as diseases, rumours, conventions, norms, standards, knowledge, technologies, or goods [18, 43, 50, 98, 110, 157, 178, 247]. From the point of view of *spatial distribution processes* there are similarities to diffusion processes in physics. An example is the dispersal of dye molecules of ink drops in a glass of water caused by the random motion of particles (cf. Sect. 10.5).

However, the spatial dispersal under *inhomogeneous* conditions such as the dissemination of information within the irregularly distributed population of a country is more complicated. This is particularly true if the information is not distributed by central media but via personal communication or telecommunication (cf. Sect. 10.5.2).

1.1.3 The Gravity Model

The gravity model can be applied to the description of *exchange processes* such as the exchange of goods, the traffic volume, or the removals ('migration') between different towns or regions (cf. e.g. [235, 305]). Although the background is completely different, the name of the gravity model stems from the mathematical similarity with the *law of gravity* which applies to the attraction between celestial bodies as earth and moon.

The gravity model bases on the observation that the frequency of the considered exchange (e.g. migration) is proportional to the size of the 'source' (to the number of inhabitants there) and to the size of the 'sink' (to the number of inhabitants of the destination town since this is a measure for the cultural infrastructure, employment facilities, etc.). The exchange frequency decreases proportionally to the reciprocal value of the *distance* between source and sink. This is, on the one hand, because the *transaction costs* (e.g. removal costs) increase with distance but, on the other hand, because the number of alternative towns grows with the distance (cf. Sect. 9.3.1).

1.1.4 The Game Theory

Whereas the up to now introduced quantitative models describe relatively simple phenomena, game theory is a convenient tool to comprehend rather complicated phenomena. Above all, this applies to the modelling of social *competition* and *cooperation* on the basis of a success or utility oriented behaviour of individuals. Important applications stem from economics (e.g. in the *oligopoly theory* [269, 280]) as well as from the social and behavioural sciences (e.g. concerning the *'prisoner's dilemma'*) [12, 13, 175, 196, 234, 252, 253, 283] but also from biology (with respect to *evolution theory* or *ecology*) [60, 61, 78, 139, 256].

It turned out that in order to understand particularly the development of cooperation among egoists, the *repeated* interaction (and thus the time dimension) is of great relevance *('shadow of the future')* [13]. To consider this, a multiple repetition of the single game and decision moves was applied for a long time *('iterated games')*. Nowadays the *game dynamical equations* find more and more resonance [137, 139, 256, 271, 303] (cf. Chap. 12). These are differential equations and thus continuous in time which at least is more realistic if there are no single, clearly defined moves at discrete times.

The game dynamical equations originate from the theory of evolution [137, 271, 303] where they celebrate great successes. Therefore, in connection with the game dynamical equations, we also speak of *evolutionary game theory*. However, in contrast to the theory of evolution where the equations can be derived from the laws of genetics (cf. [138], pp. 43ff.), no 'microscopic' explanation of the game dynamical equations could be presented for a long time. Such an explanation of the game dynamical equations on the basis of individual behavioural principles will be introduced in Sect. 12.2.4 [128].

Furthermore, in Sect. 12.4.1 a game dynamical model for the spontaneous *self-organization of behavioural conventions* in the case of competing equivalent behavioural strategies will be developed [128, 131]. A well-known example for this is the rivalry of the originally equivalent video systems VHS and BETA MAX [9, 116] which was very clearly decided by the purchasers' behaviour. Here three questions arise: Why at all can one of both equivalent alternatives win out over the other? How long does it take to gain a certain market share? Why does a certain product win out over its competitors in certain branches of industry whereas in other branches several products coexist?

The solution of these questions was found in connection with the description of a phenomenon which is known from the *behaviour of pedestrians* [123, 124]. As one can easily observe, in dense pedestrian crowds *lanes* of uniform walking directions are formed. With respect to the walking direction, these lanes move (in Germany and some other countries) almost always on the right-hand side though the left-hand side is completely equivalent [204, 205]. The reason for this is that an *avoidance manoeuvre* is successful only if two each other encountering pedestrians both swerve either to the right or to the left. Otherwise they must slow down or stop in order to avoid a collision. Therefore, it is advantageous for a pedestrian to choose the strategy which is shared by the majority. As soon as accidentally a small majority favouring a certain strategy arises it will grow on and on. Consequently, a behavioural convention originates which, in the end, is caused by social self-organization. Once established conventions are often fixed by *laws* or *norms* after some time.

Nevertheless, there are also mechanisms which hinder the triumph of one behavioural alternative. One example for these are spontaneous strategy changes which result from the occasional trying out of behavioural alternatives *(trial and error)*. These can be taken into account mathematically as well. It is expected that spontaneous strategy changes somewhat reduce the attainable majority of the developing behavioural convention. For a small frequency of spontaneous strategy changes this is actually true. However, if this frequency exceeds a critical value the formation of a behavioural convention will be suppressed and each of the two strategies will be chosen with equal probability. Whether one of many equivalent behaviours can win out at all depends on whether the advantage which results from an adaption to the behaviour of the majority exceeds the influence of spontaneous strategy changes.

In the case of the avoidance problem spontaneous strategy changes obviously play a subordinate role. Otherwise the preference for the right-hand side (in Germany) would not have developed. However, with the same probability a preference for the left-hand side could have emerged. The existing conventions can vary from one region to another. This applies, for example, to the driving direction of road traffic. Whereas in Great Britain left-hand traffic is prescribed, in the rest of Europe right-hand traffic is usual. Thus an explanation of the *symmetry breaking* regarding avoidance behaviour on the basis of the asymmetry of the human body is to be excluded.

The described model for the development of conventions is also applicable to the revolution direction of clock hands, the direction of writing, and many other examples where it is advantageous to choose the same behavioural alternative as the majority. In the case of the rivalry of the above mentioned equivalent video systems it is the *compatibility* of the recorders (concerning copying, selling or hiring of tapes) which makes it profitable to join a once accidentally arisen majority.

The above model for the self-organization of behavioural conventions can easily be extended to the case of more than two equivalent behavioural alternatives. Furthermore, situations can be described where one behavioural alternative is superior to the others. However, here the formation of a behavioural convention is trivial.

1.1.5 Decision Models

It is well-known that decisions are processes the outcome of which normally is not predictable. At best, the *probability* for a particular outcome can be estimated. A very detailed model for decision processes stems from FEGER [69] (cf. Sect. 8.2.4). According to this, decision situations are conflict situations in which one from several behavioural alternatives must be chosen. The decision process mainly consists of a collection of arguments which are in favour of one or another alternative *('anticipation of consequences')*. Here the order of arguments is quite accidental. The frequency of arguments supporting a certain alternative increases with the *preference* for it. If a certain number of arguments one after the other supports one and the same behavioural alternative the decision is made. Therefore, the probability of a decision for a certain alternative depends on the probability of arguments supporting it and thus on the preference for it.

Nevertheless, in many cases it is not the decision *process* which is of interest but only the *outcome* of the decision. The functional dependence of the decision probability for a particular behavioural alternative on the preference is described by the *multinomial logit model* [53]. In this connection, a measure for the preference is the *utility* which is expected as a consequence of a certain decision. The multinomial logit model proved itself in various areas. Its applications reach from purchase patterns up to the choice of the location of residence or of the means of transport [48, 209, 297].

The fact that a decision usually does not have a unique outcome is related to the *incompleteness of information* about the actual utility. Naturally, the estimate for the actual utility normally varies among individuals and is possibly also dependent on the mood. The more *uncertain* the knowledge about the actual utility of the possible behavioural alternatives is the more probabilistic the outcome of the decision.

If one tries to develop a *dynamic* behavioural or decision model, the problem will arise that a temporal *sequence* of decisions is to be described. In this case, the single decisions of the mentioned decision sequence will again be treated with the multinomial logit model. However, the probability of a decision now depends on the previously made decision, i.e. on the current behaviour. This probability defines the *transition probability* from one behaviour to another. It will be greater than zero if the corresponding behavioural change is connected with an *increase* of utility (cf. Sect. 9.2.1).

1.1.6 Final Remark

Up to now, the logistic equation, the gravity model, the diffusion models, the game dynamical equations, and the decision models were viewed as single theoretical concepts standing next to each other without any relations. In this book, for the first time, a general, mathematical, and dynamic model will be developed which includes all these concepts as special cases. This behavioural model represents a kind of a *metatheory* which is—at least in the above mentioned cases—empirically well-founded.

1.2 How to Describe Social Processes in a Mathematical Way

Most problems accompanying the attempts to mathematically describe social systems arise from their *complexity.* This causes an extreme *sensitivity* to all sorts of influences so that, in similar situations, social systems sometimes behave in a completely different way. An additional difficulty is the *chance affectedness (randomness, stochasticity)* of individual decisions which also influences the concrete behaviour of a social system (cf. Chap. 8).

For a long time this made the mathematical description of social systems appear generally impossible. During the last years, however, current insights in *complex systems* have shed some light on these problems. Above all, they come from the fields of *non-linear dynamics (synergetics, chaos theory)* and *statistical physics* (where numerous stochastic methods were developed for randomly behaving systems).

In Part I of this book, the most important concepts of these fields, their interrelations, and properties will be discussed in a unified and self-consistent form. On this occasion, entirely new methods are also considered. One of these is the *'path integral' solution* of the *master equation* which is applicable to problems such as *process control* and *event analysis* [51, 275].

Non-linear dynamics and statistical physics have often offered successful explanations even for non-physical phenomena, e.g. in *chemistry* [85, 181, 200, 208], in *biology* [8, 91, 92] as well as in the *economic and social sciences* [18, 42, 43, 285, 286, 289, 291, 292, 294]. For the investigation of a given complex system, the way of proceeding usually is to find appropriate stochastic equations which are afterwards further examined with the methods of non-linear dynamics. It is to be emphasized that what is done here, in contrast to the out-dated *'physicalistic'* approaches, is no *model* transfer but a *methodological* transfer. A simple model transfer by merely reinterpreting the model variables were questionable in so far as it would assume the compared systems to possess the same properties (characteristics)—at least in those aspects that are fundamental for the model. However, this *'isomorphism'* often is only insufficiently guaranteed. In contrast to this, a methodological transfer considers the system specific properties explicitly by the respective model assumptions.

1.2.1 Statistical Physics and Stochastic Methods

For some decades, *stochastic equations* have proven a versatile and useful instrument in all fields of science which deal with *fluctuations* (random variations) [117, 151, 188, 189]. Some phenomena become clearly comprehensible only with stochastic models. Above all, this applies to *phase transitions* and *critical phenomena* [145, 177, 202, 263]. Especially, the phase transitions are of particular interest because they mean a fundamental change of the system properties. An example for this is the transition from ice to water or from water to steam.

Statistical physics deals with systems consisting of a huge number of *elements (subsystems)* which possibly can be subdivided into several *types* of elements (e.g. different sorts of gas molecules). The elements of *every* type (i.e. equivalent subsystems) must be present in a huge number. Furthermore, let us assume that each element can be characterized by a number of time-dependent *variables* (quantities) which together describe its *state* and can be combined in a *state vector*. The elements have *interactions* with each other, i.e. they can change their states mutually, e.g. by exchanging energy or information. Apart from interactions there may occur also *spontaneous state changes* (which result from the properties of the elements or are induced by external influences on the system).

Even if the exact laws for the behaviour of the elements of such a system and their interactions are known, due to the huge number of elements it is impossible to exactly calculate their behaviour. In most cases, however, the point of interest is not the behaviour of the *single* elements (i.e. their temporal state changes) but rather the *distribution* of states within the system.

It turned out that for the description of the temporal evolution of the state distribution *probabilistic models (stochastic equations)* are suitable [84, 126]. Above all, this applies to the *master equation* [215] (cf. Chap. 3) but also to the **Fokker-Planck** *equation* [79, 221] which can be understood as a TAYLOR approximation of the master equation (cf. Chap. 6). A similar thing is true for the *stochastic differential equations* (**Langevin** *equations* [166]) which, in a certain respect, represent a reformulation of the FOKKER-PLANCK equations [267] (cf. Chap. 7). The LANGEVIN equations, however, characterize the behaviour of *single* elements the state changes of which, on the one hand, go back to a *systematic effect* which describes the *most probable* state change. On the other hand, they stem from *fluctuations* which reflect individual variations of the state changes. In contrast to this, the FOKKER-PLANCK equation as well as the master equation describe the temporal evolution of the state distribution of *many* subsystems (i.e. of an *'ensemble'* of elements).

1.2.1.1 The Master Equation

In order to be able to describe the time-dependent change of the *state distribution,* it is not necessary to know in detail which elements experience which state changes *(transitions).* However, we must know how many elements per time unit go over to the other possible states. Thus, it will not matter if a *particular* element is subject to a certain state change or *another one*, instead.

The temporal evolution of the state distribution is given by several quantities. One of these is the present state distribution itself because it characterizes how many elements can potentially go over from a particular state to another. The *rate* of such state changes is described by so-called *transition rates.* These are, on the one hand, given by the *frequency (per time unit)* with which spontaneous or interaction induced transitions take place *at all.* On the other hand, they are determined by the *(transition) probabilities* with which changes from the respective occupied state into the other available states occur. In order to describe a system of the considered kind mathematically, the transition rates must be determined in a *system specific* way. The resulting equation for the temporal evolution of the state distribution is a *master equation* which describes the concrete processes within the system.

Up to now, we treated the master equation in *state space* which consists of state vectors. An analogous master equation can also be formulated for *configuration space* consisting of vectors that combine the *occupation numbers* (frequencies) of all states in the *configuration* of a system. The master equation in state space then corresponds to the *approximate mean value equation* of the related master equation in configuration space.

The configurational master equation has some essential merits [133, 289]: First, it allows to use individual transition rates which depend on the occupation numbers of the states. Such cases are important for the description of *indirect interactions* which remarkably often occur in social systems. Second, the configurational master equation makes possible to gain equations which take into account *pair interactions* or even *higher order interactions.* Third, from the configurational master equation *covariance equations* can be derived which are essential for the determination of the *reliability* of the mean value equations. Finally, it allows to calculate *corrections* for the approximate mean value and covariance equations (cf. Chap. 5 and Sect. 12.4).

1.2.1.2 Application to the Description of Social Systems

Above all, it were WEIDLICH and HAAG from the University of Stuttgart in Germany who realized that the master equation is particularly suitable for the description of *social* systems [285, 286, 289, 292, 294]. The elements (subsystems) then represent individuals (human beings, persons) and the states stand for certain aspects of their behaviour (e.g. for the expression of opinions concerning a special affair). In these models, the distinguished behaviours have only to be *classified* but not to be *linearly ordered*, and they need not to form a continuous behavioural space. Thus, problems which are related to the *measurement* or *scaling* of social quantities do not arise. Nevertheless, by analyzing empirical data, the degree of dissimilarity of the different behaviours (their *'distance'*) can be determined by a method introduced in Chap. 13. From this, by means of *multidimensional scaling* [161, 301], a *behavioural space* can be constructed afterwards.

Above all, the problem is to find a suitable system specific approach for the transition rates. However, for this problem a solution was already formulated in Sect. 1.1.5. There it turned out that the transition probability with which an individual decides in favour of a new behaviour is determined by the *utility* of the behavioural change and can be described by a *multinomial logit model* (cf. Chap. 9 and [289], §3).

Since this formulation can be considered as sufficiently founded, it is not required to laboriously determine the utility of each behaviour by means of extensive surveys and suitable calibrations. It is rather simple to calculate the utilities from data material describing the temporal course of the behavioural distribution (e.g. opinion distribution) and the transition rates [294] (cf. Chap. 13).

With the empirically determined utilities, for instance, the *analysis* and *prognosis* of voting or purchasing behaviour is possible. For economy or politics it is, furthermore, of particular interest which factors contribute with which importance to the utility of certain purchasing or voting decisions. This can be determined from suitable data material via a *regression method* in which it is assumed that the utility, on the one hand, depends on the interests of purchasers and voters respectively and, on the other hand, on the various characteristics of the offered products or political concepts (cf. Sect. 13.6 and [238, 294]).

The circumstance that the utility of a behaviour (and so the probability of the corresponding behavioural change) often varies among different individuals can also be taken into account. For this purpose, the individuals of the considered social system are divided into several *(behavioural) types* ('characters'). Individuals of the same behavioural type should tend to the possible behavioural changes with similar probabilities.

In order not to have too many behavioural types and to obtain large *subpopulations* (defined by the individuals of the same type) the investigation is restricted to *simple behavioural processes* and *simple situations* (cf. Sect. 8.2.2). Then it does not matter how the single individuals behave in other situations and areas of life. It is even admissible that the modelled behaviours are influenced by other areas of behaviour as long as these influences have no *systematic* effect (with respect to the average of the individuals of a subpopulation). In the case of voting behaviour, for instance, it could suffice to determine the subpopulations by the level of income, education, and confession.

WEIDLICH and HAAG have developed master equation models for the description of a multitude of different social phenomena. Among them there are models of opinion formation, migration models, and models for the emergence of settlement structures, but also non-equilibrium models for economic decisions [289–292]. In the case of the migration models, a comparison with empirical data could be carried out. This provided very good results for the model assumptions concerning the transition rates (i.e. the frequency of behavioural changes) [294].

1.2.1.3 BOLTZMANN-Like Equations

The models of WEIDLICH and HAAG base on *indirect interactions* of individuals, i.e. the frequency of individual behavioural changes depends on the *socioconfiguration* or the behavioural distribution within the social system [289, 292]. Therefore, the behaviour of an individual has, in turn, an effect on himself/herself *(feedback effect)*.

Indirect interactions are especially suitable for the description of behavioural changes which are indirectly induced by the *media* (radio, TV, newspapers), caused by generally accessible information, or evoked by the 'socio-cultural and political climate'. In physics they play a role in *mean field theories* [24] which are, for instance, used for the description of the magnetization of iron (ISING *model:* cf. [10, 148, 304]).

Many behavioural changes, however, go back to *direct interactions*. This, for example, concerns behavioural changes which are triggered by the behaviour of other individuals. A typical case for this are opinion changes caused by discussions. The most important effects of direct interactions can already be understood by considering *pair interactions* which normally yield the largest contribution. In principle, however, simultaneous *interactions of higher order* (interactions between several individuals) can be also treated mathematically (cf. Sect. 5.4).

In order to describe pair interactions, in contrast to WEIDLICH and HAAG's master equation models, additional terms are required which take into account that the rate of the corresponding behavioural changes is proportional to the number of potential interaction partners. The resulting equations are BOLTZMANN-*like equations* (cf. Chap. 4). The fundamental significance of these becomes particularly clear in the following quotation by UHLENBECK [276]:

> The Boltzmann equation has become such a generally accepted and central part of statistical mechanics, that it almost seems blasphemy to question its validity and to seek out its limitations.

The BOLTZMANN-like equations were originally developed for the description of the behaviour of gases where the pair interactions are caused by collisions (scattering processes) of gas particles. The transferability to avoidance and deceleration manoeuvres at encounters of pedestrians is obvious (cf. [123, 125]).

Normally, however, the pair interactions in social systems are of a totally different nature. In the case of opinion formation, for instance, pair interactions are to be imagined in such a way that two individuals with certain opinions occasionally communicate with each other and possibly accept other opinions as a consequence of their exchange of ideas. So, instead of velocities this time opinions are changed. Moreover, in the case of opinion formation (but also of behavioural changes in general) there are other *kinds* of pair interactions (cf. Sects. 9.3 and 10.3). One can distinguish between *persuasion processes* in which the opinion of another person is taken over, *avoidance processes* in which a particular opinion is evaded in order to differ from the interaction partner, and *compromising processes.* With the corresponding model, e.g. *fashion cycles* can be well understood [127].

Direct and indirect interactions do not exclude each other mutually. They rather complete each other and can be integrated in a general model. The limitation to one of both interaction mechanisms could lead to the situation that the empirical data can be reproduced only insufficiently. In physics, for example, the properties of nuclei can be explained only satisfactorily by combining a mean field with direct residual interactions ('pairing effect') (cf. [278, 298]).

1.2.1.4 Social Fields and Social Forces

For the case of a *continuous* or *quasi-continuous* behavioural space, the BOLTZMANN-like equations can be transformed into BOLTZMANN-FOKKER-PLANCK *equations* by a TAYLOR approximation (cf. Sect. 6.6, [130]). These equations have a very intuitive interpretation. Accordingly, there exists a vector-like quantity which describes the *most probable behavioural change* of individuals of a certain behavioural type. This quantity can be interpreted as a *social force* because it is the drive behind behavioural changes and, furthermore, it is influenced by the behaviours of other individuals. It could also be said that the social force describes concrete *motivations to act* which are evoked in an individual as a reaction to his/her environment. In the BOLTZMANN-FOKKER-PLANCK equations, apart from social forces, there also appear so-called *diffusion coefficients.* These reflect the individual variations of behaviour (cf. Chap. 11, [130]).

In certain cases the social force can be expressed by a *social field.* This can be imagined as a dynamically changing mountain range in behavioural space where the steepness is a measure for the social force which is at work at a certain place of the behavioural space and, accordingly, affects the corresponding behaviour. The idea of a social field which describes the influence of public opinion, norms, trends, and environment on individual behaviour originally goes back to LEWIN [169]. It is also related to the mathematical formulation of indirect interactions suggested by WEIDLICH and HAAG [292].

In some contexts, the description of behavioural changes by social forces proves its extraordinary worth. For example, the distance which a pedestrian keeps to others is given by an equilibrium of two opposite forces—one force which describes the attempt to move with a certain desired speed and another, repulsive force reflecting the readiness to respect the territory which another pedestrian claims for himself/herself. If a pedestrian feels a growing pressure of time due to delays, the force describing the tendency to get ahead increases compared to the repulsive force so that he/she keeps less distance or even pushes. This effect can, interestingly enough, also be observed in queues that have come to rest though a reduction of distance does not lead to getting on faster [123, 124, 129].

The force model is also a very promising starting point for a quantitative description of *group dynamics* (cf. also p. 14). Here, for instance, attractive forces due to certain likings or common interests play a role but also repulsive forces due to conflicting goals.

1.2.2 Non-linear Dynamics

The interactions between the single elements of a system are in general *non-linear,* i.e. the effects are not simply proportional to the causes. Thus the behaviour of the system as a whole cannot simply be understood as a sum or superposition of the behaviours of the constituent elements. Especially, an element often has an indirect effect on itself *(feedback effect).*

The non-linearity of the interactions has some important consequences. On the one hand, it turns out that a few variables often determine the values of the other variables *('slaving principle')* [106, 107] so that they are sufficient to describe the behaviour of the system on their own. On the other hand, the non-linearity often causes *instabilities* which lead to *complex dynamics* or *self-organization phenomena,* i.e. to the *emergence* of new system properties. (A more detailed and very recommended discussion of this topic is given in [239].)

1.2.2.1 The 'Slaving Principle'

The dynamic processes within a considered system can be characterized by the complete set of variables which describe the time-dependent states of its subsystems. However, it is often more suitable to choose variables which reflect processes that are going on in the system at different 'speeds'. For the determination of a suitable set of variables mathematical methods have been developed (cf. Chaps. 8 and 9 of [106]).

These variables can be subdivided into three classes *('separation of time scales'):* First, variables which change approximately at a rate (on a 'time scale') which corresponds to the rate of changes of the examined processes in the system. These are denoted as *order parameters.* Second, variables which change much more slowly. They can be considered as *quasi-constant* and treated as fixed system parameters. Third, variables which change much faster than the examined processes and can be understood as *fluctuations.* Most of the time, the fast variables remain near their equilibrium values which can gradually move along with the temporal change of the order parameters.

It can be mathematically shown that the fast variables are already determined by the order parameters due to the non-linear coupling *('slaving principle').* Therefore, they can be *'quasi-adiabatically' eliminated* from the examined equations. As a consequence, from the numerous system variables only a few order parameters are required for the description of the substantial processes which means an enormous *data reduction* and *information compression.* (Cf. [106–108].)

The order parameters reflect the dynamics of the phenomena which take place on a 'macroscopic' or collective level in a self-consistent way. In general, they are described by a closed system of differential equations. Due to the information compression the dynamics of the order parameters can agree for totally different systems so that in this case one sometimes speaks of *universal laws.* These universal laws are the cause for many *analogies* between the macroscopic behaviour of different systems the elements of which do not show any similarity. (Cf. [106–108, 302].)

The method of the separation of time scales will be discussed in more detail in Sects. 3.2.3 and 10.3.4. Moreover, it is important in another context because it means that for describing social processes many conceivable influences play no essential role. So, for the modelling of behavioural changes, it will not be essential to comprehend in detail each of the different *thought processes (trains of thought)* if the thought processes are clearly faster than the behavioural changes resulting from these. On this condition, the behavioural changes can be understood as *reactions* occurring with certain probabilities to the perceptions that take place (cf. Sect. 8.2.1).

It should be emphasized that variables which change on a comparable time scale as the examined processes need not necessarily to have an influence on the latter since they possibly are not dynamically coupled with those at all. So there are several *behavioural areas* which hardly mutually influence each other and thus can be modelled separately.

1.2.2.2 Instabilities and Complex Dynamics

Non-linear equations can exhibit a much more complex behaviour than linear ones. So, instead of a unique solution, several stationary (time-independent) solutions often exist. Depending on its initial state, the system approaches one of them. A change of the *(control) parameters* which characterize the external conditions of the system, however, can lead to an *instability* of the stationary solution. As soon as a certain *critical point* is exceeded, unexpectedly we will have an entirely different system behaviour than before, e.g. another stationary state, stable *oscillations* (e.g. *limit cycles*), or even *chaos* [106] (cf. Sect. 10.3). The sudden change of the system behaviour is indicated by *critical fluctuations* (large state oscillations) and a *critical slowing down* (a very slow approach of the stable solution).

The term 'chaos' denotes a very complex dynamic behaviour which is extremely sensitive to the initial state. If the latter is varied only inconceivably little, the system may evolve completely different in the course of time *('butterfly effect')*. An example for chaos regarding opinion formation will be introduced in Sect. 10.3.4 [131]. Chaotic behaviour is probably the cause of the previously mentioned *sensitivity* and *unpredictability* of complex social processes.

1.2.2.3 Self-Organization and Emergence

When passing through *critical points* (points of instability) by changing control parameters, the system behaviour changes drastically. In this context, one often speaks of *phase transitions, self-organization processes,* or the *emergence* of new system properties. The term 'self-organization' is justified since the changed system behaviour is not determined in detail by the environmental influences (e.g. the *boundary conditions*) but bases on the autonomous dynamics of the system. (Cf. [106, 107]).

In Sect. 11.4, for example, it can be found that the opinion distribution in a subpopulation will develop completely differently in time if the control parameter 'indifference with respect to the held position' exceeds a certain critical value. Whereas below this value one sticks to the personally preferred position, above the critical value it will be abandoned in favour of new positions.

A typical phase transition in physics is the transition from gaseous to liquid state when the control parameter 'temperature' is reduced below the condensation point. Partially related with the formation of liquid drops is the grouping of individuals. Whereas, due to the respective individual behaviour, the internal structure of a group is subject to fluctuations, to the outside a group behaves as a unity. So it can be described by its own, *'macroscopic' (collective) variables* which change at a considerably slower rate than the individual behaviours.

If the interest is focused on the interactions of groups instead of on the interactions among individuals, one can find a similar phenomenon—the formation of organizations which change even slower than groups do. Finally, the organizations found a society.

Obviously, in this case the non-linear dynamics gives rise to a formation of different *levels* which are *hierarchically* ordered one below the other. A similar situation can be found in physics with regard to elementary particles. These form nuclei which, with electrons, combine to atoms. And the atoms form chemical molecules. These organize themselves, for instance, as solids which altogether yield celestial bodies. In biology comparable hierarchies can be observed.

On each lowest level we find the strongest interactions. This is obviously the reason for the circumstance that here the state changes proceed in the fastest way. If these interactions are attractive *bonds* will arise. These causes the elements to behave no longer completely individually but to form *units* (systems) which represent the elements *(subsystems)* of the next level. Since the attractive interactions are more or less 'saturated' by the bonds, the interactions *within* these units are stronger than the interactions *between* them *(boundary formation)*. The relatively weak *residual interactions* between the formed units induce their relatively slow state changes.

Consequently, a general interdependence between the interaction strength, the rate of state changes, and the formation of hierarchical levels can be found.

In this section the connections with *general system theory* [22, 33, 42, 176, 214, 232] came to light especially clearly. With the quantitative approaches discussed above we can expect that a mathematical formulation of system theoretical concepts is possible even on a much larger scale.

Chapter 2
Dynamic Decision Behavior

2.1 Introduction

Decision theory is a central field in the socio-economic sciences, as decisions determine a major part of human interactions. Therefore, decision theory is essential for the deductive derivation and microscopic understanding of the macroscopic phenomena observed in society and economics, such as

- social exchange or economic markets,
- the formation of groups, companies, institutions, or settlements,
- the dynamics of stock markets, business cycles, and other instability phenomena.

One may hope that, once the *elementary interactions* among individuals are understood by means of experimental and numerical studies, all regular phenomena should in principle be derivable from these interaction laws. This optimistic vision is motivated by the great success in the derivation of the structure, dynamic behavior, and properties of matter from elementary physical interactions. Scientists would like to understand the spatio-temporal patterns in socio-economic systems in a similar way.

This paper is an attempt to develop a consistent theoretical approach to human decision behavior (certainly an incomplete one). In Sects. 2.2.2, 2.2.3, 2.2.4 and 2.2.5, we will discuss how decisions come about, why they are so time-consuming, and what happens, if there is not enough time to complete the decision-making process. We also touch the topic of the freedom of will. Based on this, we will develop a quantitative theory for the probability of decision changes. This will take into account situations of incomplete information and limited processing capacities, thereby generalizing the concept of *homo economicus*. As a consequence, our theory implies a transitive preference scale only in special cases. In Sects. 2.2.9,

---

This chapter reprints parts of a previous publication with kind permission of the copyright owner, Springer Publishers. It is requested to cite this work as follows: D. Helbing, Dynamic decision behavior and optimal guidance through information services: Models and experiments. Pages 47–95 in: M. Schreckenberg and R. Selten (eds.) *Human Behaviour and Traffic Networks* (Springer, Berlin, 2004).

DOI 10.1007/978-3-642-11546-2_2,

2.2.10, 2.3.1, 2.3.2, 2.3.3, 2.3.4, 2.3.5 and 2.3.6, we also discuss the effects of non-linearities due to individual pair interactions. In this way, we can understand fashion cycles and chaotic decision dynamics, the self-organization of behavioral conventions, polarization phenomena, and transitions from individual to mass behavior. These phenomena can be understood by means of game-dynamical equations, which are a special case of the derived Boltzmann equations for decision changes. Other special cases are the logistic equation, the gravity model, or social force models.

Our evaluation results of a generalized day-to-day route choice experiment show (when we average over the behavior of all test persons) a non-transitive behavior, because the empirical decision probability is not monotonically increasing with the payoff or expected payoff gain. This can be incorporated into the previously developed decision theories, if learning behavior is taken into account.

It turns out that the decision dynamics is volatile and related with considerable losses in the average payoffs. We find, however, that already small differences in the way of information presentation can reach surprisingly large improvements. By far the best performance in terms of average and individual payoffs can be reached by user-specific recommendations. Taking into account the empirical compliance rates of the individuals, it is possible to solve the problem of traffic forecasts which are in harmony with the driver reactions to them. These findings are of general importance for information service providers and for the efficient distribution of scarce resources such as road capacities, time, space, money, energy, goods, or our natural environment. Nevertheless, the proposed method does not facilitate to manipulate the individuals by biased recommendations, as the compliance rate goes down accordingly: Our test persons followed the recommendations just to the degree they were useful for them to reach the user equilibrium.

2.2 Modelling Dynamic Decision Behavior

The following sections are trying to shed some new light on several old questions in the field of decision theory. Nowadays, there exists a variety of different models for the description of particular aspects of decision behavior, which have been developed by psychologists, social scientists, economists, and behavioral biologists in parallel. However, until today there is no consensus about a general and unified description of decision behavior. Therefore, the particular challenge will be the formulation of a consistent approach, which allows us to cover many aspects as special cases of one single theory.

2.2.1 Questioning Transitive Decisions and Homo Economicus

Human decision making has been subject to scientific research for a long time. In the beginning, there was a considerable progress in the interpretation and quantitative description of decision making. It was, in fact, one of the few areas in the social sciences where mathematical laws were formulated with some success. We mention the postulate that decisions are *transitive*, implying that there is a one-dimensional preference scale. In other words: If we prefer decision A to B and B to C, we will favour A compared to C. This idea was further developed with the concept of the *homo economicus*, according to which individuals would behave perfectly rational based on immediate and complete information, i.e. one would always choose—in a deterministic and predictable way—the alternative with the highest utility or pay-off. Great economic theories are based on this concept, but it is more and more questioned:

- First of all, decisions are hardly predictable. This can have several reasons: *Deterministic chaos* (like *intermittency*), *incomplete information* (i.e., 'hidden variables'), or probabilistic factors such as *fluctuations*.
- Second of all, individuals have to struggle with *imperfect information* due to *finite memory* and *limited processing capabilities*.
- Apart from this, there are *delays in information gathering*, which can cause instabilities as in other systems with delayed reaction (see [1]).
- Based on behavioral studies, scientists have also recognized that *emotions* affect the outcome of decisions (cf. the studies in *behavioral economics* [2] and *behavioral finance* [3]). For example, the decision distribution regarding emotional issues tends to be *polarized*, i.e. *bi-modal* rather than Gaussian, in contrast to unemotional or boring issues.
- The *El Farol bar problem* [4] and the *minority game* [5–7] even show that, in certain situations, there exists no rational (optimal deterministic) strategy. If all individuals had perfect information and would do the same, everyone would lose.

The facts known today call for a new theoretical approach for *boundedly rational agents* [4], but the concept of the *homo economicus* is so wide-spread, that new approaches have hard times to win through. In the beginning, this concept was a very useful tool as it allowed scientists to carry out analytical calculations and to develop economic theories. For a certain time period, this actually justified the simplifications made. Nowadays, however, classical economists are afraid of a break-down of their theories, if they would permit a questioning of the underlying assumptions. In fact, nobody knows exactly which chapters of economy would have to be rewritten, to what extent, and how it would have to be done, if these assumptions were relaxed. However, today computer simulations can complement analytical calculations where the complexity of the model is too high for obtaining rigorous results. This approach has been enormously powerful in physics and other natural sciences. Of course, computational results must always be checked for consistency, plausibility, and tested against empirical data. This requires a particular experience in modelling, programming, and data analysis, which should be a substantial part of the training of young scientists.

2.2.2 Probabilistic Decision Theories

Empirical studies clearly support that decision behavior is rather probabilistic than deterministic. In many cases, the relative frequency $p(i)$ of usage of a strategy i was found to be proportional to the number $N(i) \geq 0$ of times it was successful (which implies a *trial and error* behavior, at least in the beginning). In mathematical terms, this *law of relative effect* [8–10] reads

$$p(i) = \frac{N(i)}{\sum_{i'} N(i')} . \tag{2.1}$$

Without loss of generality, for any parameter T we can introduce a function

$$U(i) = T \ln N(i) \tag{2.2}$$

such that

$$p(i) = \frac{\mathrm{e}^{U(i)/T}}{\sum_{i'} \mathrm{e}^{U(i')/T}} . \tag{2.3}$$

The function $U(i)$ is often called *utility function*. It reflects some preference scale and the roughly logarithmic scaling of sensory stimuli, known as *Weber's* or *Fechner's law* [11]. The relation (2.3) is called the *multinomial logit model* [12, 13]. It is perhaps the most prominent example of probabilistic decision models, but there are several other ones [13, 14].

The *multinomial logit model* can be also derived in different ways. Notably enough, it resembles the *canonical distribution* of energy levels $E(i) = -k_B U(i)$ in physics [15, 16], which can be obtained by *entropy maximization* under the constraint that the average energy $k_B T$ is given [17–19]. The parameter k_B is the Boltzmann constant, and T has the meaning of the *temperature*. Therefore, the parameter T in formula (2.3) is sometimes called the *'social temperature'* (a more precise interpretation of which is given later on). The parameter T determines the sensitivity to variations of $U(i)$, specifically the sensitivity of the decision behavior on the variation of the utility. High values of T imply uniformly distributed decisions (an equi-distribution), while the limiting case $T = 0$ means that only the alternative(s) with the highest utility $U(i)$ is (are) chosen. (In non-degenerate cases, this corresponds to deterministic decision behavior.)

In the classical derivation of the multinomial logit model [12], T is a measure for the *uncertainty of information*. This derivation assumes that, due to limited information, $U(j)$ would only reflect the known part of the utility, while $\varepsilon(j)$ describes the unknown, stochastically varying part:

$$V(j) = U(j) + \varepsilon(j) . \tag{2.4}$$

The fluctuating part ε shall be extreme-value distributed, so that the maximum of two extreme-value-distributed variables is again extreme-value distributed. (One could say the extreme value distribution is the *'natural'* distribution for extreme value problems in the same way as the Gaussian distribution is the 'normal' distribution for sums of variables.) The parameter T is directly related to the variance of the extreme value distribution (which is sometimes also called Gumbel, Weibull, or Gnedenko distribution [18, 19]). If individuals choose the alternative j with the highest total utility $V(i) > V(j)$ for all $j \neq i$, the probability of selecting alternative i is again given by formula (2.3). Therefore, probabilistic decision behavior can be interpreted as effect of incomplete or uncertain information.

In the following, we will introduce several generalizations of the above multinomial logit model. By

$$p_a(i,t) = \frac{e^{U_a(i,t)/T_a(t)}}{\sum_{i'} e^{U_a(i',t)/T_a(t)}}, \qquad (2.5)$$

we take into account a possible dependence of T and $U(i)$ on the time t. Moreover, we distinguish different homogeneous subgroups a reflecting different personalities, character traits, or social backgrounds. For members of the same subgroup, the parameters T_a and utilities $U_a(i)$ are assumed to be approximately the same, while there are usually significant differences between subgroups. These originate partly from the fact that the utility $U(i)$ is composed of two parts $S_a(i)$ and $R_a(i)$ [18, 19]:

$$U_a(i,t) = S_a(i,t) + R(i,t)\,. \qquad (2.6)$$

$S_a(i,t)$ reflects the *personal preferences* or the satisfaction resulting from decision i, while $R(i,t)$ describes the *social reinforcement*, i.e. the social support or punishment an individual must expect as a consequence of decision i. It is known that individuals show a tendency to increase the *consistency* between their attitudes, behaviours, and social environment [20–24]. Therefore, three different ways of maximizing the utility are observed:

- The individual can decide for a behavior i' with $S_a(i',t) > S_a(i,t)$ instead of for behavior i.
- If, due to *social pressure* $R(i,t) < 0$, an individual takes a decision i' which does not agree with his/her attitudes, this will eventually change the assessment $S_a(i',t)$ of alternative i'. This phenomenon is known from psychology as *dissonance reduction* [24, 25]. (By the way, an attitude change does not occur in the case of a sufficiently high social reward for a behavior i' that is in disagreement with his/her attitudes!)
- The individual can also look for a social environment which has a positive attitude towards decision i: "Birds of a feather flock together."

Another generalization will be the application of the above multi-nomial logit model to *decision changes* from the present alternative i to a new one i'. That is, we will introduce the *conditional* or *transition probability*

$$p_a(i'|i,t) = \frac{\mathrm{e}^{U_a(i'|i,t)/T_a(t)}}{\sum_{i''} \mathrm{e}^{U_a(i''|i,t)/T_a(t)}}, \tag{2.7}$$

for an individual of group a to select alternative i' after i, and express it in terms of a *relative* or *conditional utility function* $U_a(i'|i,t)$ *given* an individual of group a presently pursues strategy i. The idea is that individuals try to improve their situation compared to the present one.

Let us now decompose this relative utility into a symmetric part

$$S_a(i'|i,t) = \frac{U_a(i'|i,t) + U_a(i|i',t)}{2} = S_a(i|i',t) \tag{2.8}$$

and an antisymmetric part

$$A_a(i'|i,t) = \frac{U_a(i'|i,t) - U_a(i|i',t)}{2} = -A_a(i|i',t). \tag{2.9}$$

We can, then, write

$$p_a(i'|i,t) = \frac{\mathrm{e}^{A_a(i'|i,t)/T_a(t)}}{\mathrm{e}^{-S_a(i'|i,t)/T_a(t)}} \Bigg/ \sum_{i''} \frac{\mathrm{e}^{A_a(i''|i,t)/T_a(t)}}{\mathrm{e}^{-S_a(i''|i,t)/T_a(t)}}. \tag{2.10}$$

Herein, the contribution

$$D_a(i,i',t) = \mathrm{e}^{-S_a(i'|i,t)/T_a(t)} = D_a(i',i,t) \tag{2.11}$$

can be interpreted as *effective distance* or *dissimilarity*, reflecting *transaction costs* $S_a(i'|i,t)$. Since the formula (2.10) still contains as many parameters as the conditional transition probability $p_a(i'|i,t)$, it is just another representation, but not yet a model. Possible approaches to reduce the number of parameters are, for example, the assumption of time-independent symmetric transaction costs

$$S_a(i,i',t) = S_a(i,i') = S_a(i',i) \tag{2.12}$$

(which implies time-independent effective distances for a constant parameter $T_a(t) = T_a$) and/or

$$A_a(i'|i,t) = U_a(i',t) - U_a(i,t). \tag{2.13}$$

That is, the asymmetrical part $A_a(i'|i,t)$ would describe a *utility gain*. The resulting formula for the transition probability is

$$p_a(i'|i,t) = \frac{e^{[U_a(i',t)-U'_a(i,t)]/T_a(t)}}{D_a(i,i',t)} = \underbrace{\frac{1}{D_a(i,i',t)}}_{\text{Distance factor}} \underbrace{e^{U_a(i',t)/T_a(t)}}_{\text{Pull factor}} \underbrace{e^{-U'_a(i,t)/T_a(t)}}_{\text{Push factor}} \tag{2.14}$$

with the effective distance

$$D_a(i,i',t) = e^{-S_a(i,i')/T_a(t)} \tag{2.15}$$

and the assessment

$$\begin{aligned} U'_a(i,t) &= U_a(i,t) + T_a(t) \ln \sum_{i''} \frac{e^{[U_a(i'',t)-U_a(i,t)]/T_a(t)}}{D_a(i,i'',t)} \\ &= T_a(t) \ln \sum_{i''} \frac{e^{U_a(i'',t)/T_a(t)}}{e^{-S_a(i,i'')/T_a(t)}} . \end{aligned} \tag{2.16}$$

According to Eq. (2.14), the transition probability decreases with the dissimilarity $D_a(i,i',t)$ of two alternatives, but it increases with the *pull factor* (a high utility $U_a(i',t)$ of the new alternative i') and with the *push factor* (a low assessment $U'_a(i,t)$ of the previously chosen alternative i). (For a discussion of the so-called *shadow costs* related to the difference between $U_a(i',t)$ and $U'_a(i,t)$ see [26].) In order to have uniquely defined utilities, one normally sets the average utility equal to zero or, equivalently,

$$\sum_{i'} U_a(i,t) = 0 . \tag{2.17}$$

Note that formula (2.5) is a special case of Eq. (2.14), resulting for identical effective distances $D_a(i,i',t) = D_a(t)$. The main advantage of the more general approach of the conditional decision probabilities is that they take into account transaction costs or the effective distance between two alternatives. We will see later on that this is quite important for certain applications (see Sect. 2.3).

2.2.3 Are Decisions Phase Transitions?

Decisions are *discontinuous changes* of the behavior after a period of *critical fluctuations* (repeatedly changing one's mind) and of *critical slowing down* (hesitation to take the decision). This suggests that decisions are phase transitions [18, 19].

In the following, we will develop a more detailed picture of the decision-making process based on experimental observations [27, 28] (for more details see Chap. 7,

[18, 19]). According to Feger [27, 28] decisions are *conflict situations* occurring when we can choose between several mutually exclusive *behavioral alternatives* i. At the beginning of such a conflict we estimate its *importance.* This is decisive for the time spent on decision-making.

Assume we are confronted with a complex and new situation that requires to take a decision rather than just a reflexive or automatic reaction to a standard situation (as in car driving or avoidance behavior of pedestrians). Then, a detailed consideration of the pros and cons of the different available alternatives is necessary. A large number of brain variables are involved into this consideration process, and there is some experimental evidence that they (may) show a chaotic dynamics [29, 30]. In this way, we are normally exploring a *multi-dimensional assessment space* [18, 19, 31] rather than a one-dimensional scale of options. Some areas of this assessment space are in favour of one decision, others in favour of another decision (while some may be neutral or irrelevant for the specific decision). These areas could be interpreted as *basins of attraction* of the different decision alternatives. They do not need to be connected areas, but may be *fractal sets* as well. When moving through this assessment space during the consideration phase, the 'relative sizes' (i.e. the measures) of these areas determine the occurrence probabilities of pros for the different alternatives. These can be ranked and thereby allow us to define a one-dimensional preference scale, although *the decision-making process is clearly a probabilistic process and normally not consistent with transitivity relations* (remember the multi-dimensional and possibly *fractal* assessment space). Therefore, transitivity only applies to particular (probably simple) decisions.

To be more specific, assume that the brain variables involved into the consideration process produce a random series $l_0, l_1, l_2, \dots$ of (consciously) imaginated, *anticipated consequences* of the possible decisions, and let l_k be in favour of decision $i = f(l_k)$. There is experimental evidence [27, 28] that a decision is taken if K *consecutive arguments* $l_{k'+1}, \dots, l_{k'+K}$ are in favour of the *same* decision $i = f(l_{k'+1}) = \dots = f(l_{k'+K})$. Note that the number K depends on the available *decision time* and the *importance of the decision,* which may be group-specific and time-dependent. Therefore, we replace K by $K_a(t)$ in the following. Finally note that a comparison of the consequences of alternative decisions continues even *after* a decision was made. This is experienced as *doubts* about the rightness of the decision.

2.2.4 Fast and Slow Decisions

Let $p'_a(i'|i,t)$ be the occurrence probability of pros for alternative i', if the individuum has previously chosen alternative i. We may then define *preferences* $U'_a(i'|i,t) = \ln p'_a(i'|i,t)$ such that

$$p'_a(i'|i,t) = e^{U'_a(i'|i,t)} \geq 0\,. \tag{2.18}$$

According to the above, the probability $p_a(i'|i,t)$ of deciding for alternative i' is equal to the probability of $K_a(t)$ successive favourable arguments for i', i.e.

$$p_a(i'|i,t) = \mathrm{e}^{U''_a(i'|i,t)} = [p'_a(i'|i,t)]^{K_a(t)} = \mathrm{e}^{K_a(t)U'_a(i'|i,t)} \tag{2.19}$$

with

$$U''_a(i'|i,t) = \ln p_a(i'|i,t) = K_a(t)U'_a(i'|i,t)\,. \tag{2.20}$$

Note that, in contrast to non-exponential approaches, the expression (2.19) has a invariant functional form (namely, an exponential one), which is independent of the specific value of $K_a(t)$. The requirement of having several pros before a decision is made does not only reduce the risk of accidentally choosing an alternative with small preference $U'_a(i'|i,t)$. It also *magnifies* the differences in the utilities of different alternatives i' and i'' [18, 19] because of

$$U''_a(i'|i,t) - U''_a(i''|i,t) = K_a(t)[U'_a(i'|i,t) - U'_a(i''|i,t)]\,. \tag{2.21}$$

Consequently, when $K_a(t)$ is reduced, there is a higher likelihood to decide for an alternative for which we have a lower preference. To avoid this, $K_a(t)$ is larger when the decision is important, as stated above. However, if the number of alternatives is large, the decision-making process takes particularly long (which is known as the *pain of choice*). Therefore, decision-making processes can not always be completed, especially if there is a deadline or a pressure of time. In such cases, the value of $K_a(t)$ is reduced, resulting in a different decision distribution. Because of

$$K_a(t)U'_a(i'|i,t) = U_a(i'|i,t)/T_a(t) - \ln \sum_{i''} \mathrm{e}^{U_a(i''|i,t)/T_a(t)}\,, \tag{2.22}$$

$T_a(t)$ *is basically proportional to* $1/K_a(t)$. That is, the decision distribution becomes more and more similar to a uniform distribution (equi-distribution), when the time spent on the decision is reduced. In principle, this comes close to tossing a coin, which is exactly what people tend to do when they do not have enough time to take a well-founded decision. It would, therefore, be interesting to investigate the quality of decision-making by managers, as their time budget per decision tends to be short. On the other hand, experience plays a role, as a small uncertainty $T_a(t)$ of information can compensate for a shortage of time.

2.2.5 Complete and Incomplete Decisions

There are other situations of *incomplete decision-making*, where individuals would or actually do run out of time. In such cases, it appears that a spontaneous decision is taken 'out of the stomach'. This could be the alternative which got the highest relative weight in the previous, unfinished consideration process, but it could also orient at the decisions of others based on the respective levels of sympathy or trust. We should underline three points here:

- The brain executes geometric averaging, which corresponds to an arithmetic average of the logarithms of $p'_a(i'|i,t)$ [11], i.e. of the utilities $U'_a(i'|i,t)$, so that the formula for the resulting decision probability may look similar to (2.19). The logarithmic scaling of physical stimuli, by the way, relates the decision probability to the power-law *Cobb-Douglas function* [32, 33], as is shown in [18, 19].
- *Emotionally loaded arguments* (re-)occur more frequently than others. Therefore, they have a higher impact on incomplete decision-making processes than on complete ones.
- When there is not enough time to complete a decision, imitative or avoidance behavior (see Sect. 2.2.10) play an important role.

Empirical research should focus on the investigation of incomplete decisions, as they are quite common in our everyday life.

2.2.6 The Red-Bus-Blue-Bus Problem

We have seen that the exponential approach is favourable for the derivation of the multi-nomial logit model and its properties. Modified approaches have been mainly developed because of the so-called *red-bus-blue-bus problem* [13]. It occurs when the decision alternatives are not completely exclusive, for example, when alternative busses have different colors or when alternative routes share common parts. In such cases, certain areas of the assessment space are in favour of *several* (say, j) of the alternatives. These areas are equally shared among the alternatives (i.e., divided by j), thereby reducing the resulting decision probabilities and the related preferences. We may take this into account by means of weight factors w_i (with $0 \le w_i \le 1$). If w_i^j denotes the part of the characteristics (favourable assessment space areas) that alternative i shares with $j-1$ of the I alternatives, the correct formula should be

$$w_i = \sum_{j=1}^{I} \frac{w_i^j}{j} \qquad \text{with} \qquad \sum_{j=1}^{I} w_i^j = 1\,. \tag{2.23}$$

However, there is no general and simple formula to determine w_i^j (see [34] for a related treatment of route choice behavior). The corresponding formula for the (conditional) decision probability reads

$$p_a(i'|i,t) = \frac{w_{i'} e^{U_a(i',t)/T_a(t)} / D_a(i,i',t)}{\sum_{i''} w_{i''} e^{U_a(i'',t)/T_a(t)} / D_a(i,i'',t)}\,. \tag{2.24}$$

Note that the weights $w_{i'}$ could alternatively be taken into account in the calculation of $U_a(i',t)$.

2.2.7 The Freedom of Decision-Making

A subject of particular interest in connection with decision conflicts is what we denote as the *freedom of decision-making*. One important precondition for the freedom of decision-making is a not fully externally determined outcome of decisions, i.e. it should not be predictable in a reliable way. This precondition is certainly fulfilled, although we know that certain decisions are more likely than others. The above introduced, probabilistic decision model is consistent with this. According to it, the respective decision is a result of the internal dynamics of the brain variables, which is to certain degree stochastic or chaotic. Whether this result is voluntary depends on whether the brain dynamics can be mentally controlled in a more or less arbitrary way. Recent measurements of neural activity seem to indicate that a decision is made (for example, body motion is initiated) before the conscious feeling of a free decision arises [35, 36]. This feeling could, therefore, be an interpretation or *rationalization* of our emergent behavior.

According to psychological investigations the subjectively felt freedom of decision-making increases with the *uncertainty* with respect to the final result of a decision [37, 38]. That is, the freedom of decision-making is an *entropy-like* quantity. It is greater the larger the number of alternatives is and the more equivalent they are (with regard to the related preferences). A restriction of the freedom of decision-making gives rise to *reactance* (i.e. to a kind of a *defiant reaction*). Either the limitation of freedom will be evaded if possible, or *resistance* is formed [25, 39–41].

2.2.8 Master Equation Description of Dynamic Decision Behavior

We will now discuss a stochastic description of dynamic decision behavior along the lines suggested by Weidlich [42–44]. Let us assume we have N_a individuals in group a, and a considered population of altogether

$$N = \sum_a N_a \tag{2.25}$$

persons. The so-called *occupation number* $n_i^a(t)$ shall denote how many individuals of group a pursue alternative i at time t, i.e. we have

$$\sum_i n_i^a(t) = N_a \,. \tag{2.26}$$

The *socio-configuration*

$$n = (n_1^1, n_2^1, \dots, n_i^1, \dots, n_1^2, \dots, n_i^2 \dots) \tag{2.27}$$

does, then, comprise the distribution of all N individuals over the different groups a and states i. We will represent the probability of finding the socio-configuration n at time t by $P(n,t)$. This probability is reduced by transitions to other configurations n', whose frequencies are proportional to $P(n,t)$. The proportionality factor is the *conditional probability* or *(configurational) transition probability* $P(n',t+\Delta t|n,t)$ of finding the configuration n' at time $(t+\Delta t)$, given that we have the configuration n at time t. Conversely, the probability $P(n,t)$ increases by transitions from configurations n' to n, which are proportional to the occurrence probabilities $P(n',t)$ of the socio-configurations n' and to the transition probabilities $P(n,t+\Delta t|n',t)$. The resulting balance equation governing the dynamics of the above specified *Markov chain* reads

$$P(n,t+\Delta t)-P(n,t)=\sum_{n'}P(n,t+\Delta t|n',t)P(n',t)-\sum_{n'}P(n',t+\Delta t|n,t)P(n,t) \tag{2.28}$$

or, considering the normalization

$$\sum_{n'}P(n',t+\Delta t|n,t)=1 \tag{2.29}$$

of the transition probabilities,

$$P(n,t+\Delta t)=\sum_{n'}P(n,t+\Delta t|n',t)P(n',t)\,. \tag{2.30}$$

In the continuous limit $\Delta t \to 0$, we obtain the so-called *master equation*

$$\frac{dP(n,t)}{dt}=\underbrace{\sum_{n'(\neq n)}W(n|n',t)P(n',t)}_{\text{Inflow into } n}-\underbrace{\sum_{n'(\neq n)}W(n'|n,t)P(n,t)}_{\text{Outflow from } n}\,, \tag{2.31}$$

where we have introduced the *(configurational) transition rates*

$$W(n|n',t)=\lim_{\Delta t\to 0}\frac{P(n,t+\Delta t|n',t)}{\Delta t} \qquad \text{for} \qquad n'\neq n. \tag{2.32}$$

Note that the *master equation* (2.31) assumes the *Markov property* according to which the conditional probabilities $P(n,t+\Delta t|n',t)$ depend on t and Δt only, but not on previous time steps. However, a *generalized master equation* for problems with *memory effects* exists, see [18, 19, 45]. It reads

$$\frac{dP(n,t)}{dt}=\int_{-\infty}^{t}dt'\sum_{n'(\neq n)}W_{t-t'}(n|n',t)P(n',t-t')-W_{t-t'}(n'|n,t)P(n,t-t') \tag{2.33}$$

with memory-dependent transition rates $W_{t-t'}(n'|n,t)P(n,t-t')$. For example, for an exponentially decaying memory with decay rate τ we could use the formula

$$W_{t-t'}(n'|n,t) = W(n'|n,t')\frac{1}{\tau}\exp\left(\frac{t-t'}{\tau}\right). \tag{2.34}$$

2.2.9 Mean Field Approach and Boltzmann Equation

It is often useful to consider the mean value equations for the expected values $\langle n_i^a \rangle = \sum_n n_i^a P(n,t)$, which are obtained by multiplying Eq. (2.31) with n_i^a, summing up over n, and suitably interchanging n and n':

$$\frac{d\langle n_i^a\rangle}{dt} = \sum_n n_i^a \frac{dP(n,t)}{dt} = \sum_n \sum_{n'} (n_i'^a - n_i^a) W(n'|n,t)P(n,t) = \langle m_i^a(n,t)\rangle . \tag{2.35}$$

Here, we have introduced the *first jump moments*

$$m_i^a(n,t) = \sum_{n'} (n_i'^a - n_i^a) W(n'|n,t) . \tag{2.36}$$

Let us now assume *spontaneous decisions* with transition rates $w_a(i'|i,t)$ from alternative i to i' by individuals of group a and, in addition, *pair interactions* between two individuals belonging to groups a and b, leading to a change from alternative i to i' by the a-individual and from j to j' by the b-individual with a transition rate of $w_{ab}(i',j'|i,j,t)$. Defining the resulting socio-configurations

$$n_{i\,i'}^{aa} = (n_1^1, n_2^1, \dots, n_1^a, \dots, n_{i-1}^a, n_i^a - 1, n_{i+1}^a, \dots, n_{i'-1}^a, n_{i'}^a + 1, n_{i'+1}^a, \dots) , \tag{2.37}$$

$$n_{i\,i'\,j\,j'}^{aa\,bb} = (n_1^1, \dots, n_1^a, \dots, n_i^a - 1, \dots, n_{i'}^a + 1, \dots, n_j^b - 1, \dots, n_{j'}^b + 1, \dots) , \tag{2.38}$$

the corresponding configurational transition rates are given by

$$W(n'|n,t) = \begin{cases} w_a(i'|i,t)n_i^a & \text{if } n' = n_{i\,i'}^{aa} , \\ w_{ab}(i',j'|i,j,t)n_i^a(n_j^b - \delta_{ij}^{ab}) & \text{if } n' = n_{i\,i'\,j\,j'}^{aa\,bb} , \\ 0 & \text{otherwise.} \end{cases} \tag{2.39}$$

Herein, $\delta_{ij}^{ab} = 1$ if $a = b$ and $i = j$ (to avoid self-interactions), but 0 otherwise. According to formula (2.39), the total rate of spontaneous transitions is proportional to the number n_i^a of individuals of group a who may change their previous decision i independently of each other, while the total rate of pair interactions is proportional to the number $n_i^a(n_j^b - \delta_{ij}^{ab})$ of possible interactions between a- and b-individuals pursuing alternatives i and j. Inserting Eq. (2.39) into (2.36) eventually leads to

$$m_i^a(n,t) = \sum_{i'} \Big[w_a(i|i',t) + \sum_b \sum_{j,j'} w_{ab}(i,j|i',j',t) n_{j'}^b \Big] n_{i'}^a$$
$$- \sum_{i'} \Big[w_a(i'|i,t) + \sum_b \sum_{j,j'} w_{ab}(i',j'|i,j,t) n_j^b \Big] n_i^a \,, \quad (2.40)$$

if δ_{ij}^{ab} is negligible (see, for example [18, 19]). The *mean field approach* assumes

$$\langle m_i^a(n,t) \rangle \approx m_i^a(\langle n \rangle, t) \,, \quad (2.41)$$

i.e., that the system dynamics is determined by the mean value $\langle n \rangle$, which is true for a sharply peaked, unimodal distribution $P(n,t)$. This leads to the generalized *Boltzmann equation*

$$\frac{dP_a(i,t)}{dt} = \sum_{i'} \Big[w_a(i|i',t) + \sum_b \sum_{j,j'} \tilde{w}_{ab}(i,j|i',j',t) P_b(j',t) \Big] P_a(i',t)$$
$$- \sum_{i'} \Big[w(i'|i,t) + \sum_b \sum_{j,j'} \tilde{w}_{ab}(i',j'|i,j,t) P_b(j,t) \Big] P_a(i,t) \,, \quad (2.42)$$

where we have introduced the *(expected) occurrence probabilities* $P_a(i,t) = \langle n_i^a \rangle / N_a$ of decisions i in group a and $\tilde{w}_{ab}(i',j'|i,j,t) = N_b w_{ab}(i',j'|i,j,t)$ [18, 19, 46, 47]. Note that this Boltzmann equation neglects the covariances

$$\sigma_{ij}^{ab}(t) = \big\langle (n_i^a - \langle n_i^a \rangle)(n_j^b - \langle n_j^b \rangle) \big\rangle = \langle n_i^a n_j^b \rangle - \langle n_i^a \rangle \langle n_j^b \rangle \quad (2.43)$$

and the corresponding correlations

$$r_{ij}^{ab}(t) = \sigma_{ij}^{ab}(t) \Big/ \sqrt{\sigma_{ii}^{aa}(t) \sigma_{jj}^{bb}(t)} \,. \quad (2.44)$$

For the derivation of corrected mean value equations (taking into account contributions by the covariances), see [18, 19, 48]. Without corrections, the above Boltzmann equation can be interpreted as the systematic component of a *Langevin equation* describing the *most probable* decision changes.

2.2.10 Specification of the Transition Rates of the Boltzmann Equation

In the previous section, we have derived an equation for the temporal change of the occurrence probabilities $P_a(i,t)$ of the decisions i in group a. Simplifying the above expressions, we can write

$$\frac{dP_a(i,t)}{dt} = \underbrace{\sum_{i'} w^a(i|i',t) P_a(i',t)}_{\text{Inflow into } i} - \underbrace{\sum_{i'} w^a(i'|i,t) P_a(i,t)}_{\text{Outflow from } i} \quad (2.45)$$

with the *(effective) transition rates*

$$w^a(i'|i,t) = w_a(i'|i,t) + \sum_b \sum_{j,j'} \tilde{w}_{ab}(i',j'|i,j,t) P_b(j,t) \,. \tag{2.46}$$

These have to be specified for social interactions, now. There is a detailed theory how to do this [18, 19, 46, 47, 49], but here we will only write down the finally resulting formula. Assume that $\nu^1_{ab}(t) = \nu_{ab}(t) r^1_{ab}(t)$ is the rate of *imitation processes* of an a-individual due to interactions with b-individuals, and $\nu^2_{ab}(t) = \nu_{ab}(t) r^2_{ab}(t)$ the analogous rate of *avoidance processes*. These rates are products of the *interaction rate* ν_{ab} of an a-individual with b-individuals, which depends on the *social interaction network* [50–52], and of the relative frequencies $r^1_{ab}(t)$ and $r^2_{ab}(t)$ of imitative and avoidance processes, respectively.

Now, let $p^a(i'|i,t)$ be the probability to change from alternative i to i' as discussed in Sect. 2.2.2. The effective transition rate has, then, the form

$$w^a(i'|i,t) = w_a(i'|i,t) + p^a(i'|i,t) \sum_b \left[\nu^1_{ab}(t) P_b(i',t) + \nu^2_{ab}(t) P_b(i,t) \right], \tag{2.47}$$

because the imitation rate is proportional to the occurrence probability $P_b(i',t)$ of the imitated decision i', and the avoidance rate is proportional to the occurrence probability $P_b(i,t)$ of the presently pursued alternative i [18, 19].
Finally, we can write

$$w_a(i'|i,t) = \nu^0_a(t) p_a(i'|i,t) \,, \tag{2.48}$$

where $\nu^0_a(t)$ denotes the rate of *spontaneous* decision changes. If the transition probabilities $p_a(i'|i,t)$ and $p^a(i'|i,t)$ of spontaneous and interactive decision changes are the same, the formula for the effective transition rate simplifies:

$$w^a(i'|i,t) = p_a(i'|i,t) \Big\{ \nu^0_a(t) + \sum_b \left[\nu^1_{ab}(t) P_b(i',t) + \nu^2_{ab}(t) P_b(i,t) \right] \Big\} \,. \tag{2.49}$$

It makes sense to specify $p_a(i'|i,t)$ in accordance with Eq. (2.14).

Imitation is a very common human behavior. One also speaks of *herding behavior* [53], *bandwaggon effect*, or *persuasion* [54]. Avoidance behavior is sometimes called *defiance* or *snobbish behavior.* It originates from the desire of humans to distinguish from people with different backgrounds. Note that *homo economicus* should not show avoidance or imitation behavior at all, but decide on a rational basis. Nevertheless, there are good reasons for this behavior. In many situations, we do not have enough time to collect and evaluate the information for a rational decision (see Sect. 2.2.5). (Just imagine we would really try to compare all contracts of insurance companies.) Therefore, we rely on the experience of others. The wide spreading of imitation and avoidance behavior is due to the great success of *learning by observation,* which has its roots in *evolution.* It allows us to avoid painful experiences and helps to learn faster. Therefore, we tend to imitate (successful) decisions of people who are in a similar situation as we are. Some indicator for similarity is sympathy, as we tend to like people whose background is comparable. In contrast, we may show avoidance behavior with respect to people we dislike, because we expect their decisions to be counter-productive. That is, emotions are helpful in cases where we cannot complete our decisions. Altogether, the combination of individual assessment with imitation and avoidance behavior may be viewed as *collective problem solving* [1, 55–57]. It allows us to cope with situations which one individual cannot handle on time due to the limited capacities of information collection and data processing.

2.3 Fields of Applications

The Boltzmann equation was originally developed for the description of particle collisions in gases, but the mathematically related description of social interactions has a wide range of applications. It turns out that many dynamical models that have been proposed, used, and tested in the social sciences, are special cases of the above generalized Boltzmann equation.

2.3.1 *The Logistic Equation*

Imagine a situation with one group $a = 1$ and $I = 2$ alternatives, where only spontaneous and imitative decision changes play a role. Then, we obtain an equation of the form

$$\frac{dP_1(1,t)}{dt} = C_0 + C_1 P_1(1,t) + C_2 P_1(1,t)^2 \tag{2.50}$$

with constants C_0, C_1, and C_2 given by the transition rates $w(...)$ [18, 19, 49]. Introducing the scaling $z(t) = P_1(1,t) - C$ with $C = (-C_1 - \sqrt{(C_1)^2 - 4C_0C_2})/(2C_2)$, the (initial) growth rate $r = C_1 - 2C_2C$, and the capacity $z_0 = r/C_2$, one arrives at the *logistic equation* [58, 59]

$$\frac{dz(t)}{dt} = rz(t)\big[1 - z(t)/z_0\big], \tag{2.51}$$

which describes many kinds of *limited growth processes* [60–63].

2.3.2 *The Generalized Gravity Model and Its Application to Migration*

A quite successful model to estimate the *origin-destination matrices* describing flows of goods, persons, cars, etc. between locations i and i' is the *generalized gravity model* [18, 19, 49]

$$w^a(i'|i,t)P_a(i,t) = \sum_b \nu^1_{ab}(t) \mathrm{e}^{[U_a(i',t) - U'_a(i,t)]/T_a} \frac{P_b(i',t)P_a(i,t)}{D_a(i,i')}. \tag{2.52}$$

In the case of one single population $a = 1$, we have

$$w^1(i'|i,t)P_1(i,t) \propto \frac{P_1(i',t)P_1(i,t)}{D_1(i,i')}, \tag{2.53}$$

which looks similar to the law of gravitation and explains the name of the model [64, 65]. It reflects that, for example, the person flow from place i to i' is proportional to the number of people living at location i (who can travel) and proportional to the number of people in the destination town they may meet, but the number of trips goes down with the effective distance.

Note that it was very successful to apply the above model to migration between different regions [18, 19, 49, 66, 67]. The fitted utilities even mirrored political events such as the construction of the Berlin wall. Despite of a data reduction by 87.2% corresponding to only 1.28 data values per year, the correlation with the migration data was very high, namely $r = 0.985$ [18, 19, 49].

2.3.3 Social Force Models and Opinion Formation

In this section, we will assume a continuous and m-dimensional decision space. For this reason, we will replace i by x, i' by x', and sums $\sum_{i'}$ by integrals $\int d^m x'$. Moreover, we require that decision changes mostly occur in small steps (i.e. $w^a(x'|x,t) \approx 0$ if $\|x' - x\|$ is large). Then, it is possible to derive a *Boltzmann-Fokker-Planck* equation by second order Taylor approximation of the above Boltzmann equation. This equation is equivalent to a certain *stochastic differential equation* or *Langevin equation* describing the decision changes of the single individuals α belonging to group a [18, 19, 49, 68]. It reads:

$$\frac{dx_\alpha(t)}{dt} = f_a(x_\alpha, t) + \textit{individual fluctuations}. \tag{2.54}$$

Herein, the vector

$$f_a(x,t) = \int d^m x' \, (x' - x) \, w^a(x'|x,t) \tag{2.55}$$

has the interpretation of a (non-Newtonian) *social force* [69], which determines the size and direction of the systematic part of decision changes [18, 19, 49, 68, 70]. Note that this social force does not only affect the individual behavior, but also changes with the decision distributions $P_b(x,t)$. For the effective transition rates (2.47), for example, we get

$$\begin{aligned} f_a(x,t) = \int d^m x' \, (x' - x) \Big\{ & w_a(x'|x,t) \\ & + p^a(x'|x,t) \sum_b \big[\nu^1_{ab}(t) P_b(x',t) + \nu^2_{ab}(t) P_b(x,t) \big] \Big\} \end{aligned} \tag{2.56}$$

The distributions can be expressed in terms of the individual decisions via

$$P_b(x,t) = \frac{1}{N_b}\sum_{\beta\in b}\delta\big(x - x_\beta(t)\big), \tag{2.57}$$

where $\delta(x - x_\beta)$ denotes a multivariate Gaussian distribution around x_β with a small variance. Considering this allows us to decompose the social force into components due to spontaneous decision changes and due to pair interactions [18]:

$$f_a(x_\alpha,t) = \underbrace{f_a^0(x_\alpha,t)}_{\text{Spont. force}} + \underbrace{\sum_\beta \big[f_{ab}^1(x_\alpha,x_\beta,t) + f_{ab}^2(x_\alpha,x_\beta,t)\big]}_{\text{Pair interaction forces}} \tag{2.58}$$

with the *spontaneous force*

$$f_a^0(x_\alpha,t) = \int d^m x'\,\big(x' - x_\alpha(t)\big)\,w_a\big(x'\big|x_\alpha(t),t\big), \tag{2.59}$$

the *imitation force*

$$\begin{aligned} f_{ab}^1(x_\alpha,x_\beta,t) &= \int d^m x'\,\big(x' - x_\alpha(t)\big)\,p^a\big(x'\big|x_\alpha(t),t\big)\frac{\nu_{ab}^1(t)}{N_b}\delta\big(x' - x_\beta(t)\big) \\ &= \big(x_\beta(t) - x_\alpha(t)\big)\,p^a\big(x_\beta(t)\big|x_\alpha(t),t\big)\frac{\nu_{ab}^1(t)}{N_b}, \end{aligned} \tag{2.60}$$

and the *avoidance force*

$$f_{ab}^2(x_\alpha,x_\beta,t) = \int d^m x'\,\big(x' - x_\alpha(t)\big)\,p^a\big(x'\big|x_\alpha(t),t\big)\frac{\nu_{ab}^2(t)}{N_b}\delta\big(x_\alpha(t) - x_\beta(t)\big). \tag{2.61}$$

These expressions can be further evaluated, if $w_a(x'|x,t)$ and $p^a(x'|x,t)$ are specified. We also point out that the above social force model shares some common features with the *social impact theory* [52, 71–74].

Social force models have been very successful in applications to vehicle traffic [75–77] and pedestrian flows [53, 70, 78, 79]. Here, we will discuss an application to opinion formation. Let us assume two groups a of people distributed over a *one-dimensional opinion scale* between two extreme positions regarding a certain issue. The utilities and transaction costs determining the conditional decision probabilities $p_a(x'|x,t)$ are specified as follows [18, 19, 49, 68]:

$$U_a(x,t) \propto -\left(\frac{x - x_a}{L_a}\right)^2 \quad \text{and} \quad S_a(x,x') \propto \frac{|x' - x|}{R}. \tag{2.62}$$

x_a has the meaning of the preferred position in group a, L_a is the tolerance of deviations from this position, and R the range of interaction. In the discussed examples, we neglect spontaneous transitions by setting $\nu_a^0(t) = 0$ and investigate either imitative or avoidance interactions in accordance with the interaction rates

$$\left(\nu_{ab}^1(t)\right) = \begin{pmatrix} \nu & \nu \\ \nu & \nu \end{pmatrix} \qquad \text{and} \qquad \nu_{ab}^2(t) = 0 \tag{2.63}$$

or

$$\nu_{ab}^1(t) = 0 \qquad \text{and} \qquad \left(\nu_{ab}^2(t)\right) = \begin{pmatrix} 0 & \nu \\ \nu & 0 \end{pmatrix}. \tag{2.64}$$

A selection of numerical results is presented in Figs. 2.1, 2.2 and 2.3. It is particularly interesting that, in the simulation of imitative behavior, we find a *phase transition* to *compromising behavior,* when the tolerance L_a is sufficiently large. For a more detailed discussion see [18, 19, 49, 68].

2.3.4 The Game-Dynamical Equations

The *game-dynamical equations*

$$\begin{aligned} \frac{dP_1(i,t)}{dt} &= \underbrace{\nu(t)P_1(i,t)\big[F(i,t) - \sum_{i'} F(i',t)P_1(i',t)\big]}_{\text{Selection}} \\ &\quad + \underbrace{\sum_{i'} \big[w_1(i|i')P_1(i',t) - w_1(i'|i)P_1(i,t)\big]}_{\text{Mutation}} \end{aligned} \tag{2.65}$$

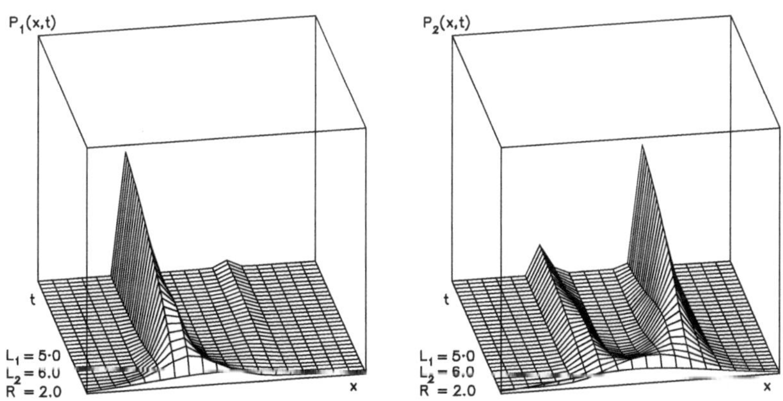

Fig. 2.1 Example of opinion formation of two groups with imitatively interacting individuals, when the tolerance L_a is small (from [18, 19, 49, 68, 80]). The opinion distributions $P_a(x,t)$ in both groups are bimodal. Most individuals decide for an opinion x close to their preferred opinion x_a, but some are convinced by the opinion preferred in the other group

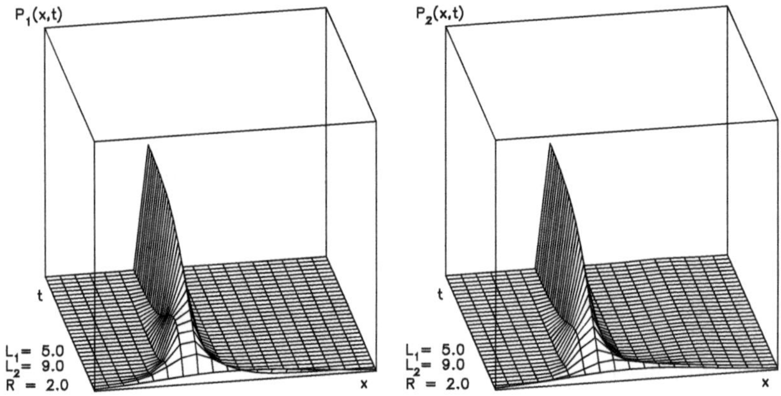

Fig. 2.2 As Fig. 2.1, but for a higher tolerance L_a (from [18, 19, 49, 68, 80]). The shape of the distributions $P_a(x, t)$ has now qualitatively changed from a bimodal to a unimodal form. This indicates a phase transition from imitative to compromising behavior at some critical value of tolerance. In both populations, the opinions are then distributed around a certain opinion x_0, which agrees in both groups, but differs from both, x_1 and x_2. Its location is closer to the preferred position x_1 in the group $a = 1$ with he smaller tolerance $L_1 < L_2$

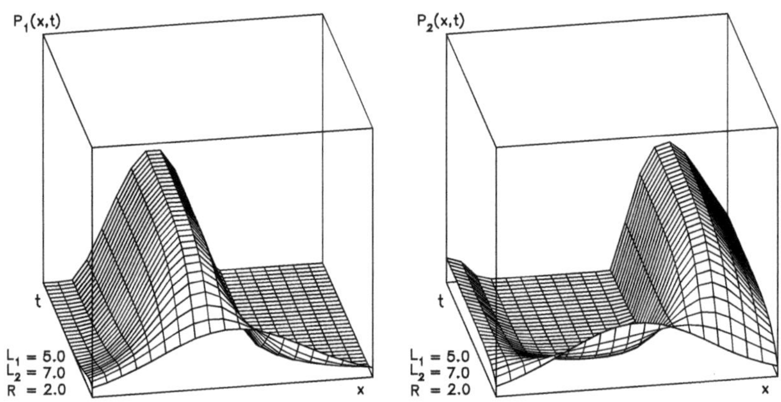

Fig. 2.3 Opinion distributions $P_a(x, t)$ for two groups a showing avoidance behavior, because the individuals in one group dislike the people of the respective other group (from [18, 19, 49, 68, 80]). As expected, there is almost no overlap between the opinion distributions $P_1(x, t)$ and $P_2(x, t)$ in the different groups. The tendency to avoid the opinions in the other group is so large that some people in group $a = 2$ even show opinions 'left' of the ones found in group $a = 1$, although these are far away from the preferred position x_2. This occurs due to their higher tolerance $L_2 > L_1$

for behavioral changes [81–84] are an adoption of *selection-mutation equations* originally developed in evolutionary biology [85, 86]. They describe the effects of spontaneous transitions (so-called *mutations*) and a *selection* of those strategies i whose *expected success* or so-called *fitness*

$$F(i, t) = \sum_{i''} P_{ii''} P_1(i'', t) \tag{2.66}$$

is higher than the average one, $\sum_{i'} F(i',t)P_1(i',t)$. Herein, $P_{ii''}$ denotes the *payoff* when strategy i is confronted with strategy i''.

The game-dynamical equations have been very successful in explaining observations in behavioral biology, sociology, and economics. However, in contrast to evolutionary biology, a 'microscopic' derivation based on individual interactions has been missing for a long time. This has been discovered in 1992 [48, 87]. Inserting $\nu_{11}^2(t) = 0$ and the expression

$$\nu_{11}^1(t)p^1(i'|i;t) = \nu(t)\max\big(F(i',t) - F(i,t), 0\big) \qquad (2.67)$$

into the Boltzmann equation (2.45) with the effective transition rates (2.47) exactly yields the game-dynamical equations because of

$$\max\big(F(i,t)-F(i',t),0\big)-\max\big(F(i',t)-F(i,t),0\big) = F(i,t)-F(i',t)\,. \qquad (2.68)$$

Formula (2.67) is nowadays called the *proportional imitation rule* [88], as it assumes that the transition probability $p^1(i'|i;t)$ is proportional to the expected gain $F(i',t) - F(i,t)$ in success, if this is positive, but zero otherwise. Note, however, that the game-dynamical equations can be also viewed as a first-order Taylor approximation of a Boltzmann-equation with an imitative transition probability of the form (2.47) with

$$D_1(i,i') = 2 \qquad \text{and} \qquad F(i,t) = \frac{U_1(i,t) + U_1'(i,t)}{2T_1(t)}\,. \qquad (2.69)$$

In the following, we will discuss some applications of the game-dynamical equations, which can, by the way, be transformed [84] into mathematically equivalent *Lotka-Volterra equations* [89–91] used to describe *predator-prey* or other (ecological) *systems* [84, 92–94].

2.3.5 Fashion Cycles and Deterministic Chaos

Now, assume one population $a = 1$, in which imitative decision changes take place between $I = 3$ kinds of *fashions* $i \in \{1,2,3\}$. If the payoff matrix is specified according to

$$\big(P_{ii''}\big) = \begin{pmatrix} 0 & -1 & 1 \\ 1 & 0 & -1 \\ -1 & 1 & 0 \end{pmatrix}, \qquad (2.70)$$

it reflects that fashion i receives negative attention by people wearing fashion $(i+1) \bmod I$ (the avantgarde), while it receives positive attention by people wearing fashion $(i-1) \bmod I$ (being behind the present fashion).

The corresponding game-dynamical equations are

$$\frac{dP_1(i,t)}{dt} = \nu P_1(i,t)\big[P_1\big((i-1) \bmod I, t\big) - P_1\big((i+1) \bmod I, t\big)\big]. \quad (2.71)$$

Apart from the *normalization condition*

$$\sum_{i=1}^{I} P_1(i,t) = 1\,, \quad (2.72)$$

these equations have an *invariant of motion:*

$$C = \prod_{i=1}^{I} P_1(i,t) = \text{const.} \quad (2.73)$$

For $I = 3$ it is, therefore, possible to calculate the exact form of the resulting *phase portraits* as a function of the initial conditions $P_1(i,0)$, despite of the non-linearity of the differential equations (2.71):

$$P_1(2,t) = \frac{1-P_1(1,t)}{2} \pm \sqrt{\left[\frac{1-P_1(1,t)}{2}\right]^2 - \frac{C}{P_1(1,t)}}\,, \quad (2.74)$$

where $P_1(3,t) = 1 - P_1(1,t) - P_1(2,t)$ and $C = P_1(1,0)P_1(2,0)P_1(3,0)$. This implies non-linear, but periodic (i.e. anharmonic) oscillations (see Fig. 2.4).

According to Eq. (2.74), there should always be the same sequence of fashions. This is, of course, not very realistic. However, for $I > 3$, we find a rather irregular sequence, as desired (see Fig. 2.5a). Note that, for certain specification of the

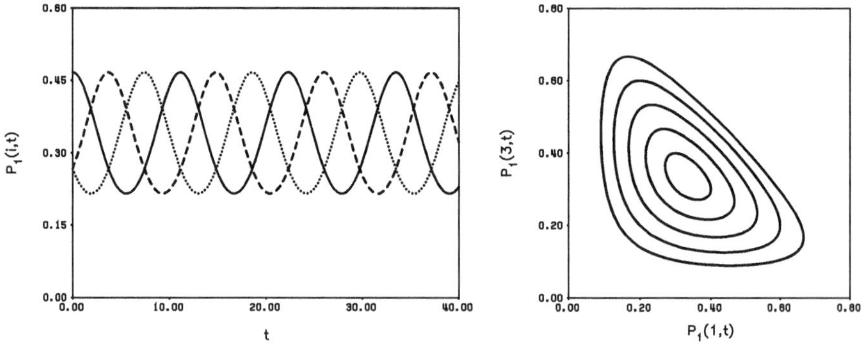

Fig. 2.4 Example of fashion cycles for the case of $I = 3$ fashions (from [18, 19, 47, 95]). We observe non-linear and regular oscillations in the plot of the time-dependent proportions $P_1(i,t)$ of the different fashions i (*left*) and in the phase portraits showing the proportion $P_1(3,t)$ of fashion $i = 3$ over the proportion $P_1(1,t)$ of fashion $i = 1$ for various initial conditions (*right*)

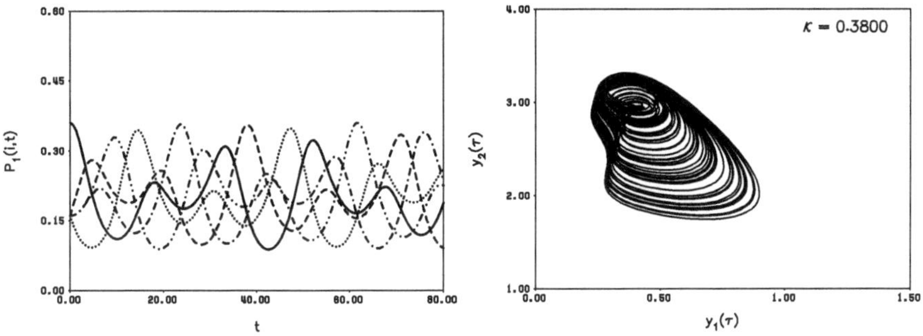

Fig. 2.5 *Left*: The non-linear oscillations for the case of $I = 5$ different fashions are irregular in the sense that they do not show short-term periodicity. *Right*: The decision dynamics can be chaotic (even if the decision dynamics is deterministic). This is illustrated by the plot of a chaotic attractor showing the result of another opinion formation model (a kind of periodically driven Brusselator) in scaled variables $y_a(\tau) = y_a^0 P_a(t)$ with $\tau = \tau_0 t$. For details see [18, 19, 46, 47], from which these plots were reproduced

payoff matrices, the equations for the most probable decision changes can show even more complex dynamical behavior such as *deterministic chaos* (see Fig. 2.5b). This implies that the decision distributions $P_a(i,t)$ would, for principal reasons, be *unpredictable* over a longer time period, even if we knew the transition rates $w^a(i'|i,t)$ exactly.

2.3.6 Polarization, Mass Psychology, and Self-Organized Behavioral Conventions

In this subsection, we will assume individuals of one group $a = 1$ that can choose between $I = 2$ *equivalent* strategies, i.e. the payoff matrix is symmetric:

$$\big(P_{ii''}\big) = \begin{pmatrix} A+B & B \\ B & A+B \end{pmatrix} . \tag{2.75}$$

An example for equivalent strategies would, for example, be the avoidance of another pedestrian on the right-hand side ($i = 1$) or on the left-hand side ($i = 2$), see Fig. 2.6.

With a constant spontaneous transition or mutation rate $w(x|x') = W$ corresponding to *trial and error behavior,* we find the specific game dynamical equation

$$\frac{dP_a(i,t)}{dt} = -2\left(P_a(i,t) - \frac{1}{2}\right)\left\{W + \nu A P_a(i,t)\big[P_a(i,t) - 1\big]\right\} . \tag{2.76}$$

For $\kappa = 1 - 4W/(\nu A) \leq 0$, the only stationary solution is $P_a(i) = 1/2$. Otherwise this solution is unstable, but there are the two stable stationary solutions $P_a(i) = (1 + \sqrt{\kappa})/2$ and $P_a(i) = (1 - \sqrt{\kappa})/2$. The finally resulting solution

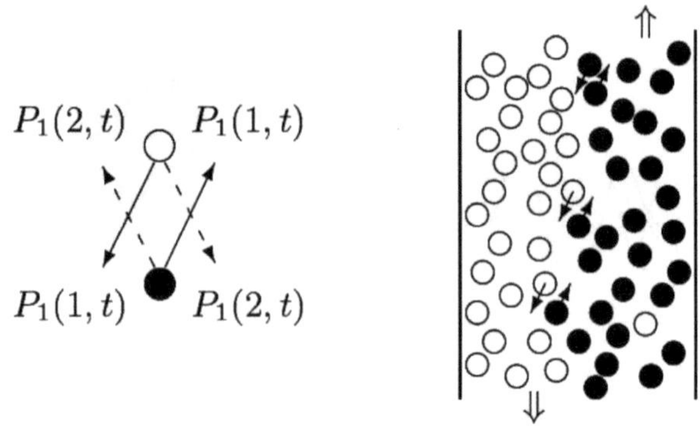

Fig. 2.6 *Left*: Illustration of the avoidance problem of two pedestrians walking in opposite directions. $P_1(1, t)$ represents the probability of deciding for an avoidance maneuver on the right-hand side, $P_1(2, t)$ the corresponding probability for the left-hand side. *Right*: Pedestrians subconsciously form lanes of uniform walking directions. In Central Europe, these lanes appear more frequently on the right-hand side, which can be interpreted as the result of a self-organized behavioral convention. (From [18, 19, 45, 46, 87, 96])

depends on the (random) initial condition (i.e., basically on initial fluctuations). Thus, we find *symmetry-breaking* or, in other words, *history-dependent behavior* [18, 19, 45, 46, 48, 70, 87, 96].

From the above, we may draw several interesting conclusions:

- If it is profitable to take the same decision as the interaction partner (i.e. $A > 0$), in each group a one of the equivalent strategies will win through, if the spontaneous transition rate W is small enough. This gives rise to a *self-organized behavioral convention.* Examples are the rotation direction of clocks, the pedestrians' asymmetric avoidance behavior [18, 19, 45, 46, 87, 96], or the triumph of VHS over Beta video [97].
- A transition from *individualistic behavior* (where people choose independently among all available alternatives) to *herding behavior* or *mass psychology* (where people tend to join the decision of the majority) occurs when the parameter κ becomes positive. This can happen, if the advantage A or the interaction rate ν increase for some reason, or if the rate W of spontaneous decision changes (i.e. the readiness to check out other alternatives) goes down.
- If we distinguish several weakly interacting groups a, for example people with separate social backgrounds living in different parts of a city or country, the alternative $i = 1$ may gain the majority in some groups, and the alternative $i = 2$ in others, if $\kappa > 0$. This corresponds to a *polarization* of society, which is common for emotional topics, possibly because of the higher interaction rate ν.

Generalizations of the above equations to $I > 2$ equivalent or several non-equivalent strategies are easily possible. In the latter case, superior strategies will tend to occur more frequently, but the polarization effect and the transition from individual to mass behavior can still occur under similar conditions as discussed above.

2.4 Summary and Outlook

In this contribution, we tried to develop a consistent theory of decision behavior. We started from socio-psychological observations regarding single decision-making processes and concluded that decisions can be interpreted as phase transitions. Moreover, a transitive preference scale was found to be restricted to particular situations. Our probabilistic approach improves the concept of *homo economicus,* as it takes into account limited or uncertain information, limited processing capabilities (e.g. incomplete decisions), and emotional aspects of decision-making. We have also considered non-linear interactions among individuals, which allowed us to understand polarized decision behavior (in particular regarding emotionally loaded issues), the self-organization of behavioral conventions, and the transition from individual to mass behavior. Several fields of application have been outlined, such as opinion formation, fashion cycles, social force models, logistic and gravity models, or dynamical game theory, which have all been special cases of the developed dynamical decision theory.

Obviously, it requires both, theoretical and experimental efforts to get ahead in decision theory. In a decade from now, the microscopic theory of human interactions will probably have been developed to a degree that allows one to systematically derive social patterns and economic dynamics on this ground. This will not only yield a deeper understanding of socio-economic systems, but also help to more efficiently distribute scarce resources such as road capacities, time, space, money, energy, goods, or our natural environment.

Acknowledgments This study was partially supported by the ALTANA-Quandt foundation. The author wants to thank Prof. Aruka, Prof. Selten, and Prof. Schreckenberg for their invitations and fruitful discussions, as well as Prof. Kondor and Dr. Schadschneider for inspiring comments.

References

1. Helbing D (2001) Traffic and related self-driven many-particle systems. Rev Mod Phys 73:1067–1141
2. Hagel JH, Roth AE (eds) (1995) The handbook of experimental economics. Princeton University, Princeton, NJ
3. Shleifer A (2000) Inefficient markets: an introduction to behavioral finance. Clarendon Lectures. Oxford University Press, Oxford
4. Arthur WB (1994) Inductive reasoning and bounded rationality. Am Econ Rev 84:406–411
5. Challet D, Zhang Y-C (1997) Emergence of cooperation and organization in an evolutionary game. Physica A 246:407ff
6. Challet D, Zhang Y-C (1998) On the minority game: analytical and numerical studies. Physica A 256:514–532
7. Challet D, Marsili M, Zhang Y-C (2000) Modeling market mechanism with minority game. Physica A 276:284–315
8. Rachlin HC (1976) Behavior and learning. Freeman, San Francisco, CA
9. Rachlin HC, Baum WM (1972) Effects of alternative reinforcement: does the source matter? J Exp Anal Behav 18:231–241

10. Schroeder SR, Holland JG (1969) Reinforcement of eye movement with concurrent schedules. J Exp Anal Behav 12:897–903
11. Fechner GT (1860) Elemente der Psychophysik. Druck und Verlag von Breitkopf und Hartel, Leipzig
12. Domencich TA, McFadden DM (1975) Urban travel demand. A behavioral analysis. North-Holland, Amsterdam, pp 61–69
13. Ortúzar JdeD, Willumsen LG (1990) Modelling transport, Chap. 7: Discrete-choice models Wiley, Chichester
14. Ben-Akiva M, McFadden DM et al (1999) Extended framework for modeling choice behavior. Marketing Lett 10:187–203
15. Reif F (1965) Fundamentals of statistical and thermal physics. McGraw-Hill, Singapore, pp 202ff, 229ff
16. Rosser WGV (1982) An introduction to statistical physics. Horwood, Chichester, pp 108f
17. Haken H (1988) Information and self-organization. Springer, Berlin
18. Helbing D (1996) Stochastische Methoden, nichtlineare Dynamik und quantitative Modelle sozialer Prozesse. Shaker, Aachen
19. Helbing D (1995) Quantitative sociodynamics (and references therein). Kluwer Academic, Dordrecht
20. Heider F (1946) Attitudes and cognitive organization. J Psychol 21:107–112
21. Heider F (1958) The psychology of interpersonal relations. Wiley, New York, NY
22. Cartwright D, Harary F (1956) Structural balance: a generalization of Heider's theory. Psychol Rev 63:277–293
23. Osgood CE, Tannenbaum PH (1955) The principle of congruity in the prediction of attitude change. Psychol Rev 62:42–55
24. Festinger L (1957) A theory of cognitive dissonance. Row & Peterson, Evanston, IL
25. Brehm SS (1976) The application of social psychology to clinical practice. Wiley, New York, NY
26. Binder J, Haag G, Koller M 2002 Modelling and simulation of migration, regional employment development and regional gross wage payment in Germany: the bottom-up approach. In: Regional Economies in Transition (Research Report 02:1), University of Trollhättan, Uddevalla
27. Feger H (ed) (1977) Studien zur intraindividuellen Konfliktforschung. Huber, Bern
28. Feger H (1978) Konflikterleben und Konfliktverhalten. Huber, Bern
29. Babloyantz A, Destexhe A (1987) Strange attractors in the human cortex. In: Rensing L, van der Heiden U, Mackey MC (eds) Temporal disorder in human oscillatory systems. Springer, Heidelberg
30. Babloyantz A, Nicolis C, Salazar M (1985) Evidence of chaotic dynamics of brain activity during the sleep cycle. Phys Lett A 111:152
31. Osgood CE, Suci GJ, Tannenbaum PH (1957) The measurement of meaning. University of Illinois Press, Urbana, IL
32. Douglas PH (1934) The theory of wages. Macmillan, New York, NY, pp 132–135
33. Nicholson W (1992) Microeconomic theory, 5th edn. Dryden Press, FortWorth, pp 90ff, 111ff
34. Ben-Akiva M, Bottom J, Ramming MS (2001) Route guidance and information systems. Int J Syst Contr Engin 215:317–324
35. Libet B, Gleason CA, Wright EW, Pearl DK (1983) Time of conscious intention to act in relation to onset of cerebral activity (readiness-potential). Brain 106:623–642
36. Haggard P, Eimer M (1999) On the relation between brain potentials and the awareness of voluntary movements. Exp Brain Res 126:128–133
37. Harvey JH (1976) Attribution of freedom. In: Harvey JH, Ickes WJ, Kidd RF (eds) New directions in attribution research, Chap. 4. Wiley, New York, NY
38. Steiner ID (1970) Perceived freedom. In: Berkowitz L (ed) Advances in experimental social psychology. Academic Press, New York, NY
39. Brehm JW (1972) Responses to loss of freedom: a theory of psychological reactance. General Learning Press, Morristown, NJ

40. Brehm JW (1966) A theory of psychological reactance. Academic Press, New York, NY
41. Wortman CB, Brehm JW (1975) Responses to uncontrollable outcomes: an integration of reactance theory and the learned helplessness model. In: Berkowitz L (ed) Advances in experimental social psychology. Academic Press, New York, NY, pp 277–336
42. Weidlich W, Haag G (1983) Concepts and models of a quantitative sociology. The dynamics of interacting populations. Springer, Berlin
43. Weidlich W (1991) Physics and social science—the approach of synergetics. Phys Rep 204: 1–163
44. Weidlich W (2000) Sociodynamics. A systematic approach to mathematical modelling in the social sciences. Harwood Academic, Amsterdam
45. Helbing D (1998) Microscopic foundation of stochastic game dynamical equations. In: Leinfellner W, Köhler E (eds) Game theory, experience, rationality. Kluwer Academic, Dordrecht, pp 211–224
46. Helbing D (1994) Amathematical model for behavioral changes by pair interactions and its relation to game theory. Angewandte Sozialforschung 18(3):117–132
47. Helbing D (1993) Stochastic and Boltzmann-like models for behavioral changes, and their relation to game theory. Physica A 193:241–258
48. Helbing D (1992) Interrelations between stochastic equations for systems with pair interactions. Physica A 181:29–52
49. Helbing D (1994) A mathematical model for the behavior of individuals in a social field. J Math Sociol 19(3):189–219
50. Burt RS (1982) Towards a structural theory of action. Network models of social structure, perception, and action. Academic Press, New York, NY
51. Braun N (1993) Socially embedded exchange. Lang, Frankfurt/Main
52. Krawiecki A, Hołyst JA, Helbing D (2002) Volatility clustering and scaling for financial time series due to attractor bubbling. Phys Rev Lett 89:158701
53. Helbing D, Farkas I, Vicsek T (2000) Simulating dynamical features of escape panic. Nature 407:487–490
54. Fishbein M, Ajzen I (1975) Belief, attitude, intention and behavior: an introduction to theory and research. Addison-Wesley, Reading, MA
55. Anderson NH (1961) Group performance in an anagram task. J Soc Psychol 55:67–75
56. Kelley HH, Thibaut JW (1969) Group problem solving. In: Lindzey G, Aronson E (eds) The handbook of social psychology, vol 4. Addison-Wesley, Reading, MA
57. Laughlin PR, Kerr NL, Davis JH, Halff HM, Marciniak KA (1975) Group size, member ability, and social decision schemes on an intellective task. J Pers Soc Psychol 31:522–535
58. Pearl R (1924) Studies in human biology. Williams & Wilkins, Baltimore, MD
59. Verhulst PF (1845) Nouveaux Mémoires de l'Académie Royale des Sciences et des Belles-Lettres de Bruxelles 18:1ff
60. Bartholomew DJ (1967) Stochastic models for social processes. Wiley, London
61. Montroll EW, Badger WW (1974) Introduction to quantitative aspects of social phenomena. Gordon and Breach, New York, NY
62. Hamblin RL, Jacobsen RB, Miller JLL (1973) A mathematical theory of social change. Wiley, New York, NY
63. Olinick M (1978) An introduction to mathematical models in the social and life sciences. Addison-Wesley, Reading, MA, pp 59–65
64. Ravenstein E (1876) The birthplaces of the people and the laws of migration. Geogr Mag III:173–177
65. Zipf GK (1946) The P1P2/D hypothesis on the intercity movement of persons. Am Sociol Rev 11:677–686
66. Weidlich W, Haag G (eds) (1988) Interregional migration. Springer, Berlin
67. Haag G (1989) Dynamic decision theory. Applications to urban and regional topics. Kluwer Academics, Dordrecht

68. Helbing D (1993) Boltzmann-like and Boltzmann-Fokker-Planck equations as a foundation of behavioral models. Physica A 196:546–573
69. Lewin K (1951) Field theory in social science. Harper and Brothers, New York, NY
70. Helbing D (1991) A mathematical model for the behavior of pedestrians. Behav Sci 36: 298–310
71. Nowak A, Szamrej J, Latané B (1990) From private attitude to public opinion: a dynamic theory of social impact. Psychol Rev 97:362–376
72. Lewenstein M, Nowak A, Latané B (1992) Statistical mechanics of social impact. Phys Rev A 45:763–776
73. Vallacher R, Nowak A (eds) (1994) Dynamic systems in social psychology. Academic Press, New York, NY
74. Kacperski K, Hołyst JA (1996) Phase transitions and hysteresis in a cellular automata-based model of opinion formation. J Stat Phys 84:169–189
75. Helbing D (1997) Verkehrsdynamik. Neue physikalische Modellierungskonzepte. Springer, Berlin
76. Helbing D, Tilch B (1998) Generalized force model of traffic dynamics. Phys Rev E 58: 133–138
77. Treiber M, Hennecke A, Helbing D (2000) Congested traffic states in empirical observations and microscopic simulations. Phys Rev E 62:1805–1824
78. Helbing D, Molnár P (1995) Social force model for pedestrian dynamics. Phys Rev E 51:4282–4286
79. Helbing D, Farkas I, Vicsek T (2000) Freezing by heating in a driven mesoscopic system. Phys Rev Lett 84:1240–1243
80. Helbing D (1994) Boltzmann-like and force models for behavioral changes. In: Bachem A, Derigs U, Jünger M, Schrader R (eds) Operations research '93. Physica, Heidelberg, pp 251–254
81. Hofbauer J, Schuster P, Sigmund K (1979) A note on evolutionarily stable strategies and game dynamics. J Theor Biol 81:609–612
82. Taylor P, Jonker L (1978) Evolutionarily stable strategies and game dynamics. Math Biosci 40:145–156
83. Zeeman EC (1980) Population dynamics from game theory. In: Nitecki A, Robinson C (eds) Global theory of dynamical systems. Springer, Berlin
84. Hofbauer J, Sigmund K (1988) The theory of evolution and dynamical systems. Cambridge University Press, Cambridge
85. Eigen M (1971) The selforganization of matter and the evolution of biological macromolecules. Naturwissenschaften 58:465
86. Fisher RA (1930) The genetical theory of natural selection. Oxford University Press, Oxford
87. Helbing D (1992) A mathematical model for behavioral changes by pair interactions. In: Haag G, Mueller U, Troitzsch KG (eds) Economic evolution and demographic change. Formal models in social sciences. Springer, Berlin, pp 330–348
88. Schlag KH (1998) Why imitate, and if so, how? A boundedly rational approach to multi-armed bandits. J Econ Theo 78(1):130–156
89. Lotka AJ (1920) Analytical note on certain rhythmic relations in organic systems. Proc Natl Acad Sci USA 6:410
90. Lotka AJ (1956) Elements of mathematical biology. Dover, New York, NY
91. Volterra V (1931) Leçons sur la Théorie Mathématique de la Lutte pour la Vie. Gauthier-Villars, Paris
92. Goel NS, Maitra SC, Montroll EW (1971) On the Volterra and other nonlinear models of interacting populations. Rev Mod Phys 43(2):231–276
93. Hallam ThG (1986) Community dynamics in a homogeneous environment. In: Hallam TG, Levin SA (eds) Mathematical ecology. Springer, Berlin
94. Peschel M, Mende W (1986) The predator-prey model. Springer, Wien

95. Helbing D (1992) A mathematical model for attitude formation by pair interactions. Behav Sci 37:190–214
96. Helbing D (1996) A stochastic behavioral model and a ‘microscopic’ foundation of evolutionary game theory. Theory Decis 40:149–179
97. Arthur WB (1989) Competing technologies, increasing returns, and lock-in by historical events. Econ J 99:116–131

Part I
Stochastic Methods and Non-linear Dynamics

Overview

Part I of this book deals with *stochastic methods* and the basic principles of *non-linear dynamics,* which will be taken up again in Part II. In this way the mathematical formalism can be developed in its full generality and the *interrelations* between the various treated equations can be particularly elaborated. Further, in the second part which focuses on applications the scope of formal aspects can be lessened so that the results will stand out more clearly.

In detail, the way of proceeding in Part I is as follows: First, in Chap. 3 the *master equations for state space* will be developed. In particular, the connection to *micro-dynamics* (to quantum mechanics) will be shown. Afterwards, a short overview of the most important properties of the master equation will be given. Especially, a new solution method will be introduced, namely the *'path integral' solution.* Since this allows the calculation of path dependent quantities, it is suitable for *process control* and *event analysis.*

In Chap. 4, by factorization of the pair distribution functions, *BOLTZMANN-like equations* will be obtained. These consider *spontaneous transitions* and *pair interactions* of subsystems but neglect interactions of higher order. *Summarizing systems of equal type* the equations simplify enormously because the number of variables is drastically reduced.

A *master equation for configuration space* (the space of occupation numbers) is derived in Chap. 5. This allows the treatment of occupation number dependent transition rates which play an essential role in social processes. The master equation in state space and the BOLTZMANN-like equations result from it as *approximate mean value equations.* However, the configurational master equation allows also the derivation of *covariance equations* and *corrected mean value equations* as well as the consideration of *higher order interactions.*

For various reasons, the derivation and solution of the covariance equations should not be omitted: First, these are necessary to determine the *time interval* for which the approximate or corrected mean value equations are valid. Second, they are required to estimate the *reliability of prognoses.* Third, they distinguish various stochastic processes which possess the same mean value equations.

By means of a TAYLOR expansion of the master equation a ***FOKKER-PLANCK*** equation will be obtained in Chap. 6. One of its advantages is that it is often easier to solve than the master equation and depends only on the first two jump moments (the drift and diffusion coefficients). Finally, for systems which consist of many subsystems the '***BOLTZMANN-FOKKER-PLANCK** equations*' result by factorization of the pair distribution functions in the FOKKER-PLANCK equation or by TAYLOR expansion of the BOLTZMANN-like equations. Since they depend non-linearly on the distribution functions, a suitable strategy for their solution seems to be a *self-consistent method* similar to that of HARTREE.

In Chap. 7 *stochastic differential equations* will be introduced. The emphasis will be on the ***LANGEVIN** equation* which, in a certain sense, is equivalent to the FOKKER-PLANCK equation. However, whereas the FOKKER-PLANCK equation describes the probability distribution of states, the LANGEVIN equation reflects the individual behaviour of single subsystems which are subject to the influence of fluctuations. Part I closes with a discussion of *phase transitions* which are characterized by a drastic change of the system behaviour. In this connection, the various *routes to chaos* will be elucidated.

Chapter 3
Master Equation in State Space

3.1 Introduction

Many systems do not change with time in a *deterministic* (unique) way. Hence for the change of a system from a certain state to one out of many possible other states often it is only *transition probabilities* that are given.[1] The reason for the *uncertainty* with respect to the state change depends on the special situation:

- It can result from *external influences* (disturbances).
- Also *internal fluctuations* are possible, e.g. in a system that consists of many subsystems (elements).
- In *quantum mechanics* the transitions are inherently non-deterministic due to HEISENBERG's *uncertainty principle* [121, 122]. However, the corresponding probabilities of their occurence can be evaluated.
- Chaotic dynamics also brings about an uncertainty due to the so-called *butterfly effect.*

If there is an uncertainty about the concrete change of a system's state, only a *stochastic description* is possible. That is, we can calculate only the *probability* $P(X, t)$ to find a state X at a future time t. The temporal change of the probability distribution is given by

$$\frac{d}{dt} P(X, t) = \text{flow into } X - \text{flow out of } X \tag{3.1}$$

where the *flow* is a measure for the number of transitions per time unit. Due to the distinction of inflow and outflow terms the construction and interpretation of master equations is particularly simple.

The 'flow into X' is the sum of all flows into state X out of the various states X'. Here the 'flow out of state X' into state X' is proportional to the probability

[1] The terms *state, system,* etc. will be specified in Table 8.1 on page 165.

D. Helbing, *Quantitative Sociodynamics*, 2nd ed.,

DOI 10.1007/978-3-642-11546-2_3,

$P(X', t)$ of state X' to occur and to the *transition rate* $w(X|X'; t)$ from X' to X, i.e. of the form

$$\text{flow out of } X' \text{ into } X = w(X|X'; t)P(X', t). \tag{3.2}$$

The transition rate $w(X|X'; t)$ denotes the *transition probability* $P(X, t + \Delta t|X', t) \geq 0$ from state X' to state X per time unit Δt:

$$w(X|X'; t) := \lim_{\Delta t \to 0} \frac{P(X, t + \Delta t|X', t)}{\Delta t} \geq 0 \qquad (X \neq X'). \tag{3.3}$$

In summary we can write

$$\text{flow into } X = \sum_{X'(\neq X)} w(X|X'; t)P(X', t). \tag{3.4a}$$

Similarly, the 'flow out of X' is just the sum of the flows out of state X into other states X'. In total analogy we get

$$\text{flow out of } X = \sum_{X'(\neq X)} w(X'|X; t)P(X, t). \tag{3.4b}$$

Inserting (3.4) into (3.1) the following stochastic equation for the temporal change of the probability distribution $P(X, t)$ is obtained:

$$\frac{d}{dt}P(X, t) = \sum_{X'(\neq X)} \Big[w(X|X'; t)P(X', t) - w(X'|X; t)P(X, t)\Big]. \tag{3.5a}$$

This equation is called the *master equation* [215]. The summation includes all states X' of the *state space* Γ. Γ is the set of all possible states of the considered system. In case of Γ being not discrete but *continuous* (and of the dimension dim $\Gamma = M$) the sum over X' must be replaced by an integral:

$$\frac{d}{dt}P(X, t) = \int d^M X' \Big[w(X|X'; t)P(X', t) - w(X'|X; t)P(X, t)\Big]. \tag{3.5b}$$

$P(X, t)$ has the meaning of a *probability density*, then.

In order to arrive at the master equation (3.5), it must be implicitly presupposed that the considered system can be described by a Markov *process* [159, 244]. In simple words, the assumed Markov *property* means that the transition rates from a state X to another state X' depend on the *previous* state of the system only. To take into account *'memory effects'* (in cases where the transition rates also depend on the former states of the system) the *generalized master equation* [156, 241]

$$\frac{d}{dt}P(X, t) = \int_{t_0}^{t} dt' \sum_{X'} \Big[w_{t-t'}(X|X'; t')P(X', t') - w_{t-t'}(X'|X; t')P(X, t')\Big] \tag{3.6}$$

is required.

Master equation (3.5) would describe a *deterministic process* if

1. the transition rates were *deterministic* which means

$$w(X'|X;t) = 0 \quad \text{for all } X \text{ except for one single state } X' \equiv F(X,t) \in \Gamma \tag{3.7}$$

   and implies

$$w(X'|X;t) \equiv w(X'|X;t)\delta_{X'F(X,t)}\,, \tag{3.8}$$

2. the state X_0 at an initial time t_0 were given:

$$P(X,t_0) = \delta_{XX_0}\,. \tag{3.9}$$

Then for all later times $t > t_0$ the state $X'(t)$ of the system would be determined in a unique way:

$$P(X,t) = \delta_{XX'(t)}\,. \tag{3.10}$$

Here

$$\delta_{xy} := \begin{cases} 1 \text{ if } x = y \\ 0 \text{ otherwise} \end{cases} \tag{3.11}$$

denotes the KRONECKER *symbol.*

3.2 Derivation

Depending on the reason for the uncertainty concerning a system, the correct derivation of the corresponding master equation varies. The simplest derivation starts directly from the MARKOV *assumption.* In the case of *external influences (disturbances)* the transition rates of the considered system depend on the probability distribution of the variables describing the external influences. For a system that consists of many elements, the fast variables can be *adiabatically eliminated.* Then the latter only appear in the form of *(internally caused) fluctuations* of the remaining slow variables.

In order to justify the master equation in an accurate way, we must consider the *microscopic dynamics* of a system. The description then starts from a *quantum theoretical* basis, namely with the *statistical operator.* A *projection* on the *relevant variables* (with respect to the calculation of expected values) leads to the NAKAJIMA-ZWANZIG *equation.* In an appropriate basis it takes on the form of a *(generalized) master equation* [76, 306]. Again the fast variables in this master equation can be adiabatically eliminated so that a master equation for the relevant macroscopic state variables remains.

Finally, a derivation of the master equation for chaotically behaving systems is possible by means of special mappings, namely so-called MARKOV *partitions* (cf. [45, 201]).

3.2.1 Derivation from the MARKOV Property

Let $P(X_n, t_n; X_{n-1}, t_{n-1}; \ldots; X_0, t_0)$ be the *joint probability* to find the considered system in states $X_0, \ldots, X_n$ at the times

$$t_0, \quad t_1 := t_0 + \Delta t, \quad \ldots, \quad t_n := t_0 + n\Delta t\,. \tag{3.12}$$

Here let the times t_i be arbitrarily close to each other, i.e. $\Delta t \to 0$. According to *probability theory* we have

$$\sum_{X_n} \sum_{X_{n-1}} \cdots \sum_{X_0} P(X_n, t_n; X_{n-1}, t_{n-1}; \ldots; X_0, t_0) = 1 \tag{3.13}$$

due to normalization. Furthermore,

$$P(X_n, t_n) := \sum_{X_{n-1}} \cdots \sum_{X_0} P(X_n, t_n; X_{n-1}, t_{n-1}; \ldots; X_0, t_0) \tag{3.14}$$

is the probability to find state X_n at the time t_n since (3.14) describes the sum over all probabilities to arrive in state X_n. Moreover, there is

$$\begin{aligned} &P(X_n, t_n; X_{n-1}, t_{n-1}; \ldots; X_0, t_0) \\ &= P(X_n, t_n | X_{n-1}, t_{n-1}; \ldots; X_0, t_0) P(X_{n-1}, t_{n-1}; \ldots; X_0, t_0) \end{aligned} \tag{3.15}$$

with

$$\sum_{X_n} P(X_n, t_n | X_{n-1}, t_{n-1}; \ldots; X_0, t_0) = 1\,. \tag{3.16}$$

Thus the probability of a system to be in states $X_0, \ldots, X_n$ at the times $t_0, \ldots, t_n$ is given as a product of the probability to be in states $X_0, \ldots, X_{n-1}$ at the times $t_0, \ldots, t_{n-1}$ and the *conditional probability* $P(X_n, t_n | X_{n-1}, t_{n-1}; \ldots; X_0, t_0)$ for a transition to state X_n provided that the system is in states $X_0, \ldots, X_{n-1}$ at the times $t_1, \ldots, t_{n-1}$.

Using the MARKOV *assumption* that the conditional probability depends only on the *preceding* moment we have

$$P(X_n, t_n | X_{n-1}, t_{n-1}; \ldots; X_0, t_0) \equiv P(X_n, t_n | X_{n-1}, t_{n-1})\,. \tag{3.17}$$

By inserting (3.17) into (3.15) and using (3.14) we obtain

$$P(X_n, t) = \sum_{X_{n-1}} P(X_n, t_n | X_{n-1}, t_{n-1}) P(X_{n-1}, t_{n-1}) \,. \tag{3.18}$$

Due to the *normalization*

$$\sum_{X'} P(X', t + \Delta t | X, t) = 1 \tag{3.19}$$

of the transition probabilities (cf. (3.16)) and with

$$X := X_n, \quad X' := X_{n-1}, \quad t := t_{n-1} \tag{3.20}$$

we get

$$\begin{aligned} \frac{P(X, t + \Delta t) - P(X, t)}{\Delta t} &= \frac{1}{\Delta t} \sum_{X'} P(X, t + \Delta t | X', t) P(X', t) \\ &\quad - \frac{1}{\Delta t} \left[\sum_{X'} P(X', t + \Delta t | X, t) \right] P(X, t) \,. \end{aligned} \tag{3.21}$$

In the limit $\Delta t \to 0$ we arrive at the *master equation*

$$\frac{d}{dt} P(X, t) = \sum_{X'} \Big[w(X|X'; t) P(X', t) - w(X'|X; t) P(X, t) \Big] \tag{3.22a}$$

with the *transition rates*

$$w(X'|X; t) := \lim_{\Delta t \to 0} \frac{P(X', t + \Delta t | X, t)}{\Delta t} \qquad (X' \neq X) \,. \tag{3.22b}$$

For $X' = X$ the transition rates $w(X'|X; t)$ can be chosen arbitrarily because the relevant terms in (3.22a) compensate each other. Often it is convenient to put

$$w(X|X; t) := 0 \,. \tag{3.23}$$

3.2.2 External Influences (Disturbances)

We will now consider a system in which the transitions between two states X and X' depend on unknown (e.g. external) quantities Q_j. Let us assume that the temporal change of the external quantities is independent of the processes within the considered system, and let $P(Q, t)$ be the probability that we have the quantity

$Q \equiv (\ldots, Q_j, \ldots)^{\text{tr}}$ at the time t. Then the *effective transition rate* of the system at the time t is

$$w(X'|X;t) := \sum_{Q,Q'} w(X',Q'|X,Q)P(Q,t) \tag{3.24}$$

Because of the dependence on $P(Q,t)$, the dynamics of the system can only be described in a *stochastic* way since, in general, it implies

$$w(X'|X;t) \neq 0 \quad \text{for various } X' \tag{3.25}$$

even in the case that the transition rates $w(X',Q'|X,Q)$ are deterministic. This is often the reason for a *time-dependence* of the effective transition rates $w(X'|X;t)$.

3.2.3 Internal Fluctuations

Now let us start from a system the dynamics of which depends on extremely many internal variables. Let us assume that we are only interested in the variables $X_{\text{s},J} \in \{X_I\}$ but not in the variables $X_{\text{f},K} \in \{X_I\}$. By means of the decomposition

$$X \equiv (X_{\text{f}}, X_{\text{s}})^{\text{tr}} \tag{3.26}$$

of the variable vector X and by use of the conditional probability $P(X_{\text{f}}|X_{\text{s}};t)$ we can achieve the representation

$$P(X,t) \equiv P(X_{\text{f}},X_{\text{s}};t) = P(X_{\text{f}}|X_{\text{s}};t)P(X_{\text{s}},t) \tag{3.27}$$

with

$$\sum_{X_{\text{f}}} P(X_{\text{f}}|X_{\text{s}};t) = 1 \qquad \text{and} \qquad \sum_{X_{\text{s}}} P(X_{\text{s}},t) = 1 \tag{3.28}$$

(cf. (3.13), (3.14), (3.15) and (3.16)). Thus the master equation takes on the form

$$\begin{aligned}
\frac{d}{dt}P(X_{\text{f}},X_{\text{s}};t) &= P(X_{\text{s}},t)\frac{d}{dt}P(X_{\text{f}}|X_{\text{s}};t) + P(X_{\text{f}}|X_{\text{s}};t)\frac{d}{dt}P(X_{\text{s}},t) \\
&= \sum_{X'_{\text{f}},X'_{\text{s}}} w(X_{\text{f}},X_{\text{s}}|X'_{\text{f}},X'_{\text{s}};t)P(X'_{\text{f}}|X'_{\text{s}};t)P(X'_{\text{s}},t) \\
&\quad - \sum_{X'_{\text{f}},X'_{\text{s}}} w(X'_{\text{f}},X'_{\text{s}}|X_{\text{f}},X_{\text{s}};t)P(X_{\text{f}}|X_{\text{s}};t)P(X_{\text{s}},t)\,.
\end{aligned} \tag{3.29}$$

Due to

$$P(X_s, t) = \sum_{X_f} P(X_f, X_s; t) \tag{3.30}$$

(cf. (3.14)) we have the relation

$$\frac{d}{dt} P(X_s, t) = \sum_{X'_s} \left[\overline{w}(X_s|X'_s; t) P(X'_s, t) - \overline{w}(X'_s|X_s; t) P(X_s, t) \right] \tag{3.31a}$$

with the *effective transition rates*

$$\overline{w}(X'_s|X_s; t) := \sum_{X_f} \sum_{X'_f} w(X'_f, X'_s|X_f, X_s; t) P(X_f|X_s; t) . \tag{3.31b}$$

Whereas $w(X'|X; t)$ may be symmetrical (cf. (3.88)), $\overline{w}(X'_s|X_s; t)$ is normally asymmetrical because of the dependence on $P(X_f|X_s; t)$.

Now let us assume that $X_{s,J}$ are just the slowly changing *macroscopic variables* and $X_{f,K}$ the fast varying *microscopic variables*. From the mathematical point of view, *fast variables* and *slow variables* differ in the order of magnitude of the transition rates (which will, in the following, be assumed *time-independent*). Concretely, let

$$w(X'_f, X'_s|X_f, X_s) =: \varepsilon(X'_f, X'_s; X_f, X_s) w(X'_f, X_s|X_f, X_s) \tag{3.32a}$$

for $X'_s \neq X_s$ and

$$\sum_{X'_s(\neq X_s)} \varepsilon(X'_f, X'_s; X_f, X_s) \ll 1 . \tag{3.32b}$$

This implies

$$w(X'_f, X'_s|X_f, X_s) \approx w(X'_f, X_s|X_f, X_s) \delta_{X_s X'_s} . \tag{3.33}$$

Consequently, the right-hand side of Eq. (3.31a) vanishes and (3.29) takes on the form

$$\begin{aligned} \frac{d}{dt} P(X_f|X_s; t) \approx & \sum_{X'_f} w(X_f, X_s|X'_f, X_s) P(X'_f|X_s; t) \\ & - \sum_{X'_f} w(X'_f, X_s|X_f, X_s) P(X_f|X_s; t) . \end{aligned} \tag{3.34}$$

For $P(X_s, t)$ not to be quasi-constant, a slower *time scale* must be chosen by means of the transformation

$$t_s := \varepsilon t \,. \tag{3.35}$$

Then, with the notations

$$P_s(X_s, t_s) := P(X_s, t_s/\varepsilon) \,, \tag{3.36a}$$

$$P_s(X_f|X_s; t_s) := P(X_f|X_s; t_s/\varepsilon) \,, \tag{3.36b}$$

and

$$\begin{aligned} w_s(X_s'|X_s; t_s) &:= \frac{\overline{w}(X_s'|X_s; t_s)}{\varepsilon} \\ &= \sum_{X_f} \sum_{X_f'} \frac{\varepsilon(X_f', X_s'; X_f, X_s)}{\varepsilon} w(X_f', X_s|X_f, X_s) P_s(X_f|X_s; t_s) \,, \end{aligned} \tag{3.37}$$

master equation (3.31a) reads

$$\frac{d}{dt_s} P_s(X_s, t_s) = \sum_{X_s'} \Big[w_s(X_s|X_s'; t_s) P_s(X_s', t_s) - w_s(X_s'|X_s; t_s) P_s(X_s, t_s) \Big] \tag{3.38}$$

in which the subscripts 's' are normally suppressed. Note that (3.38) is considerably simplified with regard to the number of variables.

Let ε be of the order of magnitude of the average of the quantities $\varepsilon(X_f', X_s'; X_f, X_s)$. Then the time scale t_s corresponds to the average 'speed of change' of the slow variables X_s whereas the fast variables X_f, compared to the time scale t_s, relax *quasi-adiabatically.* Their distribution $P(X_f|X_s; t_s) \approx P(X_f|X_s)$ is given by the slow variables X_s, i.e. the fast variables are *'slaved'* by the slow ones (cf. [107], pp. 204f.) because it is

$$\frac{d}{dt} P(X_f|X_s; t) = \frac{dt_s}{dt} \frac{d}{dt_s} P_s(X_f|X_s; t_s) = \varepsilon \frac{d}{dt_s} P_s(X_f|X_s; t_s) \approx 0 \,. \tag{3.39}$$

With (3.34) this implies

$$\begin{aligned} 0 \approx & \sum_{X_f'} w(X_f, X_s|X_f', X_s) P_s(X_f'|X_s; t_s) \\ & - \sum_{X_f'} w(X_f', X_s|X_f, X_s) P_s(X_f|X_s; t_s) \,. \end{aligned} \tag{3.40}$$

Obviously, the conditional probabilities $P_s(X_f|X_s; t_s)$ are, due to the quasi-adiabatic relaxation, almost time-independent on the time scale t_s. Equation (3.40) allows the determination of the conditional probabilities

$$P_s(X_f|X_s; t_s) \approx P_s(X_f|X_s) \tag{3.41}$$

which are necessary for the evaluation of the macroscopic transition rates

$$w_s(X'_s|X_s;t_s) \approx w_s(X'_s|X_s)\,. \tag{3.42}$$

Due to the dependence of the transition rates $w_s(X'_s|X_s)$ on $P(X_f|X_s)$, the dynamics of the variables X_s can again be described only in a stochastic way, even if the transition rates $w(X_f, X_s|X'_f, X'_s)$ are deterministic. Hence the variables X_s are subject to *fluctuations* which stem from the quasi-adiabatically eliminated variables X_f.

3.2.4 Derivation from Quantum Mechanics[2]

3.2.4.1 The Statistical Operator

Within the framework of non-relativistic quantum mechanics the microscopic dynamics of a system consisting of many subsystems is described by a *statistical operator (density matrix)* $\rho(t)$ which obeys the VON NEUMANN *equation* [66, 195]

$$\mathrm{i}\hbar\frac{\partial\rho}{\partial t} = \mathcal{H}\rho - \rho\mathcal{H}\,. \tag{3.43}$$

$\mathcal{H}$ is the so-called HAMILTON *operator* of the system describing its energy.

The statistical operator allows the evaluation of the *expected value (mean value)* $\langle\mathcal{A}\rangle$ of an *observable* (measurable quantity)

$$\mathcal{A} = \sum_X A_X|X\rangle\langle X| \tag{3.44}$$

by means of

$$\begin{aligned}\langle\mathcal{A}\rangle = \mathrm{Tr}(\mathcal{A}\rho) &\equiv \sum_Y \langle Y|\mathcal{A}\rho|Y\rangle \\ &= \sum_Y\sum_X A_X\langle Y|X\rangle\langle X|\rho|Y\rangle = \sum_X A_X\langle X|\rho|X\rangle \\ &= \sum_X A_X\rho_{XX}\,. \end{aligned}\tag{3.45}$$

Here the states $|X\rangle$ were chosen in such a way that they form an *orthonormal basis* of the HILBERT *space* belonging to the total system and that they are the eigenvectors of the observable $\mathcal{A}$ to the eigenvalues A_X. Hence we have

$$\sum_X |X\rangle\langle X| = \mathit{1}\,, \tag{3.46}$$

$$\langle X|Y\rangle = \delta_{XY}\,, \tag{3.47}$$

---

[2] This section can be omitted during a first reading.

and

$$\mathcal{A}|X\rangle = A_X|X\rangle\,. \tag{3.48}$$

X denotes a vector that is normally composed of various *quantum numbers*.

The matrix elements

$$\rho_{XY}(t) := \langle X|\rho(t)|Y\rangle \tag{3.49}$$

possess the properties

$$\rho_{XX}(t) \geq 0 \tag{3.50}$$

and

$$\sum_X \rho_{XX}(t) = 1 \tag{3.51}$$

(cf. [66]). Hence

$$P(X,t) := \rho_{XX}(t) \tag{3.52}$$

can be interpreted as the *probability* to find state $|X\rangle$ when observable $\mathcal{A}$ is measured. The formula (3.45) for the expected value $\langle\mathcal{A}\rangle$ has thus the familiar form

$$\langle\mathcal{A}\rangle = \sum_X A_X P(X,t)\,. \tag{3.53}$$

3.2.4.2 The Nakajima-Zwanzig Equation

As relation (3.45) shows, the actually *relevant part* of the statistical operator $\rho \equiv (\rho_{XY})$ is the *diagonal part*

$$\rho_1 := \left(\rho_{XX}\delta_{XY}\right). \tag{3.54}$$

Therefore, from the von Neumann equation we will derive a *reduced equation* for the smaller number of relevant variables. For that we proceed in the following way:

First, we write the von Neumann equation in the Liouville *form* [171]

$$\frac{\partial\rho}{\partial t} = -\mathrm{i}\underline{\mathcal{L}}\rho \tag{3.55a}$$

in which the linear *tetradic operator (superoperator)* $\underline{\mathcal{L}} \equiv \left(L_{XYX'Y'}\right)$ is given by

$$\underline{\mathcal{L}}\mathcal{B} := \frac{1}{\hbar}(\mathcal{H}\mathcal{B} - \mathcal{B}\mathcal{H})\,. \tag{3.55b}$$

Here $\mathcal{B}$ is an arbitrary operator which is defined in the HILBERT space of the total system. More explicitly, (3.55b) can be written in the way

$$(\underline{\mathcal{L}}\mathcal{B})_{XY} = \sum_{X'}\sum_{Y'} L_{XYX'Y'}B_{X'Y'} \tag{3.56a}$$

with

$$L_{XYX'Y'} = \frac{1}{\hbar}(H_{XX'}\delta_{YY'} - \delta_{XX'}H_{Y'Y}) . \tag{3.56b}$$

For tetradic operators we have the *multiplication rule*

$$(\underline{\mathcal{L}\mathcal{M}})_{XYX'Y'} = \sum_{Z}\sum_{Z'} L_{XYZZ'}M_{ZZ'X'Y'} \tag{3.57}$$

and the representation

$$(\underline{1})_{XYX'Y'} = \delta_{XX'}\delta_{YY'} \tag{3.58}$$

of the *identity operator* $\underline{1}$ (of *unity*).

By means of the LAPLACE *transformation* [218]

$$\mathcal{G}(u) := \int_0^\infty dt\, \mathrm{e}^{-ut}\rho(t_0+t) = \int_{t_0}^\infty dt\, \mathrm{e}^{-u(t-t_0)}\rho(t) \tag{3.59}$$

equation (3.55) can be converted into

$$u\mathcal{G}(u) - \rho(t_0) = -\mathrm{i}\underline{\mathcal{L}}\mathcal{G}(u) \tag{3.60}$$

which is formally solved by

$$\mathcal{G}(u) = (u\underline{1} + \mathrm{i}\underline{\mathcal{L}})^{-1}\rho(t_0) . \tag{3.61}$$

In order to separate out the diagonal part ρ_1 of the statistical operator ρ we decompose this into two parts

$$\rho = \rho_1 + \rho_2 \tag{3.62a}$$

with

$$\rho_1 = \underline{\mathcal{P}}\rho \tag{3.62b}$$

and

$$\rho_2 := \underline{\mathcal{Q}}\rho\,, \qquad \underline{\mathcal{Q}} := \underline{1} - \underline{\mathcal{P}} . \tag{3.62c}$$

Here the tetradic *(projection) operator* $\underline{\mathcal{P}}$ is defined by

$$P_{XYX'Y'} := \delta_{XY}\delta_{XX'}\delta_{YY'} \tag{3.63}$$

and fulfils the relations

$$\underline{\mathcal{P}}\underline{\mathcal{P}} = \underline{\mathcal{P}}, \qquad \underline{\mathcal{P}}\underline{\mathcal{Q}} = \underline{0}, \qquad \underline{\mathcal{Q}}\underline{\mathcal{Q}} = \underline{\mathcal{Q}}. \tag{3.64}$$

The decomposition of the statistical operator ρ into a relevant part ρ_1 and an irrelevant part ρ_2 also splits up the LAPLACE transform $\mathcal{G}(u)$ into two contributions

$$\mathcal{G}(u) = \mathcal{G}_1(u) + \mathcal{G}_2(u) \tag{3.65a}$$

where

$$\mathcal{G}_1(u) := \underline{\mathcal{P}}\mathcal{G}(u), \qquad \mathcal{G}_2(u) := \underline{\mathcal{Q}}\mathcal{G}(u). \tag{3.65b}$$

Applying $\underline{\mathcal{P}}$ and $\underline{\mathcal{Q}}$ to (3.60), we obtain two equations:

$$u\mathcal{G}_1(u) - \rho_1(t_0) = -\mathrm{i}\underline{\mathcal{P}}\underline{\mathcal{L}}\mathcal{G}_1(u) - \mathrm{i}\underline{\mathcal{P}}\underline{\mathcal{L}}\mathcal{G}_2(u) \tag{3.66a}$$

and

$$u\mathcal{G}_2(u) - \rho_2(t_0) = -\mathrm{i}\underline{\mathcal{Q}}\underline{\mathcal{L}}\mathcal{G}_1(u) - \mathrm{i}\underline{\mathcal{Q}}\underline{\mathcal{L}}\mathcal{G}_2(u). \tag{3.66b}$$

Solving (3.66b) with respect to $\mathcal{G}_2(u)$ we find

$$\mathcal{G}_2(u) = (u\underline{1} + \mathrm{i}\underline{\mathcal{Q}}\underline{\mathcal{L}})^{-1}\rho_2(t_0) - \mathrm{i}(u\underline{1} + \mathrm{i}\underline{\mathcal{Q}}\underline{\mathcal{L}})^{-1}\underline{\mathcal{Q}}\underline{\mathcal{L}}\mathcal{G}_1(u). \tag{3.67}$$

Inserting this into (3.66a) yields

$$\begin{aligned} u\mathcal{G}_1(u) - \rho_1(t_0) = &-\mathrm{i}\underline{\mathcal{P}}\underline{\mathcal{L}}\mathcal{G}_1(u) - \mathrm{i}\underline{\mathcal{P}}\underline{\mathcal{L}}(u\underline{1} + \mathrm{i}\underline{\mathcal{Q}}\underline{\mathcal{L}})^{-1}\rho_2(t_0) \\ &-\underline{\mathcal{P}}\underline{\mathcal{L}}(u\underline{1} + \mathrm{i}\underline{\mathcal{Q}}\underline{\mathcal{L}})^{-1}\underline{\mathcal{Q}}\underline{\mathcal{L}}\mathcal{G}_1(u). \end{aligned} \tag{3.68}$$

By means of the *inverse* LAPLACE *transformation*

$$\rho_1(t) := \frac{1}{2\pi\mathrm{i}} \int_{c-\mathrm{i}\infty}^{c+\mathrm{i}\infty} du\, \mathrm{e}^{u(t-t_0)}\mathcal{G}_1(u) \tag{3.69}$$

we finally get the NAKAJIMA-ZWANZIG *equation*

$$\frac{d}{dt}\rho_1(t) = -\mathrm{i}\underline{\mathcal{PL}}\rho_1(t) - \int_{t_0}^{t} dt'\, \underline{\mathcal{PL}}\mathrm{e}^{-\mathrm{i}\underline{\mathcal{QL}}(t-t')}\underline{\mathcal{QL}}\rho_1(t')$$
$$-\mathrm{i}\underline{\mathcal{PL}}\mathrm{e}^{-\mathrm{i}\underline{\mathcal{QL}}(t-t_0)}\rho_2(t_0) \tag{3.70}$$

by using the *convolution theorem* (cf. p. xix) and taking into account that the inverse LAPLACE transform of $[u\underline{1} + \mathrm{i}\underline{\mathcal{B}}]^{-1}$ is $\exp[-\mathrm{i}\underline{\mathcal{B}}(t - t_0)]$. If the observable $\mathcal{A}$ has been *measured* at the initial time t_0 we may assume the initial statistical operator to possess a *diagonal* form (in the basis of the eigenstates of $\mathcal{A}$), i.e.

$$\rho_1(t_0) = \rho(t_0) \qquad \text{and} \qquad \rho_2(t_0) = 0\,. \tag{3.71}$$

The last term in (3.70) vanishes, then. There remains the equation

$$\frac{d}{dt}\rho_1(t) = -\mathrm{i}\underline{\mathcal{PL}}\rho_1(t) - \int_{t_0}^{t} dt'\, \underline{\mathcal{PL}}\mathrm{e}^{-\mathrm{i}\underline{\mathcal{QL}}(t-t')}\underline{\mathcal{QL}}\rho_1(t') \tag{3.72}$$

which is a *closed equation* for the relevant part $\rho_1(t)$.

3.2.4.3 The Generalized Master Equation

In order to obtain an explicit form of the NAKAJIMA-ZWANZIG equation we decompose the HAMILTON operator

$$\mathcal{H} = \sum_X \sum_Y H_{XY}|X\rangle\langle Y| \tag{3.73}$$

into two parts:

$$\mathcal{H} = \mathcal{H}_0 + \mathcal{H}_1\,. \tag{3.74a}$$

Choosing the so-called *energy representation* defined by

$$\mathcal{H}_0 := \sum_X H_{XX}|X\rangle\langle X|\,, \quad \text{i.e.} \quad (H_0)_{XY} = H_{XX}\delta_{XY}\,, \tag{3.74b}$$

and

$$\mathcal{H}_1 := \mathcal{H} - \mathcal{H}_0\,, \tag{3.74c}$$

the LIOUVILLE-Operator splits up into two contributions

$$\underline{\mathcal{L}} = \underline{\mathcal{L}}_0 + \underline{\mathcal{L}}_1 \tag{3.75a}$$

with

$$\begin{aligned}(\underline{\mathcal{L}}_0)_{XYX'Y'} &= \frac{1}{\hbar}\Big[(\mathcal{H}_0)_{XX'}\delta_{YY'} - \delta_{XX'}(\mathcal{H}_0)_{Y'Y}\Big] \\ &= \frac{1}{\hbar}(H_{XX} - H_{YY})\delta_{XX'}\delta_{YY'}\end{aligned} \tag{3.75b}$$

and

$$(\underline{\mathcal{L}}_1)_{XYX'Y'} = \frac{1}{\hbar}\Big[(\mathcal{H}_1)_{XX'}\delta_{YY'} - \delta_{XX'}(\mathcal{H}_1)_{Y'Y}\Big]. \tag{3.75c}$$

Then we have

$$\underline{\mathcal{Q}}\underline{\mathcal{L}}_0 = \underline{0} = \underline{\mathcal{L}}_0\underline{\mathcal{Q}}\,. \tag{3.76}$$

Besides, for an arbitrary HAMILTON operators $\mathcal{H}$ we find the relation

$$\underline{\mathcal{P}\mathcal{L}\mathcal{P}} = \underline{0} \tag{3.77}$$

because of

$$(\underline{\mathcal{P}\mathcal{L}\mathcal{P}})_{XYX'Y'} = \delta_{XY} L_{XXX'X'}\delta_{X'Y'} \tag{3.78}$$

and (3.56b). Due to

$$\underline{\mathcal{P}\mathcal{L}}\rho_1 = \underline{\mathcal{P}\mathcal{L}\mathcal{P}}\rho\,, \tag{3.79}$$

the first term in (3.72) vanishes and there remains

$$\begin{aligned}\frac{d}{dt}\rho_{XX}(t) &= \int_{t_0}^{t} dt'\,[\underline{\mathcal{K}}(t-t')\rho_1(t')]_{XX} \\ &= \int_{t_0}^{t} dt' \sum_{X'} K_{XXX'X'}(t-t')\rho_{X'X'}(t')\end{aligned} \tag{3.80}$$

with

$$\begin{aligned}\underline{\mathcal{K}}(t) :=& -\underline{\mathcal{P}\mathcal{L}}\mathrm{e}^{-\mathrm{i}\underline{\mathcal{Q}\mathcal{L}}t}\underline{\mathcal{Q}\mathcal{L}} \\ =& -\underline{\mathcal{P}\mathcal{L}\mathcal{Q}}\mathrm{e}^{-\mathrm{i}\underline{\mathcal{Q}\mathcal{L}\mathcal{Q}}t}\underline{\mathcal{Q}\mathcal{L}\mathcal{P}}\,.\end{aligned} \tag{3.81}$$

These conversions base on (3.64) and the TAYLOR representation

$$\mathrm{e}^{\underline{\mathcal{B}}} = \sum_{k=0}^{\infty}\frac{\underline{\mathcal{B}}^k}{k!}\,. \tag{3.82}$$

Additionally, with

$$\sum_{X'} K_{XXX'X'}(t) = -\sum_{X'}\sum_{Y}\sum_{Y'} (\underline{\mathcal{P}\mathcal{L}} e^{-i\underline{\mathcal{Q}\mathcal{L}}t} \underline{\mathcal{Q}})_{XXYY'} L_{YY'X'X'} \tag{3.83}$$

and

$$\sum_{X'} L_{YY'X'X'} = \frac{1}{\hbar}\left(\sum_{X'} H_{YX'}\delta_{Y'X'} - \sum_{X'} \delta_{YX'} H_{X'Y'}\right) = 0 \tag{3.84}$$

we can prove the relation

$$\sum_{X'} K_{XXX'X'}(t) = 0\,. \tag{3.85}$$

This allows the representation

$$K_{XXXX}(t) = -\sum_{X'(\neq X)} K_{XXX'X'}(t) = -\sum_{X'(\neq X)} K_{X'X'XX}(t) \tag{3.86}$$

where the *symmetry* of $\underline{K}$ according to (3.81) was used for the last conversion. Defining

$$w_t(X|X') := K_{XXX'X'}(t) \tag{3.87}$$

as *transition rates,* we obtain

$$w_t(X|X') = w_t(X'|X)\,, \tag{3.88}$$

and (3.80) takes on the form of the *generalized master equation:*

$$\frac{d}{dt}P(X,t) = \int_{t_0}^{t} dt' \sum_{X'(\neq X)} \Big[w_{t-t'}(X|X')P(X',t') - w_{t-t'}(X'|X)P(X,t')\Big]. \tag{3.89}$$

For $w_{t-t'}(X'|X)$, the assumption

$$w_{t-t'}(X'|X) \sim w(X'|X)e^{-\lambda(t-t')} \tag{3.90}$$

often is a good approximation since, then, the influence of the past decreases exponentially with the distance $|t-t'|$ to the present. In the MARKOV*ian limit* $1/\lambda \to 0$ of 'short memory' we get

$$w_{t-t'}(X'|X) = w(X'|X)\delta(t'-t)\,. \tag{3.91}$$

This implies that the generalized master equation turns into the *ordinary master equation* given by (3.5a) [156]. *For reasons of clarity the next chapters exclusively treat ordinary master equations. The results for these can, without any complications, be transferred to the case of generalized master equations.* Then we arrive at *generalized* BOLTZMANN *equations* [47], *generalized* FOKKER-PLANCK *equations* [241] as well as *generalized* BOLTZMANN-FOKKER-PLANCK, *mean value,* and *covariance equations* which differ from the corresponding *ordinary* equations only by an additional time integration. This time integration allows the inclusion of *'memory effects'*.

3.2.4.4 Adiabatic Elimination of the Fast Variables

The derived master equation contains an enormous number of variables X_I of which only a few macroscopic variables $X_{s,J} \in \{X_I\}$ are of interest. Hence let us subdivide the variables $\{X_I\}$ into variables $\{X_{s,J}\}$ and $\{X_{f,K}\}$:

$$X \equiv (X_f, X_s)^{tr} . \tag{3.92}$$

Now we look for an equation for

$$P(X_s, t) := \sum_{X_f} P(X_f, X_s; t) . \tag{3.93}$$

It is reasonable to assume that the *macroscopic variables* $X_{s,J}$ change *slowly* whereas X_f summarizes the *fast* changing *microscopic variables.* As a consequence, the variables $X_{f,K}$ can, like in Sect. 3.2.3, be adiabatically eliminated so that a master equation for the macroscopic state vector X_s remains. *In contrast to the microscopic transition rates* $w(X'|X)$ *the macroscopic transition rates* $w_s(X'_s|X_s)$ *are, due to (3.37), usually asymmetrical.* For example, the *forward* and *reverse reaction rate* of a *chemical reaction* are in general different from each other (cf. [47]).

3.3 Properties

3.3.1 Normalization

From

$$\begin{aligned}
\frac{d}{dt}\sum_X P(X,t) &= \sum_{X,X'} w(X|X';t)P(X',t) - \sum_{X,X'} w(X'|X;t)P(X,t) \\
&= \sum_{X,X'} w(X|X';t)P(X',t) - \sum_{X',X} w(X|X';t)P(X',t) \\
&= 0
\end{aligned} \tag{3.94}$$

we conclude that the total probability $\sum_X P(X,t)$ of the system to be in *any* of the states $X \in \Gamma$ is constant with time. Usually the probabilities $P(X,t)$ are normalized to 1:

$$\sum_X P(X,t) \stackrel{!}{=} 1\,. \tag{3.95}$$

3.3.2 Non-negativity

If condition $P(X,t_0) \geq 0$ is fulfilled at an initial time t_0 for all states X, for all later times $t > t_0$ we have

$$P(X,t) \geq 0 \qquad \forall\, X \in \Gamma. \tag{3.96}$$

In order to prove this let us assume that a point in time $t' \geq t_0$ exists at which, for some state X'', the function $P(X,t)$ changes its sign for the *first* time, i.e.

$$P(X'',t') = 0\,, \qquad \frac{d}{dt}P(X'',t') < 0\,. \tag{3.97}$$

Then, in contrast to this, we would find

$$\begin{aligned}\frac{d}{dt}P(X'',t') &= \sum_{X'}\Big[w(X''|X';t')P(X',t') - w(X'|X'';t')\cdot 0\Big]\\ &\geq 0\end{aligned} \tag{3.98}$$

because of $w(X|X';t) \geq 0$ and $P(X',t') \geq 0$.

Therefore, together with the normalization condition, the interpretation of $P(X,t)$ as the *probability* of state X to occur at the time t is justified.

3.3.3 The Liouville Representation

Master equation (3.5) can also be written as a *linear system of differential equations*

$$\frac{d}{dt}P(X,t) = \sum_{X'} L_{XX'}(t)P(X',t) \tag{3.99}$$

or more briefly as a Liouville *equation* [171]:

$$\frac{d}{dt}P(t) = \underline{L}(t)P(t)\,. \tag{3.100a}$$

Here $\underline{L} \equiv (L_{XX'})$ is the LIOUVILLE *matrix* with the components

$$L_{XX'}(t) := w(X|X';t) - \delta_{XX'} \sum_Y w(Y|X';t) \tag{3.100b}$$

and $P(t)$ is a vector with the components $P(X,t)$. In the continuous case (3.5b) the KRONECKER symbol $\delta_{XX'}$ is to be replaced by DIRAC's *delta function* $\delta(X - X')$ (cf. p. xxv).

3.3.3.1 Time-Independent Transition Rates

In the following let the transition rates $w(X'|X;t)$ *be time-independent or of the form*

$$w(X'|X;t) := f(t) \cdot w'(X'|X) . \tag{3.101}$$

In the latter case (3.5a) can, by *time scaling*

$$\tau(t) := \int_{t_0}^{t} dt'\, f(t') , \tag{3.102}$$

be transformed into the master equation

$$\frac{d}{d\tau} P'(X,\tau) = \sum_{X'} \left[w'(X|X')P'(X',\tau) - w'(X'|X)P'(X,\tau) \right] \tag{3.103}$$

for $P'(X,\tau) := P(X,t(\tau))$ with *time-independent* transition rates $w'(X'|X)$. This follows from

$$\frac{dP(X,t)}{dt} = \frac{dP(X,t(\tau))}{d\tau}\frac{d\tau}{dt} = \frac{dP(X,t(\tau))}{d\tau} f(t) = \frac{dP'(X,\tau)}{d\tau} f(t) . \tag{3.104}$$

3.3.4 Eigenvalues

By *separation of variables,* i.e.

$$P(X,t) := P^{\lambda}(X) e^{\lambda(t-t_0)} , \tag{3.105}$$

(3.100) transforms into the *eigenvalue equation*

$$\underline{L} P^{\lambda} = \lambda P^{\lambda} \tag{3.106}$$

where P^λ denotes the eigenvector to the generally complex eigenvalue λ. It can be shown that the real parts of the eigenvalues λ are *non-positive* (cf. [107], pp. 92ff.):

$$\mathrm{Re}(\lambda) \le 0\,. \tag{3.107}$$

If there were an eigenvalue λ with $\mathrm{Re}(\lambda) > 0$, the corresponding solution would, in accordance with (3.105), exponentially grow with time and violate (3.95) from a certain time on.

It is the *stationary solution* $P(X,t) \equiv P^0(X)$ which is given by $dP(X,t)/dt = 0$ for all X and all times $t \ge t_0$ that belongs to the eigenvalue $\lambda = 0$. The *existence of at least one stationary solution* and, consequently, of the eigenvalue $\lambda = 0$ is guaranteed since the *rows* of the LIOUVILLE matrix $\underline{L} \equiv (L_{XX'})$ are *linearly dependent* (cf. p. xv) due to

$$\begin{aligned}\sum_X L_{XX'} &= \sum_X w(X|X') - \sum_X \delta_{XX'} \sum_Y w(Y|X') \\ &= \sum_X w(X|X') - \sum_Y w(Y|X') \\ &= 0\,.\end{aligned} \tag{3.108}$$

On certain conditions the stationary solution is also *unique* (cf. Sect. 3.3.5).

3.3.5 Convergence to the Stationary Solution

In order to show that $P(X,t)$ converges to a stationary solution $P^0(X)$ for time-independent transition rates, we introduce the *entropy-like* LIAPUNOV *function*

$$K(t) := \sum_X P(X,t) \ln \frac{P(X,t)}{P'(X,t)}\,. \tag{3.109}$$

Here let $P(X,t)$ and $P'(X,t)$ be two solutions of the master equation for different initial distributions $P(X,t_0)$ and $P'(X,t_0)$. Due to

$$\sum_X P(X,t) = 1\,, \qquad \sum_X P'(X,t) = 1\,, \tag{3.110}$$

and

$$\ln z \le z - 1 \tag{3.111}$$

we immediately find

$$K(t) = \sum_X P(X,t) \left(\ln \frac{P(X,t)}{P'(X,t)} + \frac{P'(X,t)}{P(X,t)} - 1 \right) \ge 0\,. \tag{3.112}$$

Moreover, we can derive

$$
\begin{aligned}
\frac{dK}{dt} &= \sum_X \left[\frac{dP(X,t)}{dt} \left(\ln \frac{P(X,t)}{P'(X,t)} + 1 \right) - \frac{dP'(X,t)}{dt} \frac{P(X,t)}{P'(X,t)} \right] \\
&= \sum_X \sum_{X'} \left[\Big(w(X|X')P(X',t) - w(X'|X)P(X,t) \Big) \left(\ln \frac{P(X,t)}{P'(X,t)} + 1 \right) \right. \\
&\qquad \left. - \Big(w(X|X')P'(X',t) - w(X'|X)P'(X,t) \Big) \frac{P(X,t)}{P'(X,t)} \right] \\
&= \sum_X \sum_{X'} w(X|X')P(X',t) \Big(\ln z - z + 1 \Big) \\
&\le 0
\end{aligned}
\tag{3.113}
$$

with

$$
z \equiv z(X, X'; t) := \frac{P(X,t)}{P'(X,t)} \frac{P'(X',t)}{P(X',t)} . \tag{3.114}
$$

Since $K(t)$ decreases monotonically due to (3.113) and has a lower bound because of (3.112), $K(t)$ approaches a *limit value* for $t \to \infty$ (cf. [207], p. 32) which is characterized by

$$
\frac{dK}{dt} = 0 . \tag{3.115}
$$

(3.115) is fulfilled if and only if $z(X, X'; t) = 1$ holds for all pairs (X, X') for which $w(X|X') \neq 0$. Now, by presupposing $w(X|X') \neq 0$ for all (X, X') with $X \neq X'$ and choosing for $P'(X, t)$ a stationary solution $P^0(X)$, we get

$$
\frac{P(X,t)}{P^0(X)} = \frac{P(X',t)}{P^0(X')} = const. = 1 . \tag{3.116}
$$

Therefore, every solution $P(X,t)$ of the master equation converges to the stationary solution $P^0(X)$ for $t \to \infty$. In particular, this means that $P^0(X)$ is the only existing stationary solution.

3.4 Solution Methods

3.4.1 Stationary Solution and Detailed Balance

In order to evaluate the stationary solution we must solve the linear system of equations

$$\begin{aligned} 0 &= \sum_{X'} L_{XX'} P^0(X') \\ &= \sum_{X'} \Big[w(X|X') P^0(X') - w(X'|X) P^0(X) \Big] . \end{aligned} \tag{3.117}$$

If, in addition to this, the condition

$$w(X|X') P^0(X') = w(X'|X) P^0(X) \qquad \forall\, X', X \in \Gamma \tag{3.118}$$

is fulfilled (i.e. the flows from states X' into states X and the inverse flows from states X into states X' cancel out each other *separately,* not only in the sum), we call this a situation of *detailed balance.* In such a case we directly obtain as the stationary solution

$$P^0(X_n) = P^0(X_0) \prod_{i=1}^{n} \frac{w(X_i|X_{i-1})}{w(X_{i-1}|X_i)} = P^0(X_0) \frac{w(\mathcal{C}_n)}{w(\mathcal{C}_n^{-1})} \tag{3.119}$$

where

$$\mathcal{C}_n := X_0 \to X_1 \to \cdots \to X_n \tag{3.120}$$

is a *path* from X_0 to $X \equiv X_n$ with non-vanishing transition rates $w(X_i|X_{i-1})$ and $w(X_{i-1}|X_i)$ $(i = 1, \ldots, n)$. $\mathcal{C}_n^{-1}$ denotes the *inverse path*

$$\mathcal{C}_n^{-1} := X_n \to \cdots \to X_1 \to X_0 \tag{3.121}$$

of $\mathcal{C}_n$. Moreover, for $n \geq 1$ the following abbreviations were used:

$$w(\mathcal{C}_n) := \prod_{i=1}^{n} w(X_i|X_{i-1}) \, , \tag{3.122a}$$

$$w(\mathcal{C}_n^{-1}) := \prod_{i=1}^{n} w(X_{i-1}|X_i) \, . \tag{3.122b}$$

$P^0(X_0)$ is obtained from the normalization condition (3.95). The result is

$$P^0(X_0) = \left[1 + \sum_{n(\neq 0)} \prod_{i=1}^{n} \frac{w(X_i|X_{i-1})}{w(X_{i-1}|X_i)} \right]^{-1} . \tag{3.123}$$

Obviously, for detailed balance the relation

$$\frac{w(\widehat{\mathcal{C}}_n)}{w(\widehat{\mathcal{C}}_n^{-1})} = 1 \tag{3.124}$$

must be fulfilled for all *cycles*

$$\widehat{\mathcal{C}}_n := X \to X_1 \to \cdots \to X_{n-1} \to X\,. \tag{3.125}$$

Conversely, detailed balance holds if condition (3.124) is satisfied for all cycles $\widehat{\mathcal{C}}_n$. Since the cycles $\widehat{\mathcal{C}}_n$ can be decomposed into *elementary cycles,* it is sufficient to prove (3.124) for these. (Cf. [294], pp. 328ff.)

An additional property of detailed balance is that the eigenvalues of the corresponding LIOUVILLE operator $\underline{L}$ are all *real* and thus *non-positive* (cf. (3.107) and [107], pp. 92ff.). This is the case because the eigenvalue equation

$$\sum_{X'} L_{XX'} P^{\lambda}(X') = \lambda P^{\lambda}(X) \tag{3.126}$$

can, with

$$\psi_{\lambda}(X) := \frac{P^{\lambda}(X)}{\sqrt{P^0(X)}}\,, \tag{3.127a}$$

be written as

$$\sum_{X'} L^{\mathrm{s}}_{XX'} \psi_{\lambda}(X') = \lambda \psi_{\lambda}(X)\,. \tag{3.127b}$$

Here

$$L^{\mathrm{s}}_{XX'} := L_{XX'} \frac{\sqrt{P^0(X')}}{\sqrt{P^0(X)}} = w(X|X') \frac{\sqrt{P^0(X')}}{\sqrt{P^0(X)}} - \delta_{XX'} \sum_{Y} w(Y|X') \tag{3.127c}$$

is a *symmetrical (self-adjoint)* matrix, i.e. its eigenvalues are *real* (cf. [20], pp. 235ff., and [165], pp. 255ff.). The symmetry results from

$$w(X|X') \frac{\sqrt{P^0(X')}}{\sqrt{P^0(X)}} = w(X|X') \frac{P^0(X')}{P^0(X)} \frac{\sqrt{P^0(X)}}{\sqrt{P^0(X')}} = w(X'|X) \frac{\sqrt{P^0(X)}}{\sqrt{P^0(X')}} \tag{3.128}$$

by using the condition of detailed balance.

If detailed balance is not fulfilled, there exist only limited possibilities for an explicit solution of the master equation. Worth mentioning are the *transition factor method* [111], the KIRCHHOFF *method* [63, 249, 287], and a few others (cf. e.g. [104]).

3.4.2 Time-Dependent Solution

According to the LIOUVILLE representation the general time-dependent solution of the master equation has the form

$$P(X,t) = \sum_{\lambda} P^{\lambda}(X)\mathrm{e}^{-\lambda(t-t_0)} \tag{3.129}$$

if the eigenvalues λ of

$$\underline{L} \equiv \Big(L_{XX'}(t)\Big) = \Big(w(X|X';t) - \delta_{XX'}\sum_{Y} w(Y|X';t)\Big) \tag{3.130}$$

are not degenerate. Here $P^{\lambda}(X)$ is the eigenfunction of $\underline{L}$ to the eigenvalue λ. If *degenerate* (multiple) *eigenvalues* occur, the solution is of the form

$$P(X,t) = \sum_{\lambda}\sum_{i=0}^{n_\lambda - 1} P_i^{\lambda}(X)p_i^{\lambda}(t)\mathrm{e}^{-\lambda(t-t_0)} \tag{3.131}$$

where n_λ is the *degree of degeneration* (the *multiplicity*) of the eigenvalue λ and $p_i^{\lambda}(t)$ is a polynom in t of degree i (cf. [28, 203]).

Formally the general solution of (3.100) reads

$$\begin{aligned} P(t) &= \mathrm{e}^{\underline{L}(t-t_0)}P(t_0) \\ &= \sum_{k=0}^{\infty} \frac{\underline{L}^k (t-t_0)^k}{k!} P(t_0) \end{aligned} \tag{3.132}$$

and is thus *uniquely* determined by the *initial distribution* $P(t_0)$. In the case of *time-dependent* transition rates we have

$$P(X,t) = \sum_{X_0} P(X,t|X_0,t_0)P(X_0,t_0)\,, \tag{3.133}$$

i.e.

$$P(t) = \underline{P}(t,t_0)P(t_0)\,. \tag{3.134a}$$

Here the *propagator* $\underline{P}(t,t_0) \equiv \Big(P(X,t|X_0,t_0)\Big)$ must fulfill the conditions

$$\frac{d}{dt}\underline{P}(t,t_0) = \underline{L}(t)\cdot\underline{P}(t,t_0) \tag{3.134b}$$

and

$$\underline{P}(t_0,t_0) = \underline{1} \tag{3.134c}$$

with the *unity matrix* $\underline{1} \equiv \left(\delta_{XX_0}\right)$. The formal solution is (as one can confirm by applying the *second fundamental theorem of calculus*)

$$\underline{P}(t, t_0) = \mathrm{T}\exp\left(\underline{L}, t, t_0\right) \tag{3.135a}$$

where

$$\begin{aligned}\mathrm{T}\exp\left(\underline{L}, t, t_0\right) := \underline{1} + \int_{t_0}^{t} dt'\, \underline{L}(t') + \int_{t_0}^{t} dt' \int_{t_0}^{t'} dt''\, \underline{L}(t')\underline{L}(t'') \\ + \int_{t_0}^{t} dt' \int_{t_0}^{t'} dt'' \int_{t_0}^{t''} dt'''\, \underline{L}(t')\underline{L}(t'')\underline{L}(t''') + \ldots\end{aligned} \tag{3.135b}$$

is the *time-ordered exponential function.* Unfortunately, this relation is not of great use because it can hardly be analysed explicitly. For a *numerical solution* of the master equation often *Monte-Carlo methods* are applied *('stochastic simulation')* [23].

3.4.3 *'Path Integral' Solution*

3.4.3.1 Path Probabilities

In the following we will introduce a new solution method for the discrete master equation which allows, in contrast to the methods that were discussed up to now, the calculation of path or process dependent quantities. For its derivation we start from equation

$$\begin{aligned}P(X, t + \Delta t) &= P(X, t) + \Delta t \sum_{X'(\neq X)} w(X|X'; t)P(X', t) \\ &\quad - \Delta t \sum_{X'(\neq X)} w(X'|X; t)P(X, t) \\ &= \sum_{X'} P(X, t + \Delta t|X', t)P(X', t)\end{aligned} \tag{3.136a}$$

which is equivalent to master equation (3.5a) in the limit $\Delta t \to 0$. Here the quantities

$$\begin{aligned}P(X, t + \Delta t|X', t) = \delta_{XX'} + \Delta t \sum_{Y(\neq X)} \left[w(X|Y; t)\delta_{X'Y}\right] \\ - \delta_{XX'}\, \Delta t \sum_{Y(\neq X)} w(Y|X; t)\end{aligned} \tag{3.136b}$$

have the meaning of *transition probabilities* (cf. (3.21), (3.22)) which define the *propagator* $\underline{P}(t + \Delta t, t) \equiv \Big(P(X, t + \Delta t|X', t)\Big)$ (cf. pp. 71f.). Obviously, the solution of (3.136) is

$$P(X,t) = \lim_{n\to\infty} \sum_{X_{n-1}} \sum_{X_{n-2}} \cdots \sum_{X_0} \left[\prod_{i=1}^{n} P(X_i, t_i|X_{i-1}, t_{i-1}) \right] P(X_0, t_0) \quad (3.137)$$

where $X_n := X$, $t_i := t_0 + i\,\Delta t$, and $\Delta t := (t - t_0)/n$. This representation is the basis of HAKEN's approximate *path integral solution* [105].

In the following we restrict ourselves to the case of time-independent transition rates $w(X'|X;t) \equiv w(X'|X)$ and try to derive a *contracted* form of (3.137). In order to achieve this simplification we utilize the circumstance that the described system normally does not change its state X_i over many time periods Δt. This means that within a time interval

$$T := t - t_0 \quad (3.138)$$

the system will change its state only for a finite number of times. So let X_0, $X_1, \dots, X_n$ be the sequence of states in which the system is found between times t_0 and t.

$$\mathcal{C}_n := X_0 \to \cdots \to X_{n-1} \to X_n \quad (3.139)$$

with $X_n \equiv X$ will be called the *path* of the system. Furthermore,

$$P(\mathcal{C}_n, T) \equiv P(X_0 \to \cdots \to X_{n-1} \to X_n, T) \quad (3.140)$$

will be denoted as the *(path) probability* with which the system takes the path $\mathcal{C}_n$ during the time interval T. Then the probability $P(X,t) = P(X, t_0 + T)$ to find state X at the time t can be expressed as a sum over all path probabilities:

$$P(X, t_0 + T) = \sum_{n=0}^{\infty} \sum_{X_{n-1}(\neq X)} \cdots \sum_{X_0(\neq X_1)} P(X_0 \to \cdots \to X_{n-1} \to X, T)\,. \quad (3.141)$$

This can be abbreviated by

$$P(X, t_0 + T) = \sum_{n=0}^{\infty} \sum_{\mathcal{C}_n} P(\mathcal{C}_n, T)\,. \quad (3.142)$$

The path integral solution (3.141) of the discrete master equation is the desired contraction of formula (3.137). It was suggested by EMPACHER [63]. In order to calculate the explicit form of the path probabilities $P(\mathcal{C}_n, T)$ we will extend an idea of WEIDLICH (cf. [289], Sect. 8.1). For this purpose, we write the master equation, after introducing the matrices

$$\underline{W} \equiv \left(W_{XX'}\right) := \left(w(X|X')\right) \quad \text{with} \quad W_{XX} := 0 \tag{3.143a}$$

and

$$\underline{D} \equiv \left(D_{XX'}\right) := \left(W_X \delta_{XX'}\right) \quad \text{with} \quad W_X := \sum_{Y(\neq X)} w(Y|X)\,, \tag{3.143b}$$

in the vectorial form

$$\frac{d}{dt} P(t) = (\underline{W} - \underline{D}) P(t) \tag{3.144}$$

(cf. (3.100)). W_X is the *total departure rate* from state X, i.e. the total rate of transitions to all other states. By applying the LAPLACE transformation

$$G(u) := \int_{t_0}^{\infty} dt\, \mathrm{e}^{-u(t-t_0)} P(t) = \int_{0}^{\infty} dT\, \mathrm{e}^{-uT} P(t_0 + T) \tag{3.145}$$

we get

$$u G(u) - P(t_0) = (\underline{W} - \underline{D}) G(u)\,, \tag{3.146}$$

i.e.

$$G(u) = (u \underline{1} - \underline{W} + \underline{D})^{-1} P(t_0)\,. \tag{3.147}$$

The desired result is finally obtained by the inverse LAPLACE transformation

$$P(t) = P(t_0 + T) = \frac{1}{2\pi \mathrm{i}} \int_{c-\mathrm{i}\infty}^{c+\mathrm{i}\infty} du\, \mathrm{e}^{uT} G(u) \tag{3.148}$$

of $G(u)$. This yields

$$\begin{aligned} P(t_0 + T) &= \frac{1}{2\pi \mathrm{i}} \int_{-c-\mathrm{i}\infty}^{-c+\mathrm{i}\infty} du\, \mathrm{e}^{-uT} (\underline{D} - u\underline{1} - \underline{W})^{-1} P(t_0) \\ &= \frac{1}{2\pi \mathrm{i}} \int_{-c-\mathrm{i}\infty}^{-c+\mathrm{i}\infty} du\, \mathrm{e}^{-uT} [\underline{1} - (\underline{D} - u\underline{1})^{-1} \underline{W}]^{-1} (\underline{D} - u\underline{1})^{-1} P(t_0)\,. \end{aligned} \tag{3.149}$$

The geometric series

$$[\underline{1} - (\underline{D} - u\underline{1})^{-1}\underline{W}]^{-1} = \sum_{n=0}^{\infty}[(\underline{D} - u\underline{1})^{-1}\underline{W}]^n \tag{3.150}$$

will be convergent if the arbitrary (but fixed) constant c is suitably chosen. Now, with

$$[(\underline{D} - u\underline{1})^{-1}\underline{W}]_{XX'} = \sum_{Y}(W_X - u)^{-1}\delta_{XY}w(Y|X') = \frac{w(X|X')}{W_X - u} \tag{3.151}$$

and (3.143a), we find

$$\sum_{n=0}^{\infty}\sum_{X_0}\left\{[(\underline{D} - u\underline{1})^{-1}\underline{W}]^n(\underline{D} - u\underline{1})^{-1}\right\}_{XX_0} P(X_0, t_0)$$
$$= \sum_{n=0}^{\infty}\sum_{X_{n-1}(\neq X)}\cdots\sum_{X_0(\neq X_1)}\frac{w(X_0 \to \cdots \to X_{n-1} \to X)}{\prod_{j=0}^{n}(W_{X_j} - u)}P(X_0, t_0) \tag{3.152a}$$

where

$$w(C_n) \equiv w(X_0 \to \cdots \to X_{n-1} \to X) := \begin{cases} \delta_{XX_0} & \text{if } n = 0 \\ \prod_{j=1}^{n} w(X_j|X_{j-1}) & \text{if } n \geq 1\,. \end{cases} \tag{3.152b}$$

Inserting (3.152) into (3.149) and comparing the result with (3.141) we gain the explicit expression for the path probabilities $P(C_n, T)$. It reads

$$P(C_n, T) = \frac{1}{2\pi\mathrm{i}}\int_{-c-\mathrm{i}\infty}^{-c+\mathrm{i}\infty} du\,\frac{\mathrm{e}^{-uT}}{\prod_{j=0}^{n}(W_{X_j} - u)}w(C_n)P(X_0, t_0)\,. \tag{3.153}$$

We can evaluate this integral by using the *residue theorem* (cf. [266], pp. 112ff.). If, for example, all rates W_{X_l} are equal ($W_{X_l} \equiv W_X$), we obtain

$$P(C_n, T) = \frac{T^n}{n!}\mathrm{e}^{-W_X T}w(C_n)P(X_0, t_0)\,. \tag{3.154a}$$

If, however, all rates W_{X_i} are different, we get

$$P(C_n, T) = \sum_{i=0}^{n} \frac{e^{-W_{X_i} T}}{\prod\limits_{\substack{j=0 \\ (j \neq i)}}^{n} (W_{X_j} - W_{X_i})} w(C_n) P(X_0, t_0) . \tag{3.154b}$$

Obviously, the expressions (3.154) grow polynomially in T for $T \approx 0$ and decrease exponentially for $T \to \infty$.

3.4.3.2 Path Occurence Times and Their Probability Distribution

In numerical evaluations of formula (3.141) by means of computers we certainly cannot sum up over infinitely many different paths. The summation has rather to be limited to those paths which provide the most significant contributions. In order to find out which paths are negligible we will calculate the probability distribution

$$P(T|C_n) := \frac{P(C_n, T)}{P(C_n)} \tag{3.155}$$

of the *occurence times*[3] T of a path C_n given that the system traverses this path *some time*. Here

$$P(C_n) := \int_0^{\infty} dT \; P(C_n, T) \tag{3.156}$$

is a normalization factor which reflects the probability with which the path C_n is traversed *at all*. By inserting (3.153) into (3.155) we find that the *occurence time distribution* $P(T|C_n)$ depends only on the traversed states $X_0, \dots, X_n$ but not on their order.

From the occurence time distribution $P(T|C_n)$ we expect that it will become maximal after a certain time interval $\widehat{T}$: For $T \ll \widehat{T}$ there is too little time for traversing the n states. In contrast, for $T \gg \widehat{T}$ the probability is high that already further states $X_{n+1}, X_{n+2}, \dots$ were traversed. More precisely, we have

$$P(T|C_n) \overset{T \to 0}{\longrightarrow} C_1(C_n) T^n \tag{3.157a}$$

and

$$P(T|C_n) \overset{T \to \infty}{\longrightarrow} C_2(C_n) e^{-W_X T} \tag{3.157b}$$

[3] Each time $t = t_0 + T$ at which the path C_n can be found is understood as an *occurence time* T of the respective path. Consequently, the occurence times of a path C_n must be distinguished from the *waiting time* which means the time period until the path C_n appears for the *first* time.

which can be derived from (3.154). Here $C_1(\mathcal{C}_n)$, $C_2(\mathcal{C}_n)$ are constants and W_X is the *smallest* of the rates $\{W_{X_0}, \ldots, W_{X_n}\}$ which were assumed to be different from each other.

For the calculation of the *mean value* $\langle T\rangle_{\mathcal{C}_n}$ and *variance*

$$\sigma_{\mathcal{C}_n} := \left\langle \left(T - \langle T\rangle_{\mathcal{C}_n}\right)^2 \right\rangle_{\mathcal{C}_n} = \langle T^2\rangle_{\mathcal{C}_n} - \left(\langle T\rangle_{\mathcal{C}_n}\right)^2 \tag{3.158}$$

of the occurence times T of path $\mathcal{C}_n$ we require the quantities

$$\langle T^k\rangle_{\mathcal{C}_n} := \int_0^\infty dT\, T^k P(T|\mathcal{C}_n)\,. \tag{3.159}$$

By means of the residue theorem (cf. [266], pp. 112ff.) we find

$$\begin{aligned}
\int_0^\infty dT\, T^k P(\mathcal{C}_n, T) &= \int_0^\infty dT\, T^k \frac{1}{2\pi \mathrm{i}} \int_{-c-\mathrm{i}\infty}^{-c+\mathrm{i}\infty} du\, \frac{\mathrm{e}^{-uT}}{\prod_{i=0}^{n}(W_{X_i} - u)} w(\mathcal{C}_n) P(X_0, t_0) \\
&= \frac{1}{2\pi \mathrm{i}} \int_{-c-\mathrm{i}\infty}^{-c+\mathrm{i}\infty} du\, \frac{w(\mathcal{C}_n) P(X_0, t_0)}{\prod_{i=0}^{n}(W_{X_i} - u)} \left(-\frac{d}{du}\right)^k \int_0^\infty dT\, \mathrm{e}^{-uT} \\
&= w(\mathcal{C}_n) P(X_0, t_0) \frac{k!}{2\pi \mathrm{i}} \int_{-c-\mathrm{i}\infty}^{-c+\mathrm{i}\infty} du\, \frac{1}{u^{k+1} \prod_{i=0}^{n}(W_{X_i} - u)} \\
&= w(\mathcal{C}_n) P(X_0, t_0) \left(\frac{d}{du}\right)^k \prod_{i=0}^{n} \frac{1}{(W_{X_i} - u)}\Bigg|_{u=0}\,.
\end{aligned} \tag{3.160}$$

Thus we obtain

$$P(\mathcal{C}_n) = w(\mathcal{C}_n) P(X_0, t_0) \prod_{i=0}^{n} \frac{1}{W_{X_i}}\,, \tag{3.161}$$

$$\langle T\rangle_{\mathcal{C}_n} := \int_0^\infty dT\, T P(T|\mathcal{C}_n) = \sum_{i=0}^{n} \frac{1}{W_{X_i}}\,, \tag{3.162}$$

and

$$\sigma_{\mathcal{C}_n} := \int_0^\infty dT \left(T - \langle T\rangle_{\mathcal{C}_n}\right)^2 P(T|\mathcal{C}_n) = \sum_{i=0}^{n} \frac{1}{(W_{X_i})^2}\,. \tag{3.163}$$

These results can also be derived from (3.154b) by means of the formulas

$$\int_0^\infty dT\ T^k \mathrm{e}^{-W_{X_i} T} = \frac{k!}{(W_{X_i})^{k+1}} \tag{3.164}$$

and

$$\sum_{i=0}^{n} \frac{1}{\prod\limits_{\substack{j=0 \\ (j \neq i)}}^{n} (W_{X_j} - W_{X_i})} = 0 \,. \tag{3.165}$$

Assuming that the rates W_{X_i} are equal ($W_{X_i} \equiv W_X$) we recognize that the *relative width*

$$\frac{\sqrt{\sigma_{\mathcal{C}_n}}}{\langle T \rangle_{\mathcal{C}_n}} = \frac{1}{\sqrt{n+1}} \tag{3.166}$$

of distribution $P(T|\mathcal{C}_n)$ becomes smaller with increasing number n of transitions.

When calculating *approximate solutions* for $P(X,t) = P(X, t_0 + T)$, only such paths $\mathcal{C}_n$ will be taken into account which fulfill the condition

$$|T - \langle T \rangle_{\mathcal{C}_n}| \leq \widehat{a}\sqrt{\sigma_{\mathcal{C}_n}} \tag{3.167}$$

with the *accuracy parameter* $\widehat{a}$. If, for example, we choose $\widehat{a} = 3$, about 99% of the distribution function $P(T|\mathcal{C}_n)$ will lie between $T_1 := \langle T \rangle_{\mathcal{C}_n} - \widehat{a}\sqrt{\sigma_{\mathcal{C}_n}}$ and $T_2 := \langle T \rangle_{\mathcal{C}_n} + \widehat{a}\sqrt{\sigma_{\mathcal{C}_n}}$. As a consequence, for any given time $t = t_0 + T$ those paths which satisfy condition (3.167) will reconstruct about 99% of the distribution function $P(X, t_0 + T)$ defined by formula (3.141).

3.4.3.3 Applications

Compared to other solution methods the advantage of the path integral solution is that it allows to determine process or path dependent quantities. For example, we can calculate the probability with which the considered system takes *'desired' paths* that consist of *selected states* only. In the same way we can evaluate the probability with which the system visits an unwanted state in the course of time and thus possibly takes a *'catastrophic' path.* These probabilities are of special interest for *prognoses* (e.g. *scenario techniques*) or for the *control* of the temporal behaviour of a technical or other system *(process control).*

Another interesting application area is the *event analysis (survival analysis)* [51, 275]. Here one investigates the distribution of *waiting times* until a certain *event* happens, i.e. a certain state is reached, and the paths taken. In the context of social sciences the events (states) describe personal circumstances. By means of

(3.141) and (3.154) it is now possible to carry out even *complex* event analyses for which the distinction of *many* different states is necessary or for which the states are not traversed in a given order (but possibly several times). For the description of the *event histories* of individuals, however, one often requires time-dependent transition rates $w(X'|X;t)$. Nevertheless, the mathematical results of this section can be expanded to such cases if the time-dependence of the transition rates has the special form (3.101).

3.5 Mean Value and Covariance Equations

In most cases the master equation is used for the delineation of the dynamics of the *individual* elements of a system and thus corresponds to a *microscopic level of description.* However, not only because the calculation of the probability distribution $P(X,t)$ is very laborious one is often merely interested in the mean value and covariance equations which are derivable from the master equation. These reflect the *collective* behaviour of the elements and, therefore, are also called *macroscopic equations.* Insofar, the master equation allows a microscopic foundation of the macroscopic dynamics.

Having defined the *mean value* $\langle f(X,t)\rangle$ of a function $f(X,t)$ as usual by

$$\langle f\rangle_t \equiv \langle f(X,t)\rangle := \sum_X f(X,t)P(X,t)\,, \tag{3.168}$$

we directly obtain, from the master equation (3.5a), the *mean value equations*

$$\begin{aligned}
\frac{d\langle X_I\rangle}{dt} &= \sum_X X_I \frac{dP(X,t)}{dt} \\
&= \sum_X \sum_{X'} \Big[X_I w(X|X';t)P(X',t) - X_I w(X'|X;t)P(X,t)\Big] && (3.169)\\
&= \sum_X \sum_{X'} (X'_I - X_I) w(X'|X;t)P(X,t) && (3.170)\\
&= \sum_X m_I(X,t)P(X,t)\,, && (3.171)
\end{aligned}$$

i.e.

$$\frac{d\langle X_I\rangle}{dt} = \langle m_I(X,t)\rangle\,. \tag{3.172}$$

In the derivation of (3.172) we interchanged the order of X and X' from (3.169) to (3.170) and introduced the quantities

$$m_I(X,t) := \sum_{X'} \Delta X'_I w(X'|X;t) \quad \text{with} \quad \Delta X'_I := X'_I - X_I \tag{3.173}$$

which are called the *first jump moments*[4] (cf. [151], pp. 130ff.). In complete analogy, for the *covariances*

$$\sigma_{IJ} \equiv \sigma_{IJ}(t) := \Big\langle (X_I - \langle X_I \rangle_t)(X_J - \langle X_J \rangle_t) \Big\rangle_t = \langle X_I X_J \rangle - \langle X_I \rangle \langle X_J \rangle \,, \tag{3.174}$$

we find the *covariance equations*

$$\frac{d\sigma_{IJ}}{dt} = \Big\langle m_{IJ}(X,t) \Big\rangle + \Big\langle (X_I - \langle X_I \rangle) m_J(X,t) \Big\rangle + \Big\langle (X_J - \langle X_J \rangle) m_I(X,t) \Big\rangle \tag{3.175}$$

with the *second jump moments*

$$m_{IJ}(X,t) := \sum_{X'} \Delta X'_I \Delta X'_J w(X'|X;t) \tag{3.176}$$

(cf. [151], pp. 130ff.).

Equations (3.172) and (3.175) are no *closed equations* because they depend on the unknown mean values $\langle m_I(X,t) \rangle$ and $\langle m_{IJ}(X,t) \rangle$ respectively. However, if the distribution $P(X,t)$ is *sharply peaked*, i.e. as long as all variances σ_{II} are *small*, a first order TAYLOR approximation can be carried out. Then, with

$$\langle f(X,t) \rangle \approx \left\langle f(\langle X \rangle, t) + \sum_J (X_J - \langle X_J \rangle) \frac{\partial f(\langle X \rangle, t)}{\partial \langle X_J \rangle} \right\rangle = f(\langle X \rangle, t) \tag{3.177}$$

and

$$\begin{aligned} &\Big\langle (X_I - \langle X_I \rangle) f(X,t) \Big\rangle \\ &\quad \approx \left\langle (X_I - \langle X_I \rangle) f(\langle X \rangle, t) + (X_I - \langle X_I \rangle) \sum_K (X_K - \langle X_K \rangle) \frac{\partial f(\langle X \rangle, t)}{\partial \langle X_K \rangle} \right\rangle \\ &\quad = \sum_K \sigma_{IK} \frac{\partial f(\langle X \rangle, t)}{\partial \langle X_K \rangle} \,, \end{aligned} \tag{3.178}$$

closed equations result, namely the *approximate mean value equations*

$$\frac{\partial \langle X_I \rangle}{\partial t} \approx m_I(\langle X \rangle, t) \tag{3.179}$$

and the *approximate covariance equations*

$$\frac{\partial \sigma_{IJ}}{\partial t} \approx m_{IJ}(\langle X \rangle, t) + \sum_K \left(\sigma_{IK} \frac{\partial m_J(\langle X \rangle, t)}{\partial \langle X_K \rangle} + \sigma_{JK} \frac{\partial m_I(\langle X \rangle, t)}{\partial \langle X_K \rangle} \right) . \tag{3.180}$$

---

[4] Some authors denote the jump moments as *derivate moments* or *transition moments*.

The assumption of small variances is often justified since in many cases the state X_0 at a initial time t_0 is known by *measurement* or *preparation*, i.e.

$$P(X, t_0) = \delta_{XX_0} .$$

Consequently, the variances are zero at the initial time t_0 and remain small at least for a certain time interval which can be determined from the approximate covariance equations (cf. Sect. 12.4.1). A considerable increase of the covariances in the course of time normally indicates a *phase transition*, i.e. a sudden change of the macroscopic system properties (cf. Sect. 12.4.1).

In the *one-dimensional* case the *formal solution* of the approximate variance equation can be written as

$$\sigma(t) \equiv \sigma(\langle X\rangle_t) = \sigma(t_0)\left[\frac{m_{(1)}(\langle X\rangle_t)}{m_{(1)}(\langle X\rangle_{t_0})}\right]^2 + \left[m_{(1)}(\langle X\rangle_t)\right]^2 \int\limits_{\langle X\rangle_{t_0}}^{\langle X\rangle_t} dX' \, \frac{m_{(2)}(X')}{\left[m_{(1)}(X')\right]^3} \tag{3.181}$$

where $\langle X\rangle_t$ satisfies the approximate mean value equation

$$\frac{d\langle X\rangle_t}{dt} = m_{(1)}(\langle X\rangle_t) \tag{3.182}$$

(cf. [103], p. 74). Here $m_{(1)}$ denotes the first and $m_{(2)}$ the second jump moment.

Chapter 4
BOLTZMANN-Like Equations

4.1 Introduction

In the following we will assume to be confronted with systems that consist of N *elements (subsystems)* α. If the dynamics of the system is mainly given by *pair interactions* of the elements (like e.g. in a rarefied gas) it is plausible to use BOLTZMANN *equations* for its description. A typical example of application is a *chemical reaction*

$$B + C \rightleftharpoons BC \tag{4.1}$$

where the different states $x_\alpha \in \{B, C\}$ distinguish whether an atom α is bounded or not (cf. [11, 167]). Apart from pair interactions, the more general 'BOLTZMANN-*like equations*' additionally take into account *spontaneous transitions* and are the main topic of Chaps. 9 and 10. Interrelations and differences between spontaneous transitions and direct pair interactions will be discussed in detail in Sect. 4.5.

The BOLTZMANN equation, like the master equation, has a structure of the form (3.1). However, the flows of the BOLTZMANN equation are proportional to the probabilities of the states of *both* interacting subsystems. They depend on the *square* of the probability distribution and thus the solution is difficult. In the case of the *gaskinetic* BOLTZMANN *equation* the *stationary* solution can be found by means of the H*-theorem.* An evaluation of the *time-dependent* solution is usually very complicated and bases, in most cases, on a recursive method.

The derivation of the BOLTZMANN-like equations can be subdivided into the following steps: First, we combine the states

$$x_\alpha = (x_{\alpha 1}, \dots, x_{\alpha i}, \dots, x_{\alpha m})^{\text{tr}} \in \gamma_\alpha \tag{4.2}$$

of all N elements α in the state vector

$$X \equiv (\dots, X_I, \dots)^{\text{tr}} := (x_1, \dots, x_\alpha, \dots, x_N)^{\text{tr}} \tag{4.3}$$

of the total system which has M components ($M := mN$, $I := (\alpha, i)^{\text{tr}}$). γ_α means the set of all possible states x of element α and will be denoted as γ_α*-space*. For

D. Helbing, *Quantitative Sociodynamics*, 2nd ed.,
DOI 10.1007/978-3-642-11546-2_4,

reasons of simplicity we have assumed by (4.2) that the state space γ_α has the same dimension m for each subsystem. This is no restriction since, if a system can be characterized by a *smaller* number $m_\alpha < m$ of variables, the surplus components $x_{\alpha j}$ with $m_\alpha < j \le m$ can be set to zero.

Next we confine ourselves to spontaneous transitions and pair interactions, i.e. simultaneous interactions between three or more subsystems will be neglected (though a treatment of these is also possible; cf. Sect. 5.4 and [40, 41, 277]). The resulting equations can be reduced to equations for the probability distributions $P_\alpha(x_\alpha, t)$ of the subsystems α but they depend on the *pair distribution functions* $P_{\alpha\beta}(x_\alpha, x_\beta; t)$. In order to obtain *closed* equations we assume the *factorization* of the pair distribution functions (i.e. the *statistical independence* of the elements). Corrections for cases where this assumption is not justified are the topic of Sect. 5.4.

Finally, if we distinguish only a few *types a* of elements, the number of equations reduces drastically. The result are the BOLTZMANN-like equations.

4.2 Derivation

First, for the derivation of the BOLTZMANN-like equations from master equation (3.5) we introduce the vectors

$$X^\alpha := (\mathbf{0}, \dots, \mathbf{0}, x_\alpha, \mathbf{0}, \dots, \mathbf{0})^{\mathrm{tr}}, \tag{4.4}$$

$$X^{\alpha_1 \dots \alpha_k} := \sum_{i=1}^{k} X^{\alpha_i}, \tag{4.5}$$

and

$$Y_{\alpha_1 \dots \alpha_k} := X - X^{\alpha_1 \dots \alpha_k} + Y^{\alpha_1 \dots \alpha_k}. \tag{4.6}$$

They allow the decomposition of the transition rates

$$w(X'|X;t) := \begin{cases} w_{\alpha_1}(Y^{\alpha_1}|X^{\alpha_1};t) & \text{if } X' = Y_{\alpha_1} \\ w_{\alpha_1\alpha_2}(Y^{\alpha_1\alpha_2}|X^{\alpha_1\alpha_2};t) & \text{if } X' = Y_{\alpha_1\alpha_2} \\ \vdots & \vdots \\ w_{\alpha_1\dots\alpha_N}(Y^{\alpha_1\dots\alpha_N}|X^{\alpha_1\dots\alpha_N};t) & \text{if } X' = Y_{\alpha_1\dots\alpha_N} \\ 0 & \text{otherwise} \end{cases} \tag{4.7}$$

into contributions

$$w_{\alpha_1\dots\alpha_k}(Y^{\alpha_1\dots\alpha_k}|X^{\alpha_1\dots\alpha_k};t) \equiv w_{\alpha_1\dots\alpha_k}(y_{\alpha_1}, \dots, y_{\alpha_k}|x_{\alpha_1}, \dots, x_{\alpha_k};t) \tag{4.8}$$

which describe the interactions between k subsystems. In order to exclude self-interactions of an element we set

$$w_{\alpha_1 \dots \alpha_k} \equiv 0 \quad \text{if two subscripts } \alpha_i, \alpha_j \text{ agree.} \tag{4.9}$$

For most *chemical reactions* [11, 167] and many other cases the contributions with $k \leq 2$ are the most important ones because simultaneous interactions between three or more elements are rare. Hence the terms with $k > 2$ will be neglected in the following. Then master equation (3.5a) obtains the form

$$\begin{aligned}
&\frac{d}{dt} P(X,t) \\
&\quad = \sum_{\beta} \sum_{Y_\beta} \Big[w_\beta(X^\beta | Y^\beta; t) P(Y_\beta, t) - w_\beta(Y^\beta | X^\beta; t) P(X,t) \Big] \\
&\quad + \frac{1}{2} \sum_{\beta,\gamma} \sum_{Y_{\beta\gamma}} \Big[w_{\beta\gamma}(X^{\beta\gamma} | Y^{\beta\gamma}; t) P(Y_{\beta\gamma}, t) - w_{\beta\gamma}(Y^{\beta\gamma} | X^{\beta\gamma}; t) P(X,t) \Big] \\
&\quad = \sum_{\beta} \sum_{y_\beta} \Big[w_\beta(x_\beta | y_\beta; t) P(Y_\beta, t) - w_\beta(y_\beta | x_\beta; t) P(X,t) \Big] \\
&\quad + \frac{1}{2} \sum_{\beta,\gamma} \sum_{y_\beta, y_\gamma} \Big[w_{\beta\gamma}(x_\beta, x_\gamma | y_\beta, y_\gamma; t) P(Y_{\beta\gamma}, t) \\
&\quad - w_{\beta\gamma}(y_\beta, y_\gamma | x_\beta, x_\gamma; t) P(X,t) \Big] .
\end{aligned} \tag{4.10}$$

The factor $1/2$ results from

$$\sum_{\substack{\alpha_1,\dots,\alpha_k \\ (\alpha_1 < \dots < \alpha_k)}} = \frac{1}{k!} \sum_{\substack{\alpha_1,\dots,\alpha_k \\ (\alpha_i \neq \alpha_j)}} \tag{4.11}$$

and ensures that identical transitions will not be counted twice. (There exist $k!$ *permutations (interchanges)* $(\beta_1, \dots, \beta_k)$ of the k elements $\alpha_i \in \{\alpha_1, \dots, \alpha_k\}$).

Now let us introduce the *multi-particle distribution functions*

$$P_{\alpha_1 \dots \alpha_k}(x_{\alpha_1}, \dots, x_{\alpha_k}; t) := \sum_{x_\beta} P_{\alpha_1 \dots \beta \dots \alpha_k}(x_{\alpha_1}, \dots, x_\beta, \dots, x_{\alpha_k}; t) \tag{4.12a}$$

for the k particles $\alpha_1, \dots, \alpha_k$ and introduce the notation

$$P_{1 \dots N}(x_1, \dots, x_N; t) \equiv P(x_1, \dots, x_N; t) . \tag{4.12b}$$

The summation of (4.10) over all $x_\beta \in \gamma_\beta$ with $\beta \neq \alpha$ (using relations which are analogous to (3.94)) leads to

$$\frac{d}{dt}P_\alpha(x_\alpha,t) = \sum_{y_\alpha}\Big[w_\alpha(x_\alpha|y_\alpha;t)P_\alpha(y_\alpha,t) - w_\alpha(y_\alpha|x_\alpha;t)P_\alpha(x_\alpha,t)\Big]$$
$$+\sum_{y_\alpha}\Big[\sum_\beta\sum_{y_\beta}\sum_{x_\beta} w_{\alpha\beta}(x_\alpha,x_\beta|y_\alpha,y_\beta;t)P_{\alpha\beta}(y_\alpha,y_\beta;t)$$
$$-\sum_\beta\sum_{y_\beta}\sum_{x_\beta} w_{\alpha\beta}(y_\alpha,y_\beta|x_\alpha,x_\beta;t)P_{\alpha\beta}(x_\alpha,x_\beta;t)\Big] \tag{4.13}$$

with

$$\sum_{x_\alpha} P_\alpha(x_\alpha,t) = 1\,. \tag{4.14}$$

Due to the dependence on the *pair distribution functions* $P_{\alpha\beta}(x_\alpha,x_\beta;t)$ Eq. (4.13) are *not closed*. The equations for the pair distribution functions which can be derived from (4.10), however, depend on the three-particle distribution functions $P_{\alpha\beta\gamma}(x_\alpha,x_\beta,x_\gamma;t)$ etc. As a consequence, a *hierarchy of non-closed equations* (cf. [170], pp. 74ff.) results so that an approximation becomes necessary. Assuming a *factorization* of the pair distribution functions of the form

$$P_{\alpha\beta}(x_\alpha,x_\beta;t) \approx P_\alpha(x_\alpha,t)P_\beta(x_\beta,t) \tag{4.15}$$

we arrive at the *closed equations*

$$\frac{d}{dt}P_\alpha(x_\alpha,t) = \sum_{y_\alpha}\Big[w_\alpha(x_\alpha|y_\alpha;t)P_\alpha(y_\alpha,t) - w_\alpha(y_\alpha|x_\alpha;t)P_\alpha(x_\alpha,t)\Big]$$
$$+\sum_{y_\alpha}\Big[\Big(\sum_\beta\sum_{y_\beta}\sum_{x_\beta} w_{\alpha\beta}(x_\alpha,x_\beta|y_\alpha,y_\beta;t)P_\beta(y_\beta,t)\Big)P_\alpha(y_\alpha,t)$$
$$-\Big(\sum_\beta\sum_{y_\beta}\sum_{x_\beta} w_{\alpha\beta}(y_\alpha,y_\beta|x_\alpha,x_\beta;t)P_\beta(x_\beta,t)\Big)P_\alpha(x_\alpha,t)\Big] \tag{4.16}$$

(cf. [170], pp. 72ff., and [25, 93, 94, 162, 224, 240]). $w_\alpha(y_\alpha|x_\alpha;t)$ describes *spontaneous* or *externally induced transitions* of element α from state x_α to state y_α. $w_{\alpha\beta}(y_\alpha,y_\beta|x_\alpha,x_\beta;t)$ reflects *pair interactions* between two systems α and β due to which their states change from x_α and x_β to y_α and y_β respectively. Introducing the *effective transition rates*

$$w^\alpha(x'|x;t) := w_\alpha(x'|x;t) + \sum_\beta\sum_{y'}\sum_{y} w_{\alpha\beta}(x',y'|x,y;t)P_\beta(y,t)\,, \tag{4.17a}$$

Eq. (4.16) can be cast into the simple form

$$\frac{d}{dt}P_\alpha(x,t) = \sum_{x'} \Big[w^\alpha(x|x';t)P_\alpha(x',t) - w^\alpha(x'|x;t)P_\alpha(x,t) \Big] . \qquad (4.17b)$$

Factorization assumption (4.15) is *exactly* valid only if two arbitrary subsystems α and β are *statistically independent.* It is *approximately* justified if the interactions are *weak* ($w_{\alpha\beta} \approx 0$) or if the correlations between two subsystems α and β due to pair interactions are soon destroyed (cf. also (5.20)). Equation (4.15) is often substantiated by the assumption of *'molecular chaos'* [154] (cf. Sect. 10.3.4). For the case that (4.15) is violated there exist methods for the calculation of corrections (cf. [25, 54]). Moreover, a generalization of the above equations to simultaneous interactions of arbitrarily many elements is possible (cf. Sect. 5.4).

4.3 Subdivision into Several Types of Subsystems

Usually the immense number N of subsystems α can be divided into A *types* a which are characterized by certain properties, e.g. by the kind of interaction. In example (4.1) one type of subsystems would consist of the molecules of sort B and a second one of the molecules of sort C.
Now let N_a be the number of elements $\alpha \in a$ (i.e. of type a). Then we have

$$\sum_{a=1}^{A} N_a = N . \qquad (4.18)$$

If subsystems $\alpha \in a$ of the same type a are not further distinguished this implies

$$\gamma_\alpha \equiv \gamma_a \quad \text{if} \quad \alpha \in a , \qquad (4.19a)$$

$$P_\alpha \equiv P_a \quad \text{if} \quad \alpha \in a , \qquad (4.19b)$$

$$w_\alpha \equiv \widehat{w}_a \quad \text{if} \quad \alpha \in a , \qquad (4.19c)$$

and

$$w_{\alpha\beta} \equiv \begin{cases} \widehat{w}_{ab} \text{ if } \alpha \neq \beta, \ \alpha \in a, \ \beta \in b \\ 0 \quad \text{if } \alpha = \beta \end{cases} \qquad (4.19d)$$

where $\widehat{w}_{..}$ are the *individual transition rates.* Furthermore, after introducing the notations

$$w_a := \widehat{w}_a \qquad (4.20a)$$

and

$$w_{ab} := \begin{cases} N_b \cdot \widehat{w}_{ab} & \text{if } b \neq a \\ (N_a - 1) \cdot \widehat{w}_{ab} & \text{if } b = a, \end{cases} \tag{4.20b}$$

from (4.14) and (4.17) we gain

$$\sum_x P_a(x,t) = 1 \tag{4.21}$$

and

$$\frac{d}{dt} P_a(x,t) = \sum_{x'} \Big[w^a(x|x';t) P_a(x',t) - w^a(x'|x;t) P_a(x,t) \Big] \tag{4.22a}$$

with the *effective transition rates*

$$w^a(x'|x;t) := w_a(x'|x;t) + \sum_b \sum_{y'} \sum_y w_{ab}(x',y'|x,y;t) P_b(y,t) \,. \tag{4.22b}$$

By distinguishing only a few types of subsystems the number of variables is considerably reduced if $A \ll N$ or $N_a \gg 1$. Thus (4.22) becomes accessible to a treatment by computers.

4.4 Properties

4.4.1 Non-negativity and Normalization

If the condition $P_a(x,t_0) \geq 0$ is fulfilled at a certain time t_0 for all states x, the *non-negativity* defined by $P_a(x,t) \geq 0$ will be guaranteed for all later times $t > t_0$. Moreover, despite the applied factorization assumption (4.15) we can derive

$$\frac{d}{dt}\left(\sum_x P_a(x,t) \right) = 0 \,, \qquad \text{i.e.} \qquad \sum_x P_a(x,t) = const. \tag{4.23}$$

which allows the *normalization* of distribution $P_a(x,t)$. The corresponding proofs are analogous to those in Sects. 3.3.1 and 3.3.2.

4.4.2 The Gaskinetic BOLTZMANN Equation

The BOLTZMANN equation for *fairly rare gases* is a special case of Eqs. (4.22a,). It results, if one combines in the state

$$x := (r, v) \qquad (\text{or } y := (s, w)) \tag{4.24}$$

the *place* r (or s) and the *velocity* v (or w) of a gas particle. By types a we distinguish different *sorts* of particles. The *spontaneous state changes* are given by the *motion*

$$\frac{dr}{dt} = v \tag{4.25}$$

and the *acceleration*

$$\frac{dv}{dt} = \frac{d^2r}{dt^2} := \frac{F_a(t)}{m_a} \tag{4.26}$$

where $F_a(t)$ is the *force* exerted on a particle of *mass* m_a. They are described by the transition rates

$$\widehat{w}_a(r', v'|r, v; t) := \lim_{\Delta t \to 0} \frac{1}{\Delta t} \delta\Big(r' - \big(r + v\Delta t\big)\Big) \delta\left(v' - \left(v + \frac{F_a(t)}{m_a}\Delta t\right)\right) . \tag{4.27}$$

In the Kramers-Moyal *representation* (cf. Sect. 6.2) they contribute to Eq. (4.22) with the *divergence terms*

$$-\nabla_r\Big[vP_a(r, v; t)\Big] - \nabla_v\left[\frac{F_a(t)}{m_a}P_a(r, v; t)\right] . \tag{4.28}$$

Additionally, state changes result from *scattering processes* of particles (from *direct pair interactions*). These are described by the transition rates

$$\widehat{w}_{ab}(x', y'|x, y; t) := \widehat{w}'_{ab}(v', w'|v, w)\delta(r' - r)\delta(s' - s)\delta(r - s) \tag{4.29}$$

which reflect the *collisions* of two particles at location $r = s$ and the resulting velocity changes.

With respect to the *densities*

$$\rho(a, r, v; t) := N_a P_a(r, v; t) \tag{4.30}$$

where N_a denotes the number of particles of sort a within the considered volume V, we have the *gaskinetic* Boltzmann *equation* [26, 154, 163]

$$\begin{aligned}
&\frac{\partial}{\partial t}\rho(a, r, v; t) \\
&= -\nabla_r\Big[v\rho(a, r, v; t)\Big] - \nabla_v\left[\frac{F_a(t)}{m_a}\rho(a, r, v; t)\right] \\
&\quad + \sum_b \int d^3v' \int d^3w \int d^3w'\, \widehat{w}'_{ab}(v, w|v', w')\rho(a, r, v'; t)\rho(b, r, w'; t) \\
&\quad - \sum_b \int d^3v' \int d^3w \int d^3w' \widehat{w}'_{ab}(v', w'|v, w)\rho(a, r, v; t)\rho(b, r, w; t) .
\end{aligned} \tag{4.31}$$

In most cases the pair interaction rates $\widehat{w}'_{ab}$ are expressed by the *differential scattering cross section*

$$d^6\sigma'_{ab} \equiv d^6\sigma'_{ab}(v', w'|v, w) := \frac{\widehat{w}'_{ab}(v', w'|v, w)d^3v'd^3w'}{\|v - w\|} . \tag{4.32}$$

Furthermore we have

$$\widehat{w}'_{ab}(v, w|v', w') = \widehat{w}'_{ab}(v', w'|v, w) \tag{4.33}$$

since the pair interaction rates $\widehat{w}_{ab}$ are *symmetrical* (cf. (3.88)):

$$\widehat{w}_{ab}(x, y|x', y') = \widehat{w}_{ab}(x', y'|x, y) . \tag{4.34}$$

(This symmetry is a consequence of the *time reversal invariance* and the *isotropy of space;* cf. [154], pp. 69ff.) Consequently, the gaskinetic BOLTZMANN equation (4.31) can be written in the form

$$\begin{aligned} \frac{d}{dt}\rho(a, r, v; t) = \sum_b \int d^3w \int d^6\sigma'_{ab}\, \|v - w\| \Big[\rho(a, r, v'; t)\rho(b, r, w'; t) \\ -\rho(a, r, v; t)\rho(b, r, w; t)\Big] . \end{aligned} \tag{4.35}$$

Here

$$\begin{aligned} \frac{d}{dt}\rho(a, r, v; t) :&= \frac{\partial}{\partial t}\rho(a, r, v; t) + \nabla_r \cdot \Big[v\rho(a, r, v; t)\Big] + \nabla_v \cdot \left[\frac{F_a(t)}{m_a}\rho(a, r, v; t)\right] \\ &= \frac{\partial}{\partial t}\rho(a, r, v; t) + v \cdot \nabla_r \rho(a, r, v; t) + \frac{F_a(t)}{m_a} \cdot \nabla_v \rho(a, r, v; t) \end{aligned} \tag{4.36}$$

is the *total (substantial) time derivative* which describes the temporal change of the density $\rho(a, r, v; t)$ in a *transformed coordinate system* which moves *along* with velocity $(v, F_a/m_a)^{\text{tr}}$. Obviously, $d\rho(a, r, v; t)/dt$ is only given by the scattering processes. In this sense, BOLTZMANN equation (4.35) is already determined by *pair interactions*. In contrast to this, the terms due to *spontaneous transitions* usually *cannot* be eliminated from the general Eq. (4.22): In the case of *non-deterministic* transition rates $w_a(x'|x; t)$ the KRAMERS-MOYAL expansion leads not only to divergence terms but also to further contributions which cannot be eliminated by a transformation.

In the following, Eq. (4.22) will be called '**BOLTZMANN** *equations*' if these are solely given by *pair interactions*. If, in addition to pair interactions, *spontaneous transitions* play a role, it is reasonable to speak of '**BOLTZMANN**-*like equations*'.

4.4.3 The H-Theorem for the Gaskinetic BOLTZMANN Equation

For the gaskinetic BOLTZMANN equation (4.31) there exists, analogous to Sect. 3.3.5, a LIAPUNOV *function*

$$H(t) := \sum_a \int d^6x\, \rho(a, x; t) \ln \rho(a, x; t) \tag{4.37}$$

with the property

$$\frac{dH}{dt} \leq 0\,. \tag{4.38}$$

Again, this causes the system to approach an *equilibrium.* Except for an *affine transformation,* $H(t)$ is identical with BOLTZMANN's *entropy function*

$$S(t) := -\mathrm{k} \sum_a \int d^6x\, P(a, x; t) \log_2 P(a, x; t) \tag{4.39}$$

with

$$P(a, x; t) := \frac{N_a}{N} P_a(x, t) = \frac{\rho(a, x; t)}{N}\,. \tag{4.40}$$

For the proof of (4.38) we first show

$$\begin{aligned}
\frac{dH}{dt} &= \sum_a \int d^6x \left[\frac{\partial\rho(a, x; t)}{\partial t} \ln \rho(a, x; t) + \frac{\partial\rho(a, x; t)}{\partial t}\right] \\
&= -\sum_a \int d^6x \ln \rho(a, x; t) \left\{\nabla_r \Big[v\rho(a, x; t)\Big] + \nabla_v \left[\frac{F_a(t)}{m_a}\rho(a, x; t)\right]\right\} \\
&\quad + \sum_{a,b} \iiiint \ln \rho(a, x; t) \Big[\widehat{w}_{ab}(x, y|x', y')\rho(a, x'; t)\rho(b, y'; t) \\
&\quad - \widehat{w}_{ab}(x', y'|x, y)\rho(a, x; t)\rho(b, y; t)\Big] d^6x\, d^6x'\, d^6y\, d^6y' \qquad (4.41) \\
&= -\sum_a \int d^6x \left\{\nabla_r \Big[v\Big(\rho(a, x; t) \ln \rho(a, x; t) - \rho(a, x; t)\Big)\Big]\right. \\
&\quad \left. + \nabla_v \left[\frac{F_a(t)}{m_a}\Big(\rho(a, x; t) \ln \rho(a, x; t) - \rho(a, x; t)\Big)\right]\right\} \\
&\quad + \sum_{a,b} \iiiint \ln \rho(a, x; t) \Big[\widehat{w}_{ab}(x, y|x', y')\rho(a, x'; t)\rho(b, y'; t) \\
&\quad - \widehat{w}_{ab}(x', y'|x, y)\rho(a, x; t)\rho(b, y; t)\Big] d^6x\, d^6x'\, d^6y\, d^6y'\,. \qquad (4.42)
\end{aligned}$$

The term

$$\sum_a \int d^6x \, \frac{\partial \rho(a,x;t)}{\partial t} = \sum_a \frac{\partial}{\partial t} \int d^6x \, \rho(a,x;t) = \sum_a \frac{\partial N_a}{\partial t} = \frac{\partial N}{\partial t} \tag{4.43}$$

vanishes on the condition that the considered system is a *closed system* which implies a *conservation of the number of particles.* The divergence terms can be transformed into surface integrals by using GAUSS' *divergence theorem* (cf. (6.16)). Due to

$$\lim_{\|v\| \to \infty} \rho(a,r,v;t) = 0\,, \tag{4.44}$$

however, these provide no contributions. Thus the following remains:

$$\begin{aligned} \frac{dH}{dt} &= \sum_{a,b} \iiiint \ln \rho(a,x;t) \Big[\widehat{w}_{ab}(x,y|x',y') \rho(a,x';t) \rho(b,y';t) \\ &\quad - \widehat{w}_{ab}(x',y'|x,y) \rho(a,x;t) \rho(b,y;t) \Big] d^6x \, d^6x' \, d^6y \, d^6y' &(4.45)\\ &= \sum_{a,b} \iiiint \Big[\ln \rho(a,x;t) - \ln \rho(a,x';t) \Big] \\ &\quad \times \widehat{w}_{ab}(x,y|x',y') \rho(a,x';t) \rho(b,y';t) d^6x \, d^6x' \, d^6y \, d^6y' &(4.46)\\ &= \frac{1}{2} \sum_{a,b} \iiiint \Big[\ln \rho(a,x;t) - \ln \rho(a,x';t) + \ln \rho(b,y;t) - \ln \rho(b,y';t) \Big] \\ &\quad \times \widehat{w}_{ab}(x,y|x',y') \rho(a,x';t) \rho(b,y';t) d^6x \, d^6x' \, d^6y \, d^6y'\,. &(4.47) \end{aligned}$$

From (4.45) we come to (4.46) by renaming the variables of the second term in accordance with $x \leftrightarrow x'$ and $y \leftrightarrow y'$. The identity of (4.46) and (4.47) results by interchanging the variables of two interacting particles α, β and taking into account

$$\widehat{w}_{ba}(y,x|y',x') = w_{\beta\alpha}(y,x|y',x') = w_{\alpha\beta}(x,y|x',y') = \widehat{w}_{ab}(x,y|x',y')\,. \tag{4.48}$$

On the condition

$$\int d^6x' \int d^6y' \, \widehat{w}_{ab}(x,y|x',y') = \int d^6x' \int d^6y' \, \widehat{w}_{ab}(x',y'|x,y) \tag{4.49}$$

which is fulfilled due to (4.34) we further have

$$\begin{aligned}
&\frac{1}{2}\sum_b \iiiint \widehat{w}_{ab}(x, y|x', y')\Big[\rho(a, x'; t)\rho(b, y; t) \\
&\qquad - \rho(a, x; t)\rho(b, y; t)\Big] d^6x\, d^6x'\, d^6y\, d^6y' \\
&= \frac{1}{2}\sum_b \iiiint \Big[\widehat{w}_{ab}(x, y|x', y')\rho(a, x'; t)\rho(b, y'; t) \\
&\qquad - \widehat{w}_{ab}(x', y'|x, y)\rho(a, x; t)\rho(b, y; t)\Big] d^6x\, d^6x'\, d^6y\, d^6y' \\
&= 0\,.
\end{aligned} \tag{4.50}$$

This can immediately be seen by renaming the variables in the second term in accordance with $x \leftrightarrow x'$ and $y \leftrightarrow y'$.

Now, formally adding in the vanishing term (4.50) to (4.47) and introducing the abbreviations

$$z_{ab} \equiv z_{ab}(x, y; x', y'; t) := \frac{\rho(a, x'; t)\rho(b, y'; t)}{\rho(a, x; t)\rho(b, y; t)}\,, \tag{4.51a}$$

$$z'_{ab} \equiv z'_{ab}(v, w; v', w'; t) := \frac{\rho(a, r, v'; t)\rho(b, r, w'; t)}{\rho(a, r, v; t)\rho(b, r, w; t)}\,, \tag{4.51b}$$

we find the relation

$$\begin{aligned}
\frac{dH}{dt} &= -\frac{1}{2}\sum_{a,b} \iiiint \rho(a, x; t)\rho(b, y; t)\Big(z_{ab}\ln z_{ab} - z_{ab} + 1\Big) \\
&\quad \times \widehat{w}_{ab}(x, y|x', y') d^6x\, d^6x'\, d^6y\, d^6y' \\
&= -\frac{1}{2}\sum_{a,b} \iiiiint \rho(a, r, v; t)\rho(b, r, w; t)\Big(z'_{ab}\ln z'_{ab} - z'_{ab} + 1\Big) \\
&\quad \times \widehat{w}'_{ab}(v, w|v', w') d^3v\, d^3v'\, d^3w\, d^3w'\, d^3r \\
&\le 0
\end{aligned} \tag{4.52}$$

because of

$$z'_{ab}\ln z'_{ab} - z'_{ab} + 1 \ge 0 \qquad \text{for} \qquad z'_{ab} > 0\,. \tag{4.53}$$

Consequently, a system described by the gaskinetic BOLTZMANN equation (4.35) moves towards an *equilibrium*. However, without the validity of relation (4.49) a BOLTZMANN equation can also show an *oscillatory* or *chaotic* behaviour (cf. Sects. 10.3.2 and 10.3.4).

4.4.4 Solution of the Gaskinetic BOLTZMANN Equation

According to (4.52), the relation $dH/dt = 0$ presupposes that the *equilibrium condition* $z'_{ab} \stackrel{!}{=} 1$ is fulfilled. This implies the condition

$$\psi_a(v) + \psi_b(w) \stackrel{!}{=} \psi_a(v') + \psi_b(w') \tag{4.54a}$$

with

$$\psi_a(v) := \ln \rho(a, r, v) \,. \tag{4.54b}$$

Obviously, (4.54) is satisfied by the *collisional invariants*

$$\text{mass} \quad m_a \,, \tag{4.55a}$$

$$\text{momentum} \quad m_a v_i \,, \tag{4.55b}$$

$$\text{energy} \quad \frac{m_a}{2}\|v\|^2 = \frac{m_a}{2}\sum_i (v_i)^2 \,, \tag{4.55c}$$

and *linear combinations* of these. Inserting

$$\psi_a(v) := c_a^1 m_a + \sum_i c_{a,i}^2 m_a v_i + c_a^3 \frac{m_a}{2} \sum_i (v_i)^2 \tag{4.56}$$

into (4.54) and representing the density of particles of sort a at location r by $\rho(a, r)$ we obtain

$$\rho(a, r, v) = \rho(a, r) P_a(v) \tag{4.57a}$$

with the MAXWELL-BOLTZMANN *distribution*

$$P_a(v) := \prod_i \left(\frac{1}{\sqrt{2\pi\sigma_a}} \mathrm{e}^{-(v_i - \langle v_i \rangle_a)^2/(2\sigma_a)} \right) = \frac{1}{(2\pi\sigma_a)^{3/2}} \mathrm{e}^{-(v - \langle v \rangle_a)^2/(2\sigma_a)} \,. \tag{4.57b}$$

Here $\langle v \rangle_a$ has the meaning of the mean velocity of the particles of sort a. σ_a is related to their *absolute temperature* T'_a via the BOLTZMANN *constant* k:

$$\sigma_a = \mathrm{k} T'_a \,. \tag{4.58}$$

For the determination of the *time-dependent* solution of the gaskinetic BOLTZMANN equation there exists a (rather complicated) recursive method which was developed by CHAPMAN [38] and ENSKOG [64] (cf. [170], pp. 187ff.).

4.5 Comparison of Spontaneous Transitions and Direct Interactions

In the following we will elaborate the interrelations and differences between spontaneous transitions and direct interactions.

4.5.1 Transitions Induced by Interactions

Let us assume that we have $A' \equiv A + B$ types a of elements and that the elements of all types are able to change their states only by *pair interactions*. Provided the dynamics of all A' types is known, it is described by the BOLTZMANN equations

$$\begin{aligned}\frac{d}{dt} P_a(x,t) &= \sum_{b=1}^{A'} \sum_{x'} \sum_{y} \sum_{y'} w_{ab}(x, y|x', y'; t) P_b(y', t) P_a(x', t) \\ &- \sum_{b=1}^{A'} \sum_{x'} \sum_{y} \sum_{y'} w_{ab}(x', y'|x, y; t) P_b(y, t) P_a(x, t) \end{aligned} \quad (4.59)$$

($a \in \{1, \dots, A+B\}$). If, however, only the dynamics of the A types $a \in \{1, \dots, A\}$ is known, it obeys the BOLTZMANN-like equations

$$\frac{d}{dt} P_a(x,t) = \sum_{x'} \Big[w^a(x|x'; t) P_a(x', t) - w^a(x'|x; t) P_a(x, t) \Big] \quad (4.60a)$$

with

$$w^a(x'|x; t) := w_a(x'|x; t) + \sum_{b=1}^{A} \sum_{y} \sum_{y'} w_{ab}(x', y'|x, y; t) P_b(y, t) . \quad (4.60b)$$

The connection with (4.59) is given by

$$w_a(x'|x; t) := \sum_{b=A+1}^{A+B} \sum_{y} \sum_{y'} w_{ab}(x', y'|x, y; t) P_b(y, t) . \quad (4.60c)$$

So the transition rates w_a may describe transitions which are *induced* by unknown pair interactions.

4.5.2 Exponential Function and Logistic Equation

Let us consider the case of one type a of subsystems ($A = 1$) which can change between $S = 2$ states $x \equiv x \in \{1, 2\}$. Then there is

$$P(2,t) = 1 - P(1,t) \tag{4.61}$$

where subscript a was omitted. If only *spontaneous transitions* occur and the transition rates are time-independent, the corresponding BOLTZMANN-like equation is the *master equation*

$$\begin{aligned} \frac{d}{dt}P(1,t) &= w(1|2)P(2,t) - w(2|1)P(1,t) \\ &= w(1|2) - \Big[w(1|2) + w(2|1)\Big]P(1,t)\,. \end{aligned} \tag{4.62}$$

Its solution is the *exponential function*

$$P(1,t) = P^0(1) + \Big[P(1,t_0) - P^0(1)\Big]\mathrm{e}^{-\lambda(t-t_0)} \tag{4.63a}$$

with

$$P^0(1) := \frac{w(1|2)}{w(1|2) + w(2|1)} \leq 1 \tag{4.63b}$$

and

$$\lambda := w(1|2) + w(2|1)\,. \tag{4.63c}$$

If, in contrast to this, only *direct pair interactions* take place and if the transition rates are again time-independent, the corresponding BOLTZMANN-like equation is the BOLTZMANN *equation*

$$\begin{aligned} \frac{d}{dt}P(1,t) = &\Big[w(1,1|2,2)P(2,t)P(2,t) + w(1,2|2,2)P(2,t)P(2,t)\Big] \\ &- \Big[w(2,1|1,2)P(2,t)P(1,t) + w(2,2|1,2)P(2,t)P(1,t)\Big] \\ &+ \Big[w(1,1|2,1)P(1,t)P(2,t) + w(1,2|2,1)P(1,t)P(2,t)\Big] \\ &- \Big[w(2,1|1,1)P(1,t)P(1,t) + w(2,2|1,1)P(1,t)P(1,t)\Big] \\ \equiv &-C_1[P(1,t)]^2 + C_2P(1,t) + C_3 \end{aligned} \tag{4.64}$$

where C_1, C_2, and C_3 are constants that are determined by the pair interaction rates w together with (4.61). After introducing

$$Q(1,t) := P(1,t) - C \qquad \text{and} \qquad C := \frac{C_2 - \sqrt{(C_2)^2 + 4C_1C_3}}{2C_1} \tag{4.65}$$

we can transform (4.64) into the *logistic equation*

$$\frac{d}{dt}Q(1,t) = rQ(1,t)\left(1 - \frac{Q(1,t)}{Q^0(1)}\right) \tag{4.66}$$

with

$$r := C_2 - 2C_1C = \sqrt{(C_2)^2 + 4C_1C_3} \tag{4.67a}$$

and

$$Q^0(1) := \frac{r}{C_1} = \frac{\sqrt{(C_2)^2 + 4C_1C_3}}{C_1}. \tag{4.67b}$$

(4.66) has the solution

$$Q(1,t) = \frac{Q^0(1)}{1 + \frac{Q^0(1) - Q(1,t_0)}{Q(1,t_0)}\mathrm{e}^{-r(t-t_0)}}. \tag{4.68}$$

The exponential solution of (4.62) and the logistic solution of (4.64) are compared with each other in Fig. 4.1.

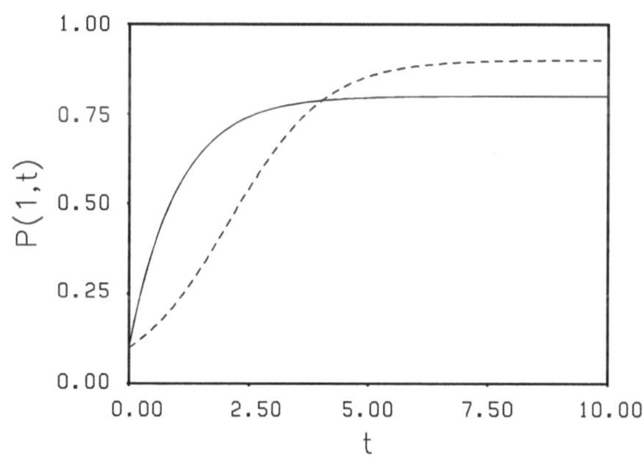

Fig. 4.1 Comparison of an exponential solution (—) and a solution of the *logistic equation* (– –)

4.5.3 Stationary and Oscillatory Solutions

In the case of exclusively spontaneous transitions with time-independent transition rates ($w^a(x'|x;t) \equiv w_a(x'|x)$) the BOLTZMANN-like equations for the temporal evolution of the distributions $P_a(x,t)$ have the form of master equations (cf. (4.62)). According to the discussion in Sect. 3.3.5, $P_a(x,t)$ converges in the course of time to a *unique* stationary *equilibrium distribution* $P_a^0(x)$.

In contrast to this, there exist BOLTZMANN equations for pair interactions which have *various* stationary solutions, *oscillatory solutions,* or *chaotic solutions* even in the case of time-independent pair interaction rates (cf. Sects. 10.3.2 and 10.3.4). However, this presupposes that condition (4.49) is violated so that the H*-theorem* does not hold.

Chapter 5
Master Equation in Configuration Space

5.1 Introduction

In many cases we are not interested in the states x_α of the single elements α of a system but only in the *number* n_x^a of subsystems of type a which are in state x (e.g. the number of molecules of sort B which are in a bounded or an unbounded state). Since the states of the subsystems α change *stochastically,* the *occupation numbers* n_x^a change stochastically, too. Consequently, the occupation numbers must be described by a probability distribution. Its temporal evolution is given by a master equation in *configuration space* (the space of occupation numbers). When considering only spontaneous state changes we find, as *approximate mean value equation* of the *configurational master equation,* the master equation in state space from Chap. 3. Taking into account also pair interactions, as approximate mean value equations we obtain BOLTZMANN-like equations, instead. A generalization to *interactions of higher order* (between three or more systems) is possible without any problems. Furthermore, the configurational master equation allows a calculation of corrections of the approximate mean value and covariance equations.

The covariance equations are important for three reasons: First, they reflect the stochastic nature of the temporal changes and distinguish different systems with the same approximate mean value equations. Second, they are necessary for the determination of the *reliability of prognoses* on the basis of mean value equations and for the evaluation of the *time interval* after which the approximate mean value equations become useless. Third, they facilitate *corrections* to the approximate mean value equations. This will be illustrated in Sect. 12.4.1 by concrete examples.

The configurational master equation admits the introduction of *occupation number dependent* transition rates which are necessary for the description of *indirect interactions.* Such models are especially useful for *mean field approaches* or the description of *social* processes (cf. Sects. 10.2 and 12.4).

Finally, the differences and interrelations between direct and indirect interactions will be elaborated in particular.

D. Helbing, *Quantitative Sociodynamics*, 2nd ed.,
DOI 10.1007/978-3-642-11546-2_5,

5.2 Transitions in Configuration Space

The following considerations take up the discussion from Chap. 4 about systems which consist of many elements α of a few types a. It will be assumed that the subsystems can be only in one of S states $x \in \gamma_a := \{x_1, \dots, x_S\}$ though a generalization to infinite or continuous state spaces γ_a is also possible. In addition, the following relations can, by exchanging S for $\widehat{S}_a$, easily be extended to the case of state spaces γ_a which consist of different numbers $\widehat{S}_a$ of states.

The vector

$$n := (n^1_{x_1}, \dots, n^1_{x_S}, \dots, n^A_{x_1}, \dots, n^A_{x_S})^{\text{tr}} \equiv (n^1, \dots, n^A)^{\text{tr}} \tag{5.1}$$

of the *occupation numbers* n^a_x will be denoted as the *configuration* of the total system and represents the distribution of the elements over the states. Then the *configuration space* Γ' consists of the vectors n which fulfill the relation

$$\sum_x n^a_x \equiv \sum_{i=1}^{S} n^a_{x_i} \overset{!}{=} N_a \qquad (n^a_x \in \{0, 1, \dots, N_a\}) . \tag{5.2}$$

Here N_a is the number of subsystems of type a. *Since we will neglect birth and death processes of subsystems in the following, N_a is time independent.* (For a discussion of birth and death processes cf. [84], pp. 262ff., and [292], pp. 86ff.)

5.2.1 Spontaneous Transitions

If we restrict ourselves temporarily to the consideration of spontaneous transitions (so that interactions between the elements are excluded: $w_{ab} \equiv 0$), the single subsystems can in general be regarded as *statistically independent* of each other. Then the probability distribution $P(n, t)$ of the configuration n is given by

$$P(n, t) = \prod_{a=1}^{A} P^a(n^a, t) \tag{5.3a}$$

with

$$n^a := (n^a_{x_1}, \dots, n^a_{x_S})^{\text{tr}} \tag{5.3b}$$

and the *multinomial distribution* [81, 82]

$$P^a(n^a, t) := \frac{N_a!}{n^a_{x_1}! \cdot \dots \cdot n^a_{x_S}!} P_a(x_1, t)^{n^a_{x_1}} \cdot \dots \cdot P_a(x_S, t)^{n^a_{x_S}} . \tag{5.3c}$$

$P(n,t)$ is the probability with which a certain distribution of the elements over the states occurs. Evaluating the time derivative of (5.3) and using (4.22), we obtain for the temporal change of $P(n,t)$ a *master equation in configuration space (configurational master equation)*. It reads

$$\frac{d}{dt}P(n,t) = \sum_{n'}\Big[w(n|n';t)P(n',t) - w(n'|n;t)P(n,t)\Big] \tag{5.4a}$$

with

$$w(n'|n;t) := \begin{cases} n_x^a w_a(x'|x;t) & \text{if } n' = n_{x'x}^{a\,a} \\ 0 & \text{otherwise}, \end{cases} \tag{5.4b}$$

$$n_{x'x}^{a\,a} := (n_{x_1}^1, \ldots, n_{x_S}^1, \ldots, (n_{x'}^a + 1), \ldots, (n_x^a - 1), \ldots, n_{x_1}^A, \ldots, n_{x_S}^A)^{\text{tr}}, \tag{5.5}$$

and

$$P(n,t) \geq 0, \qquad \sum_n P(n,t) = 1. \tag{5.6}$$

According to this, the *configurational transition rate* $n_x^a w_a(x'|x;t) = n_x^a \widehat{w}_a(x'|x;t)$ is, on the one hand, proportional to the number n_x^a of the subsystems which can leave state x. On the other hand, it is proportional to the *individual transition rate* $\widehat{w}_a(x'|x;t)$.

5.2.2 Pair Interactions

In order to consider, apart from spontaneous transitions, also pair interactions we apply the method from Sect. 4.2: We decompose the total transition rate $w(n'|n;t)$ into contributions which describe interactions between k subsystems. Afterwards we again neglect interactions of higher order ($k > 2$). Then master equation (5.4) becomes

$$\begin{aligned}\frac{d}{dt}P(n,t) = &\sum_{a,x,a',x'}\Big[w(n|n_{x'x}^{a'a};t)P(n_{x'x}^{a'a},t) - w(n_{x'x}^{a'a}|n;t)P(n,t)\Big] \\ &+\frac{1}{2}\sum_{a,x,a',x'}\sum_{b,y,b',y'}\Big[w(n|n_{x'y'xy}^{a'b'ab};t)P(n_{x'y'xy}^{a'b'ab},t) \\ &\qquad - w(n_{x'y'xy}^{a'b'ab}|n;t)P(n,t)\Big]\end{aligned} \tag{5.7}$$

with

$$n_{x'x}^{a'a} := (\ldots, (n_{x'}^{a'} + 1), \ldots, (n_x^a - 1), \ldots)^{\text{tr}} \tag{5.8a}$$

and

$$n^{a'b'ab}_{x'y'xy} := (\ldots, (n^{a'}_{x'}+1), \ldots, (n^{a}_{x}-1), \ldots, (n^{b'}_{y'}+1), \ldots, (n^{b}_{y}-1), \ldots)^{\text{tr}} . \quad (5.8b)$$

If the type a of an element does not change (or at least very slowly), the transition rates will normally have the form

$$w(n^{a'a}_{x'x}|n;t) := \delta_{aa'}\widehat{w}_a(x'|x;t)n^a_x = \delta_{aa'}w_a(x'|x;t)n^a_x \quad (5.9a)$$

and

$$\begin{aligned} w(n^{a'b'ab}_{x'y'xy}|n;t) &:= \delta_{aa'}\delta_{bb'}\widehat{w}_{ab}(x',y'|x,y;t)n^a_x n^b_y \\ &\overset{N_a \gg 1}{\approx} \delta_{aa'}\delta_{bb'}w_{ab}(x',y'|x,y;t)n^a_x \frac{n^b_y}{N_b} , \end{aligned} \quad (5.9b)$$

i.e.

$$w(n'|n;t) := \begin{cases} \widehat{w}_a(x'|x;t)n^a_x & \text{if } n' = n^{a\,a}_{x'x} \\ \widehat{w}_{ab}(x',y'|x,y;t)n^a_x n^b_y & \text{if } n' = n^{a\,b\,ab}_{x'y'xy} \\ 0 & \text{otherwise.} \end{cases} \quad (5.10)$$

Here the configurational transition rate due to spontaneous transitions is again proportional to the individual transition rate $\widehat{w}_a(x'|x;t)$ and to the number n^a_x of the elements which can leave state x. In contrast, the configurational transition rate which goes back to pair interactions is, on the one hand, proportional to the *individual pair interaction rate* $\widehat{w}_{ab}(x',y'|x,y;t)$. On the other hand, it is proportional to the number $n^a_x n^b_y$ of possible pair interactions between elements of type a which are in state x and elements of type b which are in state y. Therefore, in order to be precise, *in the case* $n' = n^{a\,a\,aa}_{x'y'xx}$ *the configurational transition rate* $\widehat{w}_{aa}(x',y'|x,x;t)n^a_x n^a_x$ *must be replaced by* $\widehat{w}_{aa}(x',y'|x,x;t)n^a_x(n^a_x-1)$. *Otherwise we would admit pair interactions of an element with itself (which was explicitly excluded by (4.9)). Nevertheless, the form (5.10) of the configurational transition rates can be used as an approximation if*

- $\sum_{y'} \widehat{w}_{aa}(x',y'|x,x;t) \ll \widehat{w}_a(x'|x;t)$ *is fulfilled or*
- $n^a_x \gg 1$ *is valid for all configurations* n *for which* $P(n,t)$ *is not negligible.*

Whereas the first case mainly applies to systems with weak internal interactions ($\widehat{w}_{ab} \approx 0$) *the second case is mostly given for systems which consist of many elements* ($N \gg 1$). *In the following the validity of approximation (5.10) will always be presupposed.*

5.3 Mean Value and Covariance Equations

With respect to the configurational master equation, the mean value and covariance equations once again are of special interest. In analogy to (3.168), the mean value $\langle f(n,t)\rangle$ of a function $f(n,t)$ is defined by

$$\langle f\rangle_t \equiv \langle f(n,t)\rangle := \sum_n f(n,t)P(n,t)\,. \tag{5.11}$$

Due to the formal relation between the master equation (3.5) in state space and the configurational master equation (5.4), the *mean value equations* have the form

$$\frac{d}{dt}\langle n_x^a\rangle = \langle m_x^a(n,t)\rangle \tag{5.12}$$

with the *first jump moments*

$$\begin{aligned} m_x^a(n,t) &:= \sum_{n'} (\Delta n')_x^a w(n'|n;t) \\ &= \sum_b \sum_{y,y'} (\Delta n_{yy'}^{bb})_x^a w(n_{yy'}^{bb}|n;t) \\ &+ \frac{1}{2}\sum_b \sum_{y,y'} \sum_c \sum_{z,z'} (\Delta n_{y'z'yz}^{b\,c\,bc})_x^a w(n_{y'z'yz}^{b\,c\,bc}|n;t)\,. \end{aligned} \tag{5.13}$$

Since the individual pair interaction rates are independent of the arbitrary numbering α of the elements, we have

$$\begin{aligned} \widehat{w}_{ba}(y',x'|y,x;t) &= w_{\beta\alpha}(y',x'|y,x;t) \\ &= w_{\alpha\beta}(x',y'|x,y;t) = \widehat{w}_{ab}(x',y'|x,y;t) \end{aligned} \tag{5.14}$$

and an explicit evaluation of (5.13) leads (after a substitution of y by x') to

$$m_x^a(n,t) = \sum_{x'} \left[w^a(x|x';n;t)n_{x'}^a - w^a(x'|x;n;t)n_x^a \right] \tag{5.15a}$$

with the *effective transition rates*

$$\begin{aligned} w^a(x'|x;n;t) &:= \widehat{w}_a(x'|x;t) + \sum_b \sum_{y'} \sum_y \widehat{w}_{ab}(x',y'|x,y;t) n_y^b \\ &\overset{N_a\gg 1}{=} w_a(x'|x;t) + \sum_b \sum_{y'} \sum_y w_{ab}(x',y'|x,y;t)\frac{n_y^b}{N_b}\,. \end{aligned} \tag{5.15b}$$

Hence, the *approximate mean value equations*

$$\frac{d}{dt}\frac{\langle n_x^a\rangle}{N_a} \approx m_x^a(\langle n\rangle, t) \tag{5.16}$$

can be written in the form

$$\frac{d}{dt}\frac{\langle n_x^a\rangle}{N_a} \approx \sum_{x'}\left[\overline{w}^a(x|x';t)\frac{\langle n_{x'}^a\rangle}{N_a} - \overline{w}^a(x'|x;t)\frac{\langle n_x^a\rangle}{N_a}\right] \tag{5.17a}$$

with the *mean transition rates*

$$\begin{aligned}\overline{w}^a(x'|x;t) &:= \widehat{w}_a(x'|x;t) + \sum_b\sum_{y'}\sum_y \widehat{w}_{ab}(x',y'|x,y;t)\langle n_y^b\rangle \\ &\stackrel{N_a\gg 1}{=} w_a(x'|x;t) + \sum_b\sum_{y'}\sum_y w_{ab}(x',y'|x,y;t)\frac{\langle n_y^b\rangle}{N_b}\end{aligned} \tag{5.17b}$$

(cf. [150]). For $N_a \gg 1$ the approximate equations for the temporal change of the *expected proportions* $\langle n_x^a\rangle_t/N_a$ agree with the BOLTZMANN-like equations (4.22), i.e. we have

$$\frac{\langle n_x^a\rangle_t}{N_a} \stackrel{N_a\gg 1}{=} P_a(x,t) \tag{5.17c}$$

and the *mean* transition rates (5.17b) correspond to the *effective* transition rates (4.22b):

$$\overline{w}^a(x'|x;t) \stackrel{N_a\gg 1}{=} w^a(x'|x;t)\,. \tag{5.17d}$$

In the following we will always assume $N_a \gg 1$, i.e. the existence of a system that consists of many elements.

The mean value equations (5.17) are *exact* equations if the configurational transition rates $w(n'|n;t)$ depend *linearly* on the occupation numbers (if interactions do not play any role). Otherwise they presuppose the validity of the condition

$$\langle n_x^a n_y^b\rangle \approx \langle n_x^a\rangle\langle n_y^b\rangle \tag{5.18}$$

which obviously corresponds to *factorization assumption* (4.15). As a consequence, the applicability of the mean value equations assumes the absolute values of the covariances

$$\sigma_{x\,x'}^{ab} = \langle n_x^a n_{x'}^b\rangle - \langle n_x^a\rangle\langle n_{x'}^b\rangle \tag{5.19}$$

to be small so that the condition

$$\left|\sigma^{ab}_{x\,x'}\right| \overset{!}{\ll} \langle n^a_x\rangle\langle n^b_{x'}\rangle \tag{5.20}$$

is fulfilled. Equation (5.20) must be checked only for the *variances* $\sigma^{aa}_{x\,x}$ because the CAUCHY-SCHWARZ *inequality* (cf. p. xxvii) together with the *triangle inequality* (cf. p. xxvii) imply the relation

$$\left|\sigma^{ab}_{x\,x'}\right| \leq \sqrt{\sigma^{aa}_{x\,x}\sigma^{bb}_{x'x'}}\,. \tag{5.21}$$

In any case we require *covariance equations.* Analogously to (3.180), the *approximate covariance equations* read

$$\frac{\partial\sigma^{ab}_{x\,x'}}{\partial t} \approx m^{ab}_{xx'}(\langle n\rangle,t) + \sum_{c,y}\left(\sigma^{ac}_{x\,y}\frac{\partial m^b_{x'}(\langle n\rangle,t)}{\partial\langle n^c_y\rangle} + \sigma^{bc}_{x'y}\frac{\partial m^a_x(\langle n\rangle,t)}{\partial\langle n^c_y\rangle}\right). \tag{5.22}$$

Here

$$\begin{aligned} m^{ab}_{xx'}(n,t) &:= \sum_{n'}(\Delta n')^a_x(\Delta n')^b_{x'}w(n'|n;t) \\ &= \sum_c\sum_{y,y'}(\Delta n^{c\,c}_{y\,y'})^a_x(\Delta n^{c\,c}_{y\,y'})^b_{x'}w(n^{c\,c}_{y\,y'}|n;t) \\ &\quad + \frac{1}{2}\sum_c\sum_{y,y'}\sum_d\sum_{z,z'}(\Delta n^{c\;d\,c\,d}_{y'z'y\,z})^a_x(\Delta n^{c\;d\,c\,d}_{y'z'y\,z})^b_{x'}w(n^{c\;d\,c\,d}_{y'z'y\,z}|n;t) \end{aligned} \tag{5.23}$$

are the *second jump moments* which have the explicit form

$$\begin{aligned} &m^{ab}_{xx'}(n,t) \\ &\quad = \delta_{ab}\left\{\delta_{xx'}\sum_y\left[w^a(x|y;n;t)n^a_y + w^a(y|x;n;t)n^a_x\right]\right. \\ &\quad\quad \left. -\left[w^a(x|x';n;t)n^a_{x'} + w^a(x'|x;n;t)n^a_x\right]\right\} \qquad (5.24\text{a}) \\ &\quad\quad +\sum_{y'}\sum_y\left[\widehat{w}_{ab}(x,x'|y,y';t)n^a_y n^b_{y'} + \widehat{w}_{ab}(y,y'|x,x';t)n^a_x n^b_{x'}\right] \\ &\quad\quad -\sum_{y'}\sum_y\left[\widehat{w}_{ab}(y,x'|x,y';t)n^a_x n^b_{y'} + \widehat{w}_{ab}(x,y'|y,x';t)n^a_y n^b_{x'}\right]. \qquad (5.24\text{b}) \end{aligned}$$

An elegant derivation of the mean value and covariance equations is also possible by means of the *operators* $\mathcal{A}^{a'a}_{x'x}$ which are defined by

$$\mathcal{A}^{a'a}_{x'x}f(n,t) := f\left(\mathcal{A}^{a'a}_{x'x}n,t\right) := f(n^{a'a}_{x'x},t) \tag{5.25}$$

(cf. [289, 294]). This bases on the following representation of configurational master equation (5.7):

$$
\begin{aligned}
\frac{d}{dt}P(n,t) = & \sum_{a,x,a',x'} \left[w(n|n^{aa'}_{xx'};t)P(n^{aa'}_{xx'},t) - w(n^{a'a}_{x'x}|n;t)P(n,t) \right] \\
& + \frac{1}{2} \sum_{a,x,a',x'} \sum_{b,y,b',y'} \left[w(n|n^{aba'b'}_{xyx'y'};t)P(n^{aba'b'}_{xyx'y'},t) \right. \\
& \left. - w(n^{a'b'ab}_{x'y'xy}|n;t)P(n,t) \right] \\
= & \sum_{a,x,a',x'} \left(\mathcal{A}^{aa'}_{xx'} - 1 \right) w(n^{a'a}_{x'x}|n;t)P(n,t) \\
& + \frac{1}{2} \sum_{a,x,a',x'} \sum_{b,y,b',y'} \left(\mathcal{A}^{aa'}_{xx'}\mathcal{A}^{bb'}_{yy'} - 1 \right) w(n^{a'b'ab}_{x'y'xy}|n;t)P(n,t) \,. \qquad (5.26)
\end{aligned}
$$

Moreover, the *moments*

$$
\left\langle (n^1_{x_1})^{l^1_1} \cdot \ldots \cdot (n^A_{x_S})^{l^A_S} \right\rangle \qquad (5.27)
$$

can be calculated via the *characteristic function (*FOURIER *transform)*

$$
F(\omega;t) := \sum_n P(n;t)\exp(\mathrm{i}\omega \cdot n) \qquad (5.28)
$$

and the *generating function (*LAPLACE *transform)*

$$
G(U;t) := \sum_n P(n;t) \prod_{a,i} (U^a_i)^{n^a_{x_i}} \,. \qquad (5.29)
$$

This methods utilize the relations

$$
\left\langle (n^1_{x_1})^{l^1_1} \cdot \ldots \cdot (n^A_{x_S})^{l^A_S} \right\rangle = \frac{1}{\mathrm{i}^{l^1_1+\cdots+l^A_S}} \frac{\partial^{l^1_1}}{\partial(\omega^1_1)^{l^1_1}} \cdots \frac{\partial^{l^A_S}}{\partial(\omega^A_S)^{l^A_S}} F(\omega;t) \bigg|_{\omega^1_1=\cdots=\omega^A_S=0} \qquad (5.30)
$$

and

$$
\left\langle (n^1_{x_1})^{l^1_1} \cdot \ldots \cdot (n^A_{x_S})^{l^A_S} \right\rangle = \left(U^1_1 \frac{\partial}{\partial U^1_1} \right)^{l^1_1} \cdots \left(U^A_S \frac{\partial}{\partial U^A_S} \right)^{l^A_S} G(U,t) \bigg|_{U^1_1=\cdots=U^A_S=1} \,. \qquad (5.31)
$$

5.4 Corrections and Higher Order Interactions

According to Sect. 5.3, the exact equations

$$\frac{d\langle n_x^a\rangle}{dt} = \langle m_x^a(n,t)\rangle \tag{5.32}$$

for the *first moments* $\langle n_x^a\rangle$ (i.e. the mean value equations) depend, because of relations (5.15), on the *second moments* $\langle n_x^a n_y^b\rangle$. However, the exact equations

$$\frac{d\langle n_x^a n_{x'}^b\rangle}{dt} = \langle m_{xx'}^{ab}(n,t)\rangle + \langle n_x^a m_{x'}^b(n,t)\rangle + \langle n_{x'}^b m_x^a(n,t)\rangle \tag{5.33}$$

for the temporal evolution of the second moments $\langle n_x^a n_{x'}^b\rangle$ depend, due to the relations (5.24), on the *third moments* $\langle n_x^a n_{x'}^b n_{x''}^c\rangle$ etc. Like in Sect. 4.2, we face the problem that the exact moment equations represent a *hierarchy of non-closed equations.* Hence suitable approximations for the moment equations must be found.

For the discussion of this problem let us consider the general case that interactions between up to k elements have to be taken into account. Then in master equation (5.4) we must insert the transition rates

$$w(n'|n;t) := \begin{cases} \widehat{w}_{a_1}(x_1'|x_1;t)n_{x_1}^{a_1} & \text{if } n' = n_{x_1' x_1}^{a_1 a_1} \\ \widehat{w}_{a_1 a_2}(x_1',x_2'|x_1,x_2;t)n_{x_1}^{a_1}n_{x_2}^{a_2} & \text{if } n' = n_{x_1' x_2' x_1 x_2}^{a_1 a_2 a_1 a_2} \\ \vdots & \vdots \\ \widehat{w}_{a_1\dots a_k}(x_1',\dots,x_k'|x_1,\dots,x_k;t) & \\ \quad \times n_{x_1}^{a_1}\cdot\ldots\cdot n_{x_k}^{a_k} & \text{if } n' = n_{x_1'\dots x_k' x_1\dots x_k}^{a_1\dots a_k a_1\dots a_k} \\ 0 & \text{otherwise} \end{cases} \tag{5.34}$$

where

$$n_{x_1'\dots x_k' x_1\dots x_k}^{a_1\dots a_k a_1\dots a_k} := (\dots,(n_{x_1'}^{a_1}+1),\dots,(n_{x_1}^{a_1}-1),\dots,(n_{x_k'}^{a_k}+1),\dots,(n_{x_k}^{a_k}-1),\dots)^{\text{tr}}. \tag{5.35}$$

Consequently, the temporal evolutions of the *lth moments* $\langle n_{x_1}^{a_1}\cdot\ldots\cdot n_{x_l}^{a_l}\rangle$ obey the equations

$$\begin{aligned} &\frac{d}{dt}\langle n_{x_1}^{a_1}\cdot\ldots\cdot n_{x_l}^{a_l}\rangle \\ &= \sum_n\sum_{n'}\Big[(n')_{x_1}^{a_1}\cdot\ldots\cdot(n')_{x_l}^{a_l} - n_{x_1}^{a_1}\cdot\ldots\cdot n_{x_l}^{a_l}\Big]w(n'|n;t)P(n,t) \\ &= \sum_n\sum_{n'}\Big[\Big((\Delta n')_{x_1}^{a_1} + n_{x_1}^{a_1}\Big)\cdot\ldots\cdot\Big((\Delta n')_{x_l}^{a_l} + n_{x_l}^{a_l}\Big) \\ &\quad - n_{x_1}^{a_1}\cdot\ldots\cdot n_{x_l}^{a_l}\Big]w(n'|n;t)P(n,t) \end{aligned} \tag{5.36}$$

(cf. (3.170)), i.e.

$$\frac{d}{dt}\langle n_{x_1}^{a_1}\cdot\ldots\cdot n_{x_l}^{a_l}\rangle = \sum_{m=1}^{l}\sum_{\mathcal{M}_m}\left\langle m_{x_{i_1}\ldots x_{i_m}}^{a_{i_1}\ldots a_{i_m}}\cdot n_{x_{i_{m+1}}}^{a_{i_{m+1}}}\cdot\ldots\cdot n_{x_{i_l}}^{a_{i_l}}\right\rangle . \tag{5.37}$$

Here

$$m_{x_{i_1}\ldots x_{i_m}}^{a_{i_1}\ldots a_{i_m}}(n,t) := \sum_{n'}(\Delta n')_{x_{i_1}}^{a_{i_1}}\cdot\ldots\cdot(\Delta n')_{x_{i_m}}^{a_{i_m}} w(n'|n;t) \tag{5.38}$$

are the *mth jump moments* and $\sum_{\mathcal{M}_m}$ means the summation over all subsets $\{i_1,\ldots,i_m\}$ consisting of m elements of the set $\{1,\ldots,l\}\equiv\{i_1,\ldots,i_l\}$. *If the transition rates $w(n'|n;t)\equiv w(n'|n)$ are time-independent, the moments are expected to approach a unique solution in the limit $t\to\infty$ since $P(n,t)$ converges to a unique equilibrium distribution* (cf. Sect. 3.3.5 and Fig. 5.1).

Due to (5.34) and $(\Delta n')_x^a \in \{0,\pm 1,\ldots,\pm k\}$, the mth jump moments are polynomials of *order* k in the occupation numbers $n_{x_{i_n}}^{a_{i_n}}$. Thus Eq. (5.37) for the lth moments depend on the first up to the $(l-1+k)$th moments. Consequently, the exact moment Eq. (5.37) will represent a closed system of equations only if $k=1$ (if interactions between elements can be neglected). In the case of interactions between $k>1$ subsystems we must choose an appropriate approximation of (5.37) or of

$$\begin{aligned} n_{x_1}^{a_1}\cdot\ldots\cdot n_{x_n}^{a_n} &= \left[\left(n_{x_1}^{a_1}-\langle n_{x_1}^{a_1}\rangle\right)+\langle n_{x_1}^{a_1}\rangle\right]\cdot\ldots\cdot\left[\left(n_{x_n}^{a_n}-\langle n_{x_n}^{a_n}\rangle\right)+\langle n_{x_n}^{a_n}\rangle\right] \qquad (5.39)\\ &= \sum_{m=0}^{n}\sum_{\mathcal{M}_m}\left(n_{x_{i_1}}^{a_{i_1}}-\langle n_{x_{i_1}}^{a_{i_1}}\rangle\right)\cdot\ldots\cdot\left(n_{x_{i_m}}^{a_{i_m}}-\langle n_{x_{i_m}}^{a_{i_m}}\rangle\right)\\ &\quad\times\left\langle n_{x_{i_{m+1}}}^{a_{i_{m+1}}}\right\rangle\cdot\ldots\cdot\left\langle n_{x_{i_n}}^{a_{i_n}}\right\rangle . \qquad (5.40)\end{aligned}$$

(5.40) is the *exact* TAYLOR expansion of (5.39) (cf. [270] and [100], pp. 340ff.). If we are interested in the first up to the jth moments ($1\le l\le j$, $j\ge 1$) there are essentially two ways of getting *approximate moment equations:*

(a) The substitution of all jump moments $m_{x_{i_1}\ldots x_{i_m}}^{a_{i_1}\ldots a_{i_m}}$ by TAYLOR approximations of first order:

$$\begin{aligned} m_{x_{i_1}\ldots x_{i_m}}^{a_{i_1}\ldots a_{i_m}}(n,t) &\approx m_{x_{i_1}\ldots x_{i_m}}^{a_{i_1}\ldots a_{i_m}}(\langle n\rangle,t)\\ &\quad+\sum_{a,x}(n_x^a-\langle n_x^a\rangle)\frac{\partial}{\partial\langle n_x^a\rangle}m_{x_{i_1}\ldots x_{i_m}}^{a_{i_1}\ldots a_{i_m}}(\langle n\rangle,t) . \end{aligned}\tag{5.41}$$

In this case the equations for the lth moments depend only on the first up to the lth moments but each equation is an approximation. This method was applied in Sects. 3.5 and 5.3. A comparison with (5.40) shows that it corresponds to the assumption

$$\frac{\left|\left\langle (n_{x_{i_1}}^{a_{i_1}} - \langle n_{x_{i_1}}^{a_{i_1}} \rangle) \cdot \ldots \cdot (n_{x_{i_m}}^{a_{i_m}} - \langle n_{x_{i_m}}^{a_{i_m}} \rangle) \right\rangle\right|}{\left\langle n_{x_{i_1}}^{a_{i_1}} \right\rangle \cdot \ldots \cdot \left\langle n_{x_{i_m}}^{a_{i_m}} \right\rangle} \ll 1 \quad \text{for} \quad 1 < m \leq k \tag{5.42}$$

which can be checked by evaluating the first up to the kth moments. Hence one should choose $j \geq k$.

(b) The substitution of all expressions $\left(m_{x_{i_1} \ldots x_{i_m}}^{a_{i_1} \ldots a_{i_m}} \cdot n_{x_{i_{m+1}}}^{a_{i_{m+1}}} \cdot \ldots \cdot n_{x_{i_l}}^{a_{i_l}} \right)$ by TAYLOR approximations of jth order: In this case the equations for the first up to the $(j-k+1)$th moments are exact (cf. (5.40)). However, the equation for an arbitrary lth moment depends on the first up to the $(l-1+k)$th moments the higher ones of which are determined by approximate equations. Method (b) corresponds to the assumption

$$\frac{\left|\left\langle (n_{x_{i_1}}^{a_{i_1}} - \langle n_{x_{i_1}}^{a_{i_1}} \rangle) \cdot \ldots \cdot (n_{x_{i_m}}^{a_{i_m}} - \langle n_{x_{i_m}}^{a_{i_m}} \rangle) \right\rangle\right|}{\left\langle n_{x_{i_1}}^{a_{i_1}} \right\rangle \cdot \ldots \cdot \left\langle n_{x_{i_m}}^{a_{i_m}} \right\rangle} \ll 1 \quad \text{for} \quad j < m \leq l-1+k \tag{5.43}$$

with $1 \leq l \leq j$. Equation (5.43) is fulfilled for a certain time interval which, in contrast to method (a), cannot be determined by solving the approximate moment equations. Nevertheless, (5.43) is normally guaranteed for large m if the absolute values of the covariances are small in the sense of

$$\left|\sigma_{x\,x'}^{ab}\right| < \epsilon \langle n_x^a \rangle \langle n_{x'}^b \rangle \qquad \text{with} \qquad 0 \leq \epsilon \ll 1 \tag{5.44}$$

(which is often the case in the *thermodynamic limit* $N \gg 1$). Then, where $P(n,t)$ is not negligible, the absolute value of

$$\left(\frac{n_{x_{i_1}}^{a_{i_1}}}{\langle n_{x_{i_1}}^{a_{i_1}} \rangle} - 1 \right) \cdot \ldots \cdot \left(\frac{n_{x_{i_m}}^{a_{i_m}}}{\langle n_{x_{i_m}}^{a_{i_m}} \rangle} - 1 \right) \tag{5.45}$$

is normally small since

$$\frac{\left|\left\langle (n_{x_{i_1}}^{a_{i_1}} - \langle n_{x_{i_1}}^{a_{i_1}} \rangle) \cdot \ldots \cdot (n_{x_{i_m}}^{a_{i_m}} - \langle n_{x_{i_m}}^{a_{i_m}} \rangle) \right\rangle\right|}{\left\langle n_{x_{i_1}}^{a_{i_1}} \right\rangle \cdot \ldots \cdot \left\langle n_{x_{i_m}}^{a_{i_m}} \right\rangle} \sim \varepsilon^{m/2} \,. \tag{5.46}$$

For $j = 1$ (i.e. a restriction to mean value equations) methods (a) and (b) lead to the same approximate equations. However, as soon as we are also interested in

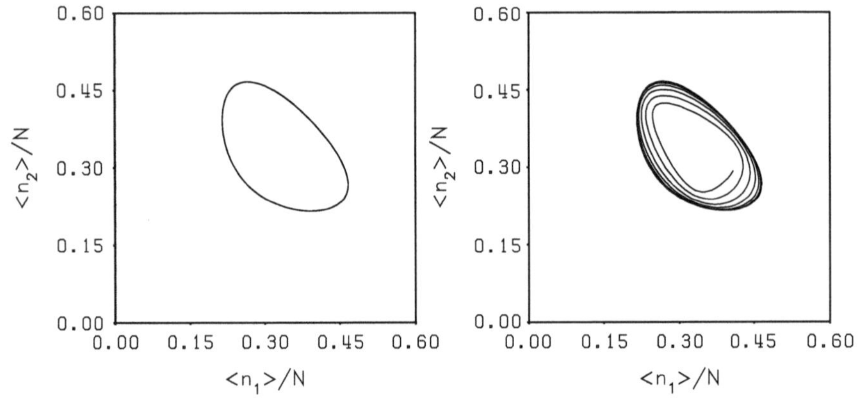

Fig. 5.1 *Left*: Phase portrait of a periodic solution of a special *approximate* mean value equation (cf. Sect. 10.3.2). *Right*: The solution of the corresponding *corrected* mean value equation moves in a spiral way towards a stationary point as expected according to Sect. 3.3.5

moments of higher order ($j \geq 2$) we can expect better results from method (b) than from (a) because method (b) bases on an approximation of higher order (cf. Fig. 5.1 and Sect. 12.4.1).

For $k = 2$ (i.e. including pair interactions) and $j = 2$, for example, method (b) would provide the *corrected mean value equations*

$$\frac{\partial \langle n_x^a \rangle}{\partial t} \approx m_x^a(\langle n \rangle, t) + \frac{1}{2} \sum_{b,y} \sum_{c,y'} \sigma_{yy'}^{bc} \frac{\partial^2 m_x^a(\langle n \rangle, t)}{\partial \langle n_y^b \rangle \partial \langle n_{y'}^c \rangle} \tag{5.47}$$

(cf. [151], pp. 130ff.) which, due to (5.15), contain terms of the thus form $\langle n_x^a n_{x'}^b \rangle$ and depend on the covariances $\sigma_{x\,x'}^{ab} = \langle n_x^a n_{x'}^b \rangle - \langle n_x^a \rangle \langle n_{x'}^b \rangle$. The *corrected covariance equations* according to method b) were given by

$$\begin{aligned} \frac{\partial \sigma_{x\,x'}^{ab}}{\partial t} \approx\; & m_{x\,x'}^{ab}(\langle n \rangle, t) + \frac{1}{2} \sum_{c,y} \sum_{d,y'} \sigma_{y\,y'}^{cd} \frac{\partial^2 m_{x\,x'}^{ab}(\langle n \rangle, t)}{\partial \langle n_y^c \rangle \partial \langle n_{y'}^d \rangle} \\ & + \sum_{c,y} \left(\sigma_{x\,y}^{ac} \frac{\partial m_{x'}^b(\langle n \rangle, t)}{\partial \langle n_y^c \rangle} + \sigma_{x'\,y}^{b\,c} \frac{\partial m_x^a(\langle n \rangle, t)}{\partial \langle n_y^c \rangle} \right) \end{aligned} \tag{5.48}$$

and would use (5.24) as well as the approximations

$$\langle n_{x1}^{a_1} n_{x2}^{a_2} n_{x3}^{a_3} \rangle \approx \langle n_{x1}^{a_1} \rangle \langle n_{x2}^{a_2} \rangle \langle n_{x3}^{a_3} \rangle + \sigma_{x1\,x2}^{a_1 a_2} \langle n_{x3}^{a_3} \rangle + \sigma_{x1\,x3}^{a_1 a_3} \langle n_{x2}^{a_2} \rangle + \sigma_{x2\,x3}^{a_2 a_3} \langle n_{x1}^{a_1} \rangle . \tag{5.49}$$

5.5 Indirect Interactions and Mean Field Approaches

In cases of *indirect interactions* not only the *configurational* transition rates but also the *individual* transition or interaction rates depend on the occupation numbers n_x^a of the states x. Restricting ourselves to spontaneous transitions and pair interactions ($k \le 2$), we must insert into the master equation

$$\frac{d}{dt}P(n,t) = \sum_{n'} \Big[w(n|n';t)P(n',t) - w(n'|n;t)P(n,t) \Big] \tag{5.50a}$$

the *generalized transition rates*

$$w(n'|n;t) = \begin{cases} \widehat{w}_a(x'|x;n;t)n_x^a & \text{if } n' = n_{x'x}^{a'a} \\ \widehat{w}_{ab}(x',y'|x,y;n;t)n_x^a n_y^b & \text{if } n' = n_{x'y'xy}^{a'b'ab} \\ 0 & \text{otherwise.} \end{cases} \tag{5.50b}$$

A derivation of the corresponding approximate and corrected mean value and covariance equations is directly possible with the TAYLOR approximations from Sect. 5.4 (cf. Sect. 12.4).

The generalization to occupation number dependent transition rates is particularly important for the description of *social* processes where the behaviour of the considered individuals can depend on the (socio)configuration n [289, 292]. However, it is also relevant for *mean field approaches* [24] which are applied in physics (e.g. in the ISING *model* of ferromagnetism [10, 148, 304]). Mean field approaches essentially correspond to the approximate mean value equations (5.16) or their stationary solution.

5.6 Comparison of Direct and Indirect Interactions

In the following we will elaborate the interrelations and differences between direct and indirect interactions in more detail.

5.6.1 Differences Concerning the Covariance Equations

First, let us compare the mean value and covariance equations of two different configurational master equations: In the first case the transition rates are assumed to have the form

$$w(n'|n;t) := \begin{cases} \widehat{w}_{ab}(x',y'|x,y;t)n_x^a n_y^b & \text{if } n' = n_{x'y'xy}^{a\,b\,ab} \\ 0 & \text{otherwise} \end{cases} \tag{5.51}$$

which exclusively describes *direct* pair interactions. For the second case let us choose

$$w(n'|n;t) := \begin{cases} \sum_b \sum_y \sum_{y'} \widehat{w}_{ab}(x',y'|x,y;t) n_y^b n_x^a & \text{if } n' = n_{x'x}^{a\ a} \\ 0 & \text{otherwise} \end{cases} \tag{5.52}$$

which means configuration dependent individual transition rates

$$\widehat{w}_a(x'|x;n;t) := \sum_b \sum_y \sum_{y'} \widehat{w}_{ab}(x',y'|x,y;t) n_y^b \tag{5.53}$$

and, therefore, *indirect* interactions.

A comparison of both cases with respect to (5.12) and (5.15) shows that the mean value equations of the corresponding master equations (5.50) are completely identical. However, a difference appears concerning the covariance equations: In the case (5.52) of indirect interactions only the terms (5.24a) of the second jump moments contribute to the covariance equations whereas in the case (5.51) of direct interactions additionally the terms (5.24b) have to be taken into account. Hence *we cannot manage without the concept of direct interactions.*

In summary, by means of the covariance equations we can distinguish various stochastic processes which possess identical mean value equations.

5.6.2 Differences Concerning the Mean Value Equations

In the last section it became clear that for each model which bases on direct interactions there exists a model which bases on indirect interactions and has the same mean value equations. Now it will be demonstrated that the opposite is not true. For this purpose, let us consider

$$\begin{aligned} \widehat{w}(x'|x;n;t) &:= \nu e^{\kappa(n_{x'} - n_x)} \\ &= \nu \sum_{l=0}^{\infty} \frac{1}{l!} \Big[\kappa(n_{x'} - n_x)\Big]^l \\ &= \nu \sum_{l=0}^{\infty} \frac{\kappa^l}{l!} \sum_{k=0}^{l} \frac{l!}{(l-k)!k!} (n_{x'})^{l-k} (-n_x)^k \end{aligned} \tag{5.54}$$

which can be transformed into

$$\widehat{w}(x'|x;n;t) = \sum_{k=0}^{\infty} (n_x)^k \widehat{w}_{(k)} \tag{5.55a}$$

with

$$\widehat{w}_{(k)} := \nu(-1)^k \sum_{l=k}^{\infty} \frac{\kappa^l}{k!(l-k)!}(n_{x'})^{l-k} . \tag{5.55b}$$

The transition rates $\widehat{w}(x'|x; n; t)$ could be understood as a result of direct interactions only if the functions $\widehat{w}_{(k)}$ were interpretable as transition rates of direct interactions between k elements. Because of $\widehat{w}_{(k)} < 0$ for odd k and the dependence of $\widehat{w}_{(k)}$ on the occupation numbers $n_{x'}$ this is not possible. Therefore, *we cannot manage without the concept of indirect interactions.*

Chapter 6
The FOKKER-PLANCK Equation

6.1 Introduction

From the master equation we can derive a FOKKER-PLANCK equation by means of a second order TAYLOR approximation. The FOKKER-PLANCK equation is a *linear partial differential equation of second order* so that, not least thanks to the analogy to the SCHRÖDINGER *equation* [250], there exist many solution methods for it [27, 83, 84, 86, 95, 96, 241, 242, 259]. In contrast to the master equation the FOKKER-PLANCK equation takes into account only the first two jump moments. The mean value and covariance equations, however, agree with those of the master equation.

From the FOKKER-PLANCK equation, by factorizing the pair distribution function, we can gain '**BOLTZMANN-FOKKER-PLANCK** *equations'* which are also derivable from the BOLTZMANN-like equations by means of a second order TAYLOR approximation. An application of these new equations is discussed in Chap. 11. Since the BOLTZMANN-FOKKER-PLANCK equations depend *non-linearly* on the probability distributions their solution is complicated. However, there exists a recursive solution strategy that is related to HARTREE'S *self-consistent field method* for quantum mechanical many-particle systems [114] (which is applied to the evaluation of *atomic* and *molecular orbitals*).

6.2 Derivation

The continuous master equation (3.5b) can, by writing X' in the first term as $X - Y$ and in the second term as $X + Y$, be transformed into

$$\frac{d}{dt}P(X,t) = \int d^M Y\Big[w[Y|X-Y;t]P(X-Y,t) - w[Y|X;t]P(X,t)\Big] \quad (6.1a)$$

with the notation

$$w[Y|X;t] := w(X+Y|X,t)\,. \quad (6.1b)$$

D. Helbing, *Quantitative Sociodynamics*, 2nd ed.,
DOI 10.1007/978-3-642-11546-2_6,

For the right-hand side of (6.1a) a TAYLOR *approximation* [270] can be carried out (cf. [100], pp. 550ff.) which, when applied to the master equation, is denoted as the KRAMERS-MOYAL *expansion* [84, 160, 193]. Introducing the *lth jump moments*

$$m_{l_1\dots l_M}(X,t) := \int d^M Y\, Y_1^{l_1}\cdot\ldots\cdot Y_M^{l_M}\, w[Y|X;t] \tag{6.2a}$$

with

$$l := l_1 + \cdots + l_M \tag{6.2b}$$

we obtain

$$\begin{aligned}
&\frac{\partial}{\partial t}P(X,t)\\
&= \int d^M Y \Bigg\{ \Bigg[\sum_{l=0}^{\infty} \sum_{\substack{l_1,\ldots,l_M\\(l_1+\cdots+l_M=l)}} \frac{(-Y_1)^{l_1}\ldots(-Y_M)^{l_M}}{l_1!\ldots l_M!}\\
&\quad \times \frac{\partial^{l_1}}{\partial X_1^{l_1}}\cdots\frac{\partial^{l_M}}{\partial X_M^{l_M}}\Big(w[Y|X;t]P(X,t)\Big)\Bigg] - w[Y|X;t]P(X,t)\Bigg\}\\
&= \sum_{l=1}^{\infty}\frac{(-1)^l}{l!} \sum_{\substack{l_1,\ldots,l_M\\(l_1+\cdots+l_M=l)}} \frac{l!}{l_1!\ldots l_M!}\frac{\partial^{l_1}}{\partial X_1^{l_1}}\cdots\frac{\partial^{l_M}}{\partial X_M^{l_M}}\Big[m_{l_1\ldots l_M}(X,t)P(X,t)\Big].
\end{aligned} \tag{6.2c}$$

If series (6.2) is convergent, (6.2) and (6.1) are completely equivalent formulations. In particular, the transition rates $w[Y|X;t]$ of the master equation can be uniquely reconstructed from the corresponding jump moments (6.2a): $w[Y|X;t]$ is the inverse FOURIER transform

$$w[Y|X;t] = \frac{1}{(2\pi)^M}\int d^M\omega\, \exp(-\mathrm{i}\omega\cdot Y)\Theta(\omega,X;t) \tag{6.3a}$$

of the *charakteristic function*

$$\Theta(\omega,X;t) := \int d^M Y\, \exp(\mathrm{i}\omega\cdot Y) w[Y|X;t]\,. \tag{6.3b}$$

In addition, because of

$$m_{l_1\ldots l_M}(X,t) = \frac{1}{\mathrm{i}^{l_1+\cdots+l_M}}\frac{\partial^{l_1}}{\partial\omega_1^{l_1}}\cdots\frac{\partial^{l_M}}{\partial\omega_M^{l_M}}\Theta(\omega,X;t)\Bigg|_{\omega_1=\cdots=\omega_M=0}, \tag{6.3c}$$

the TAYLOR representation of the characteristic function is determined by the jump moments $m_{l_1 \dots l_M}$ (cf. [241], pp. 16ff.):

$$\Theta(\omega, X; t) = \sum_{l=0}^{\infty} \sum_{\substack{l_1,\dots,l_M \\ (l_1+\dots+l_M=l)}} m_{l_1\dots l_M}(X,t) \frac{(\mathrm{i}\omega_1)^{l_1}}{l_1!} \cdot \dots \cdot \frac{(\mathrm{i}\omega_M)^{l_M}}{l_M!} . \tag{6.3d}$$

In order to analytically and numerically cope with the KRAMERS-MOYAL expansion (6.2), it is desirable to approximate this by l terms of the infinite series. Since $P(X, t)$ has the meaning of a probability density, our demand is

$$P(X,t) \overset{!}{\geq} 0 \qquad \forall X, t . \tag{6.4}$$

According to the PAWULA *theorem* [216, 241] this condition is only fulfilled for $l \leq 2$. In the second order approximation ($l = 2$), (6.2) takes on the form of the FOKKER-PLANCK *equation* [84]:

$$\begin{aligned} \frac{\partial}{\partial t} P(X,t) = &- \sum_I \frac{\partial}{\partial X_I} \Big[m_I(X,t) P(X,t) \Big] \\ &+ \frac{1}{2} \sum_{I,J} \frac{\partial}{\partial X_I} \frac{\partial}{\partial X_J} \Big[m_{IJ}(X,t) P(X,t) \Big] \end{aligned} \tag{6.5}$$

where

$$m_I := m_{l_1\dots l_M} \quad \text{with} \quad l_I = 1,\ l_J = 0 \text{ for } J \neq I \tag{6.6a}$$

and

$$m_{IJ} := \begin{cases} m_{l_1\dots l_M} & \text{with} \quad l_I = 1 = l_J,\ l_K = 0 \text{ for } K \neq I, J \text{ if } J \neq I \\ m_{l_1\dots l_M} & \text{with} \quad l_I = 2,\ l_K = 0 \text{ for } K \neq I \qquad\qquad \text{if } J = I . \end{cases} \tag{6.6b}$$

The *configurational* FOKKER-PLANCK *equation* corresponding to the configurational master equation (5.50) can be obtained analogously. It reads

$$\begin{aligned} \frac{\partial}{\partial t} P(n,t) = &- \sum_{a,x} \frac{\partial}{\partial n_x^a} \Big[m_x^a(n,t) P(n,t) \Big] \\ &+ \frac{1}{2} \sum_{a,x} \sum_{b,x'} \frac{\partial}{\partial n_x^a} \frac{\partial}{\partial n_{x'}^b} \Big[m_{xx'}^{ab}(n,t) P(n,t) \Big] \end{aligned} \tag{6.7}$$

if n is formally treated as a continuous quantity. Combining the subscripts x and superscripts a in a vectorial subscript

$$I := (a, x)^{\mathrm{tr}} \tag{6.8}$$

and replacing $n_x^a \equiv n_I$ by X_I as well as n by X, (6.7) has the same form as (6.5) with $M = AS$. In the following, notation (6.5) will therefore also be used for the configurational FOKKER-PLANCK equation (6.7).

For the second order TAYLOR approximation to be applicable, the *expansion parameters* Y_I have, on average, to be small compared to the *system size*

$$\Omega := \max_I \max_{X',X} |X'_I - X_I| \,. \tag{6.9}$$

The decisive quantity for this is the *smallness parameter* δ defined by

$$\begin{aligned} \delta^2 &:= \max_I \max_X \frac{m_{II}(X,t)}{\Omega^2 \int d^M Y' \, w[Y'|X;t]} \\ &= \max_I \max_X \int d^M Y \left(\frac{Y_I}{\Omega}\right)^2 \frac{w[Y|X;t]}{\int d^M Y' \, w[Y'|X;t]} \end{aligned} \tag{6.10}$$

since it is a measure for the *relative average jump distance.* Transforming master equation (6.1) into *scaled co-ordinates*

$$Z := \varepsilon X \qquad \text{with} \qquad \varepsilon := \frac{1}{\Omega} \,, \tag{6.11}$$

the corresponding KRAMERS-MOYAL expansion takes on the form

$$\begin{aligned} \frac{\partial}{\partial t} P(Z,t) = -\sum_{l=1}^{\infty} \frac{(-\varepsilon)^{l-1}}{l!} & \sum_{\substack{l_1,\ldots,l_M \\ (l_1+\cdots+l_M=l)}} \frac{l!}{l_1!\ldots l_M!} \\ & \times \frac{\partial^{l_1}}{\partial Z_1^{l_1}} \cdots \frac{\partial^{l_M}}{\partial Z_M^{l_M}} \Big[M_{l_1\ldots l_M}(Z,t) P(Z,t) \Big] \end{aligned} \tag{6.12a}$$

with the *scaled jump moments*

$$M_{l_1\ldots l_M}(Z,t) := \varepsilon \, m_{l_1\ldots l_M}\left(\frac{Z}{\varepsilon}, t\right) . \tag{6.12b}$$

Therefore, with regard to the order of magnitude, we find

$$\begin{aligned} \varepsilon^{l-1} |M_{l_1\ldots l_M}(Z,t)| \;&=\; \frac{1}{\Omega^l} |m_{l_1\ldots l_M}(Z/\varepsilon, t)| \\ \le \int d^M Y \left| \left(\frac{Y_1}{\Omega}\right)^{l_1} \cdot \ldots \cdot \left(\frac{Y_M}{\Omega}\right)^{l_M} \right| & \frac{w[Y|X;t]}{\int d^M Y' \, w[Y'|X;t]} \int d^M Y' \, w[Y'|X;t] \\ \sim \delta^l \int d^M Y' \, w[Y'|X;t] \,. & \end{aligned} \tag{6.13}$$

According to this, in the case $\delta \ll 1$ the scaled jump moments become smaller and smaller for increasing values of $l = l_1 + \cdots + l_M$. Consequently, they are most often neglected for $l > 2$. However, we must not neglect the second jump moments if we do not want to suppress the effect of the fluctuations. Hence as a suitable approximation we obtain the FOKKER-PLANCK equation

$$\frac{\partial}{\partial t} P(Z,t) = -\sum_I \frac{\partial}{\partial Z_I}\Big[k_I(Z,t)P(Z,t)\Big] + \frac{\varepsilon}{2}\sum_{I,J} \frac{\partial}{\partial Z_I}\frac{\partial}{\partial Z_J}\Big[q_{IJ}(Z,t)P(Z,t)\Big] \tag{6.14a}$$

with the *scaled* first jump moments

$$k_I(Z,t) := M_I(Z,t) := \varepsilon m_I\left(\frac{Z}{\varepsilon}, t\right) \tag{6.14b}$$

and the *scaled* second jump moments

$$q_{IJ}(Z,t) := M_{IJ}(Z,t) := \varepsilon m_{IJ}\left(\frac{Z}{\varepsilon}, t\right). \tag{6.14c}$$

6.3 Properties

In the following we will briefly treat the properties of FOKKER-PLANCK equation (6.14).

6.3.1 The Continuity Equation

Equation (6.14) can be written in the form of the *continuity equation*

$$\frac{\partial}{\partial t} P(Z,t) = -\nabla j(Z,t). \tag{6.15a}$$

Here

$$j_I(Z,t) := k_I(Z,t)P(Z,t) - \frac{\varepsilon}{2}\sum_J \frac{\partial}{\partial Z_J}\Big[q_{IJ}(Z,t)P(Z,t)\Big] \tag{6.15b}$$

is the component I of the *probability current density* j.

6.3.2 Normalization

In the following we will occasionally require GAUSS' *divergence theorem* which facilitates the transformation of a *volume integral* over the divergence $\nabla j(Z,t)$ of a *vector field* $j(Z,t)$ into a *surface integral* (cf. [246], pp. 1048ff.):

$$\int_V d^M Z \, \nabla j(Z,t) = \int_{\partial V} dO(Z) \cdot j(Z,t) \,. \tag{6.16}$$

Here ∂V means the *surface* of *volume* V. $dO(Z)$ denotes vectors of magnitude dO which are perpendicular to the differential surface elements of ∂V of area dO and point to the outside. Thus we integrate over the *component*

$$\frac{dO}{dO} \cdot j(Z,t) \tag{6.17}$$

of $j(Z,t)$ that is perpendicular to the surface.
GAUSS' divergence theorem and continuity equation (6.15) imply

$$\begin{aligned} \frac{\partial}{\partial t} \int_V d^M Z \, P(Z,t) &= -\int_V d^M Z \, \nabla j(Z,t) \\ &= -\int_{\partial V} dO \cdot j(Z,t) \\ &= 0 \end{aligned} \tag{6.18}$$

if the probability current j vanishes on the surface ∂V of the system. Hence, for the possibility of *normalization*

$$\int_V d^M Z P(Z,t) \stackrel{!}{=} 1 \tag{6.19}$$

we must demand the *boundary condition*

$$j(Z,t)\Big|_{Z \in \partial V} \stackrel{!}{=} 0 \,. \tag{6.20}$$

In order to guarantee this we will assume that the relations

$$\frac{\partial}{\partial Z_J} P(Z,t)\Big|_{Z \in \partial V} = 0 \tag{6.21a}$$

and

$$P(Z,t)\Big|_{Z \in \partial V} = 0 \tag{6.21b}$$

are valid, i.e. that the probability distribution $P(Z,t)$ decreases fast enough in the neighbourhood of the surface ∂V and vanishes on it. This is most often the case.

6.3.3 The LIOUVILLE Representation

The FOKKER-PLANCK equation can be represented in the form

$$\frac{\partial}{\partial t} P(Z,t) = \mathcal{L}(Z,t) P(Z,t) \tag{6.22a}$$

with the LIOUVILLE *operator*

$$\mathcal{L}(Z,t) := -\sum_I \frac{\partial}{\partial Z_I} k_I(Z,t) + \frac{\varepsilon}{2} \sum_{I,J} \frac{\partial}{\partial Z_I} \frac{\partial}{\partial Z_J} q_{IJ}(Z,t) \,. \tag{6.22b}$$

Since, due to (6.3), every FOKKER-PLANCK equation is equivalent to a master equation (6.1), many properties of the FOKKER-PLANCK equation directly correspond to those of the master equation (cf. Sect. 3.3). Therefore, only a short review of these will be given in the following.

6.3.4 Non-negativity

If condition $P(Z,t_0) \ge 0$ is fulfilled for all states Z at an initial time t_0, we have

$$P(Z,t) \ge 0 \tag{6.23}$$

for all later times $t > t_0$.

6.3.5 Eigenvalues

The eigenvalues λ of the LIOUVILLE operator have the property that their real part is *non-positive:*

$$\mathrm{Re}(\lambda) \le 0 \tag{6.24}$$

(cf. [241], pp. 104f., 143ff.). Since the FOKKER-PLANCK equation possesses the eigenvalue $\lambda = 0$, the *existence* of a *stationary solution* $P^0(Z)$ is always guaranteed.

6.3.6 Convergence to the Stationary Solution

For the *entropy-like* LIAPUNOV *function*

$$K(t) := \int d^M Z \, P(Z,t) \ln \frac{P(Z,t)}{P'(Z,t)} \tag{6.25}$$

(where $P(Z,t)$ and $P'(Z,t)$ are two solutions of the FOKKER-PLANCK equation (6.14) for different initial distributions) one can show

$$K(t) \geq 0\,, \tag{6.26a}$$

$$\frac{\partial K(t)}{\partial t} \leq 0 \tag{6.26b}$$

in the case of time-independent jump moments. If $P'(Z,t) := P^0(Z)$ is a *stationary* solution and if $q_{IJ}(Z) > 0$ and $P^0(Z) > 0$ holds for almost every Z (i.e. with probability 1), every solution $P(Z,t)$ of the FOKKER-PLANCK equation will converge to $P^0(Z)$ in the course of time and the stationary solution $P^0(Z)$ is *unique.* (Cf. [84], pp. 61ff.)

6.4 Solution Methods

6.4.1 Stationary Solution

The stationary solution $P^0(Z)$ of the FOKKER-PLANCK equation in several dimensions can be explicitly found only under certain circumstances. First of all, the jump moments $k_I(Z,t)$ and $q_{IJ}(Z,t)$ must be time-independent. Furthermore, it will be assumed that in the stationary case not only the divergence $\nabla j(Z)$ of the probability current $j(Z)$ vanishes but also the probability current itself:

$$j(Z) \equiv \mathbf{0}\,. \tag{6.27}$$

Then (6.27) can, by means of the product rule (cf. p. xxviii), be brought into the form

$$\frac{\varepsilon}{2}\sum_J q_{IJ}(Z)\frac{\partial P^0(Z)}{\partial Z_J} = P^0(Z)\left[k_I(Z) - \frac{\varepsilon}{2}\sum_J \frac{\partial}{\partial Z_J} q_{IJ}(Z)\right]. \tag{6.28}$$

If the matrix $\underline{q}(Z) \equiv (q_{IJ}(Z))$ has the *inverse* $\underline{q}^{-1}(Z) \equiv \left(q^{-1}_{IJ}(Z)\right)$ we obtain

$$\begin{aligned}
\frac{\partial}{\partial Z_I}[\ln P^0(Z)] &= \frac{\partial P^0(Z)/\partial Z_I}{P^0(Z)} \\
&= \sum_K q^{-1}_{IK}(Z)\left[\frac{2}{\varepsilon}k_K(Z) - \sum_J \frac{\partial}{\partial Z_J} q_{KJ}(Z)\right] \\
&=: f_I(Z)\,.
\end{aligned} \tag{6.29}$$

Since the left-hand side of (6.29) is a *gradient,* (6.29) can only be fulfilled on the *integrability conditions*

$$\frac{\partial f_I}{\partial Z_J} \stackrel{!}{=} \frac{\partial f_J}{\partial Z_I} \qquad \forall\, I, J\,. \tag{6.30}$$

However, if these are satisfied, the stationary solution can be written in the form

$$P^0(Z) = P^0(Z_0)\mathrm{e}^{-\phi(Z)} \tag{6.31a}$$

with the *potential*

$$\phi(Z) := -\int_{Z_0}^{Z} dZ' \cdot f(Z') \tag{6.31b}$$

(cf. [84], pp. 106f.). This implies

$$f_I(Z) = -\frac{\partial}{\partial Z_I}\phi(Z)\,. \tag{6.32}$$

In the one-dimensional case integrability condition (6.30) is automatically fulfilled. The FOKKER-PLANCK equation then possesses the stationary solution

$$P^0(Z) = P^0(Z_0)\frac{q(Z_0)}{q(Z)} \exp\left(\frac{2}{\varepsilon}\int_{Z_0}^{Z} dZ' \frac{k(Z')}{q(Z')}\right)\,. \tag{6.33}$$

6.4.2 Path Integral Solution

For the time-dependent solution of the FOKKER-PLANCK equation a *path integral representation* exists which is related to the quantum mechanical path integral solution of the SCHRÖDINGER equation [73, 75]. This reads

$$P(Z,t) = \lim_{n\to\infty}\int d^M Z_{n-1}\ldots\int d^M Z_0\left[\prod_{i=1}^{n} P(Z_i,t_i|Z_{i-1},t_{i-1})\right] P(Z_0,t_0) \tag{6.34a}$$

with

$$Z_n \equiv Z, \qquad t_i := t_0 + i\,\Delta t, \qquad \Delta t := \frac{t-t_0}{n}\,, \tag{6.34b}$$

and the *conditional probability*

$$P(Z_{i+1},t_{i+1}|Z_i,t_i) := \frac{1}{\sqrt{(2\pi)^M \varepsilon\,\Delta t\,|\underline{q}|}} \times \exp\left(-\frac{1}{2\varepsilon\Delta t}\Big[Z_{i+1}-Z_i-k(Z_i,t_i)\Big]^{\mathrm{tr}}\underline{q}^{-1}(Z_i,t_i)\Big[Z_{i+1}-Z_i-k(Z_i,t_i)\Big]\right) \tag{6.34c}$$

(cf. [241], pp. 73ff.). Here $\underline{q}^{-1}$ denotes the *inverse* of $\underline{q} \equiv (q_{IJ})$ and $|\underline{q}|$ its *determinant.*

6.4.3 Interrelation with the SCHRÖDINGER Equation[1]

Since quantum mechanics provides many solution methods for the SCHRÖDINGER equation, it is sometimes useful that the FOKKER-PLANCK equation can, on certain conditions, be transformed into a SCHRÖDINGER equation.

Let us assume the existence of a stationary solution of the form (6.31) which fulfils integrability conditions (6.30) and makes the probability current (6.15b) vanish. Then, like in Sect. 3.4.1, the FOKKER-PLANCK equation

$$\frac{\partial}{\partial t} P(Z,t) = \mathcal{L}(Z) P(Z,t) \tag{6.35}$$

(cf. (6.22)) can, by means of the transformation

$$\psi(Z,t) := \frac{P(Z,t)}{\sqrt{P_0(Z)}} = \mathrm{e}^{\phi(Z)/2} P(Z,t)\,, \tag{6.36a}$$

be converted into the equation

$$\frac{\partial}{\partial t} \psi(Z,t) = \mathcal{H} \psi(Z,t) \tag{6.36b}$$

with the HERMITian *(self-adjoint) operator*

$$\mathcal{H} := \mathrm{e}^{\phi/2} \mathcal{L} \mathrm{e}^{-\phi/2}\,. \tag{6.36c}$$

$\mathcal{H}$ can be written in the form

$$\mathcal{H} = \frac{\partial}{\partial Z_I} \varepsilon q_{IJ} \frac{\partial}{\partial Z_J} - V(Z) \tag{6.36d}$$

with the potential

$$\begin{aligned} V(Z) &:= \mathrm{e}^{\phi/2} \left(\frac{\partial}{\partial Z_I} \varepsilon q_{IJ} \frac{\partial}{\partial Z_J} \mathrm{e}^{-\phi/2} \right) \\ &= \frac{\varepsilon}{4} q_{IJ} \frac{\partial \phi}{\partial Z_I} \frac{\partial \phi}{\partial Z_J} - \frac{\varepsilon}{2} \left(\frac{\partial}{\partial Z_I} q_{IJ} \frac{\partial \phi}{\partial Z_J} \right) \end{aligned} \tag{6.36e}$$

(cf. [241], pp. 139ff., and [242]). Now, after formally setting

$$\tau := \frac{t}{\mathrm{i}\hbar} = -\frac{\mathrm{i}}{\hbar} t \tag{6.37a}$$

---

[1] This section can be omitted during a first reading.

and

$$|\psi(\tau)\rangle := \psi(Z, \mathrm{i}\hbar\tau)\,, \tag{6.37b}$$

(6.36) takes on the form

$$\mathrm{i}\hbar\frac{\partial}{\partial\tau}|\psi(\tau)\rangle = \mathcal{H}|\psi(\tau)\rangle \tag{6.37c}$$

of a SCHRÖDINGER *equation* [250] with the *'mass tensor'*

$$\frac{\hbar^2}{2\varepsilon}q_{IJ}^{-1}\,. \tag{6.38}$$

6.5 Mean Value and Covariance Equations

First, let us define the mean value of a function in the same way as in (3.168). Then, by multiplying the FOKKER-PLANCK equation (6.14) by Z_K and $Z_K Z_L$ and integrating over Z we obtain

$$\frac{d\langle Z_K\rangle}{dt} = \langle k_K(Z,t)\rangle \tag{6.39}$$

and

$$\frac{d\langle Z_L Z_K\rangle}{dt} = \langle Z_K k_L(Z,t)\rangle + \langle Z_L k_K(Z,t)\rangle + \varepsilon\langle q_{KL}(Z,t)\rangle\,. \tag{6.40}$$

Here we applied the relations

$$\sum_I Z_K\frac{\partial}{\partial Z_I}k_I = \sum_I \frac{\partial}{\partial Z_I}\Big(Z_K k_I\Big) - k_K\,, \tag{6.41a}$$

$$\sum_{I,J} Z_K\frac{\partial}{\partial Z_I}\frac{\partial}{\partial Z_J}q_{IJ} = \sum_{I,J}\frac{\partial}{\partial Z_I}\frac{\partial}{\partial Z_J}\Big(Z_K q_{IJ}\Big) - \sum_I\frac{\partial}{\partial Z_I}\Big(q_{KI}+q_{IK}\Big)\,, \tag{6.41b}$$

$$\sum_I Z_K Z_L\frac{\partial}{\partial Z_I}k_I = \sum_I\frac{\partial}{\partial Z_I}\Big(Z_K Z_L k_I\Big) - \Big(Z_K k_L + Z_L k_K\Big)\,, \tag{6.41c}$$

$$\sum_{I,J} Z_K Z_L\frac{\partial}{\partial Z_I}\frac{\partial}{\partial Z_J}q_{IJ} = \sum_{I,J}\frac{\partial}{\partial Z_I}\frac{\partial}{\partial Z_J}\Big(Z_K Z_L q_{IJ}\Big) + \Big(q_{KL}+q_{LK}\Big) - \sum_I\frac{\partial}{\partial Z_I}\Big[Z_L\Big(q_{IK}+q_{KI}\Big) + Z_K\Big(q_{IL}+q_{LI}\Big)\Big]. \tag{6.41d}$$

Apart from this, we used GAUSS' divergence theorem and neglected the surface terms because of the assumed vanishing of $P(Z,t)$ on the surface ∂V.

Finally, for the covariances

$$\sigma'_{KL}(t) := \langle Z_K Z_L \rangle - \langle Z_K \rangle \langle Z_L \rangle \tag{6.42}$$

we find the equations

$$\frac{d\sigma'_{KL}}{dt} = \Big\langle \varepsilon q_{KL}(Z,t) \Big\rangle + \Big\langle (Z_K - \langle Z_K \rangle) k_L(Z,t) \Big\rangle + \Big\langle (Z_L - \langle Z_L \rangle) k_K(Z,t) \Big\rangle . \tag{6.43}$$

If the mean value equations (6.39) and the covariance equations (6.43) are subject to an inverse transformation into the co-ordinates $X = Z/\varepsilon$, they read

$$\frac{d\langle X_I \rangle}{dt} = \langle m_I(X,t) \rangle \tag{6.44}$$

and

$$\frac{d\sigma_{IJ}}{dt} = \Big\langle m_{IJ}(X,t) \Big\rangle + \Big\langle (X_I - \langle X_I \rangle) m_J(X,t) \Big\rangle + \Big\langle (X_J - \langle X_J \rangle) m_I(X,t) \Big\rangle . \tag{6.45}$$

A comparison with (3.172) and (3.175) shows that these equations completely agree with the mean value and covariance equations of master equation (3.5).

During the evaluation of (6.39) and (6.43) we will, in general, apply the same approximations as in Sect. 3.5 in order to obtain *closed equations*. By means of a method analogous to the *projector formalism* of Sect. 3.2.4.2 one can also derive *exact closed mean value equations* from (6.22) which, however, like (3.72), contain a time integral (i.e. they are non-local in time: cf. [292], pp. 73ff.).

6.5.1 Interpretation of the Jump Moments

The interpretation of the first and second jump moments follows directly from mean value Equation (6.44) and covariance Equation (6.45). If X is the state of the considered system at the time t this implies

$$P(X',t) = \delta_{X'X} \qquad \text{and} \qquad \langle f(X',t) \rangle_t = f(X,t) \tag{6.46}$$

for an arbitrary function $f(X,t)$. Consequently, the following holds:

$$\begin{aligned} m_I(X,t) &= \langle m_I(X',t)\rangle_t \\ &= \frac{d}{dt}\langle X'_I\rangle_t \\ &= \lim_{\Delta t\to 0}\frac{1}{\Delta t}\Big(\langle X'_I\rangle_{t+\Delta t} - \langle X'_I\rangle_t\Big) \\ &= \lim_{\Delta t\to 0}\frac{1}{\Delta t}\langle X'_I - X_I\rangle_{t+\Delta t} \end{aligned} \tag{6.47a}$$

and

$$\begin{aligned} m_{IJ}(X,t) &= \frac{d}{dt}\sigma_{IJ}(t) + 0 + 0 = \frac{\partial\langle X'_I X'_J\rangle_t}{\partial t} - \frac{\partial(\langle X'_I\rangle_t\langle X'_J\rangle_t)}{\partial t} \\ &= \frac{\partial\langle X'_I X'_J\rangle_t}{\partial t} - \langle X'_I\rangle_t\frac{\partial\langle X'_J\rangle_t}{\partial t} - \langle X'_J\rangle_t\frac{\partial\langle X'_I\rangle_t}{\partial t} \\ &= \lim_{\Delta t\to 0}\frac{1}{\Delta t}\Big(\langle X'_I X'_J\rangle_{t+\Delta t} - X_I X_J\Big) \\ &\quad - \lim_{\Delta t\to 0} X_I\frac{1}{\Delta t}\Big(\langle X'_J\rangle_{t+\Delta t} - X_J\Big) \\ &\quad - \lim_{\Delta t\to 0} X_J\frac{1}{\Delta t}\Big(\langle X'_I\rangle_{t+\Delta t} - X_I\Big) \\ &= \lim_{\Delta t\to 0}\frac{1}{\Delta t}\Big\langle (X'_I - X_I)(X'_J - X_J)\Big\rangle_{t+\Delta t} \end{aligned} \tag{6.47b}$$

with

$$\Big\langle f(X',t+\Delta t)\Big\rangle_{t+\Delta t} = \sum_{X'} f(X',t+\Delta t)P(X',t+\Delta t)\,. \tag{6.48}$$

That is, m_I determines the *systematic change* of the mean value $\langle X_I\rangle$ (the *drift*) and m_{IJ} the *change of the covariance* σ_{IJ} (the *diffusion,* i.e. the *broadening* of the distribution $P(X,t)$ by fluctuations). Therefore, we call the first jump moments m_I the *drift coefficients* and the second jump moments m_{IJ} the *diffusion coefficients* [151, 241, 292].

6.6 BOLTZMANN-FOKKER-PLANCK Equations

In the following we will start with a FOKKER-PLANCK equation for a system that consists of N subsystems α and proceed in a similar way as in Chap. 4. Again we combine the states x_α of the N subsystems α in the vectors

$$X := (x_1,\ldots,x_\alpha,\ldots,x_N)^{\mathrm{tr}}\,, \tag{6.49a}$$

$$x_\alpha := (x_{\alpha 1},\ldots,x_{\alpha i},\ldots,x_{\alpha m})^{\mathrm{tr}} \in \gamma_\alpha\,. \tag{6.49b}$$

Then in (6.5) we have $I := (\alpha, i)^{tr}$ and $M = m \cdot N$.

Now let us define the vectors

$$X^{\alpha} := (\mathbf{0}, \ldots, \mathbf{0}, x_{\alpha}, \mathbf{0}, \ldots, \mathbf{0})^{tr} \tag{6.50}$$

and

$$X^{\alpha_1 \ldots \alpha_k} := \sum_{i=1}^{k} X^{\alpha_i} . \tag{6.51}$$

The transition rates $w[Y|X;t]$ can be written in the form

$$w[Y|X;t] := \begin{cases} w_{\alpha_1}[Y^{\alpha_1}|X^{\alpha_1};t] & \text{if } Y = Y^{\alpha_1} \\ w_{\alpha_1\alpha_2}[Y^{\alpha_1\alpha_2}|X^{\alpha_1\alpha_2};t] & \text{if } Y = Y^{\alpha_1\alpha_2} \\ \vdots & \vdots \\ w_{\alpha_1\ldots\alpha_N}[Y^{\alpha_1\ldots\alpha_N}|X^{\alpha_1\ldots\alpha_N};t] & \text{if } Y = Y^{\alpha_1\ldots\alpha_N} \\ 0 & \text{otherwise} \end{cases} \tag{6.52}$$

in which they are decomposed into contributions

$$w_{\alpha_1\ldots\alpha_k}[Y^{\alpha_1\ldots\alpha_k}|X^{\alpha_1\ldots\alpha_k};t] \tag{6.53}$$

that describe the interactions between k subsystems. The restriction

$$w_{\alpha_1\ldots\alpha_k} \equiv 0 \qquad \text{if two subscripts } \alpha_i, \alpha_j \text{ are identical} \tag{6.54}$$

guarantees that self-interactions are excluded.

Furthermore, only the terms $k \leq 2$ will be taken into account which is well justified if the transitions of the subsystems are essentially due to spontaneous transitions w_{α} and pair interactions $w_{\alpha\beta}$. Then, as jump moments of the corresponding FOKKER-PLANCK equation

$$\frac{\partial}{\partial t}P(X,t) = -\sum_{\alpha,i}\frac{\partial}{\partial x_{\alpha i}}\Big[m_{\alpha i}(X,t)P(X,t)\Big]$$
$$+\frac{1}{2}\sum_{\alpha,i}\sum_{\beta,j}\frac{\partial}{\partial x_{\alpha i}}\frac{\partial}{\partial x_{\beta j}}\Big[m_{\alpha i\beta j}(X,t)P(X,t)\Big] \tag{6.55a}$$

result

$$\begin{aligned}
m_{\alpha i}(X,t) &:= \int d^M Y\, y_{\alpha i} w[Y|X;t] \\
&= \sum_{\beta} \int d^M Y\, y_{\alpha i} w_\beta[Y^\beta|X^\beta;t] \\
&\quad + \frac{1}{2}\sum_{\beta,\gamma} \int d^M Y\, y_{\alpha i} w_{\beta\gamma}[Y^{\beta\gamma}|X^{\beta\gamma};t] \\
&= \int d^M Y\, y_{\alpha i} w_\alpha[Y^\alpha|X^\alpha;t] \\
&\quad + \frac{1}{2}\left(\sum_{\gamma} \int d^M Y\, y_{\alpha i} w_{\alpha\gamma}[Y^{\alpha\gamma}|X^{\alpha\gamma};t]\right. \\
&\quad \left. + \sum_{\beta} \int d^M Y\, y_{\alpha i} w_{\beta\alpha}[Y^{\beta\alpha}|X^{\beta\alpha};t]\right) \\
&= m^\alpha_{\alpha i}(X^\alpha,t) + \sum_\beta m^{\alpha\beta}_{\alpha i}(X^{\alpha\beta},t)
\end{aligned} \tag{6.55b}$$

and

$$\begin{aligned}
m_{\alpha i\beta j}(X,t) &:= \int d^M Y\, y_{\alpha i} y_{\beta j} w[Y|X;t] \\
&= \sum_{\beta} \int d^M Y\, y_{\alpha i} y_{\beta j} w_\gamma[Y^\gamma|X^\gamma;t] \\
&\quad + \frac{1}{2}\sum_{\gamma,\delta} \int d^M Y\, y_{\alpha i} y_{\beta j} w_{\gamma\delta}[Y^{\gamma\delta}|X^{\gamma\delta};t] \\
&= \delta_{\alpha\beta}\left[m^\alpha_{\alpha i\alpha j}(X^\alpha,t) + \sum_\gamma m^{\alpha\gamma}_{\alpha i\alpha j}(X^{\alpha\gamma},t)\right] \\
&\quad + m^{\alpha\beta}_{\alpha i\beta j}(X^{\alpha\beta},t)\,.
\end{aligned} \tag{6.55c}$$

Here the following notations were introduced:

$$m^\gamma_{\alpha i}(X^\gamma,t) := \int d^M Y\, y_{\alpha i} w_\gamma[Y^\gamma|X^\gamma;t]\,, \tag{6.56a}$$

$$m^{\gamma\delta}_{\alpha i}(X^{\gamma\delta},t) := \int d^M Y\, y_{\alpha i} w_{\gamma\delta}[Y^{\gamma\delta}|X^{\gamma\delta};t]\,, \tag{6.56b}$$

$$m^\gamma_{\alpha i\beta j}(X^\gamma,t) := \int d^M Y\, y_{\alpha i} y_{\beta j} w_\gamma[Y^\gamma|X^\gamma;t]\,, \tag{6.56c}$$

$$m^{\gamma\delta}_{\alpha i\beta j}(X^{\gamma\delta},t) := \int d^M Y\, y_{\alpha i} y_{\beta j} w_{\gamma\delta}[Y^{\gamma\delta}|X^{\gamma\delta};t]\,. \tag{6.56d}$$

Now we integrate FOKKER-PLANCK equation (6.55) in an analogous way as (4.12) over all variables x_β with $\beta \neq \alpha$ and again assume the factorization

$$P_{\alpha\beta}(x_\alpha, x_\beta; t) := P_\alpha(x_\alpha, t) P_\beta(x_\beta, t) \tag{6.57}$$

of the pair distribution function $P_{\alpha\beta}(x_\alpha, x_\beta; t)$. Then we obtain, apart from

$$\int d^m x_\alpha \, P_\alpha(x_\alpha, t) = 1\,, \tag{6.58}$$

the equations

$$\begin{aligned} \frac{\partial}{\partial t} P_\alpha(x_\alpha, t) = &- \sum_i \frac{\partial}{\partial x_{\alpha i}} \Big[K_{\alpha i}(x_\alpha, t) P_\alpha(x_\alpha, t) \Big] \\ &+ \frac{1}{2} \sum_{i,j} \frac{\partial}{\partial x_{\alpha i}} \frac{\partial}{\partial x_{\alpha j}} \Big[Q_{\alpha ij}(x_\alpha, t) P_\alpha(x_\alpha, t) \Big]. \end{aligned} \tag{6.59a}$$

The *effective drift coefficients*

$$K_{\alpha i}(x_\alpha, t) := m^\alpha_{\alpha i}(x_\alpha, t) + \sum_\beta \int d^m x_\beta \, m^{\alpha\beta}_{\alpha i}(x_\alpha, x_\beta, t) P_\beta(x_\beta, t) \tag{6.59b}$$

and the *effective diffusion coefficients*

$$Q_{\alpha ij}(x_\alpha, t) := m^\alpha_{\alpha i \alpha j}(x_\alpha, t) + \sum_\beta \int d^m x_\beta \, m^{\alpha\beta}_{\alpha i \alpha j}(x_\alpha, x_\beta, t) P_\beta(x_\beta, t) \tag{6.59c}$$

make use of the conventions

$$m^\alpha_{\cdot\cdot}(x_\alpha, t) \equiv m^\alpha_{\cdot\cdot}(X^\alpha, t)\,, \tag{6.60a}$$

$$m^{\alpha\beta}_{\cdot\cdot}(x_\alpha, x_\beta, t) \equiv m^{\alpha\beta}_{\cdot\cdot}(X^{\alpha\beta}, t) \tag{6.60b}$$

and take into account, each with its last term, the interactions between the subsystems β. Moreover, we applied GAUSS' divergence theorem in order to convert the terms

$$\sum_j \int d^m x_\beta \frac{\partial}{\partial x_{\beta j}} \left(\frac{\partial}{\partial x_{\alpha i}} \Big[m_{\alpha i \beta j}(X, t) P(X, t) \Big] \right) \tag{6.61}$$

for $\beta \neq \alpha$ into vanishing surface integrals (cf. Sect. 6.3.2). Finally, we used the relation

$$m^{\gamma\delta}_{\cdot\cdot} \equiv 0 \qquad \text{for} \qquad \gamma = \delta \tag{6.62}$$

which follows from (6.54).

Distinguishing only A different *types* a of elements consisting of N_a subsystems (cf. Sect. 4.3) implies

$$\gamma_\alpha \equiv \gamma_a \quad \text{if} \quad \alpha \in a\,, \tag{6.63a}$$

$$P_\alpha \equiv P_a \quad \text{if} \quad \alpha \in a\,, \tag{6.63b}$$

$$\int d^m x\, P_a(x,t) = 1\,, \tag{6.63c}$$

$$m^\alpha_{..} \equiv \widehat{m}^a_{..} \quad \text{if} \quad \alpha \in a\,, \tag{6.63d}$$

and

$$m^{\alpha\beta}_{..} \equiv \begin{cases} \widehat{m}^{ab}_{..} \text{ if } \alpha \neq \beta,\ \alpha \in a,\ \beta \in b \\ 0 \quad \text{ if } \alpha = \beta. \end{cases} \tag{6.63e}$$

So we obtain (with respect to the considerably reduced number of variables) equations which are significantly simplified:

$$\begin{aligned} \frac{\partial}{\partial t} P_a(x,t) = &- \sum_i \frac{\partial}{\partial x_i}\Big[K_{ai}(x,t) P_a(x,t)\Big] \\ &+ \frac{1}{2}\sum_{i,j} \frac{\partial}{\partial x_i}\frac{\partial}{\partial x_j}\Big[Q_{aij}(x,t) P_a(x,t)\Big] \end{aligned} \tag{6.64a}$$

with the effective drift coefficients

$$\begin{aligned} K_{ai}(x,t) &:= m^a_{ai}(x,t) + \sum_b \int d^m y\, m^{ab}_{ai}(x,y,t) P_b(y,t) \\ &= \int d^m x'\, \Delta x'_i w^a(x'|x;t)\,, \end{aligned} \tag{6.64b}$$

the effective diffusion coefficients

$$\begin{aligned} Q_{aij}(x,t) &:= m^a_{aiaj}(x,t) + \sum_b \int d^m y\, m^{ab}_{aiaj}(x,y,t) P_b(y,t) \\ &= \int d^m x'\, \Delta x'_i \Delta x'_j w^a(x'|x;t)\,, \end{aligned} \tag{6.64c}$$

and

$$m^a_{..} := \widehat{m}^a_{..}\,, \tag{6.65a}$$

$$m^{ab}_{..} := \begin{cases} N_b \cdot \widehat{m}^{ab}_{..} & \text{if } b \neq a \\ (N_a - 1)\cdot \widehat{m}^{ab}_{..} & \text{if } b = a\,. \end{cases} \tag{6.65b}$$

The result (6.64) could also (and even very easily) have been obtained from the BOLTZMANN-like equations (4.22) by means of a second order TAYLOR approximation (cf. Sect. 11.2 and [184]). Thus it is suggested to denote Eqs. (6.64) as the '*BOLTZMANN-FOKKER-PLANCK equations*'.

6.6.1 Self-Consistent Solution

Due to factorization assumption (6.57) Eqs. (6.64) are *closed* equations, i.e. they contain no functions (like, for instance, pair distribution functions $P_{\alpha\beta}(x_\alpha, x_\beta; t)$) which cannot be determined by the others. However, since they are *non-linear* in $P_a(x,t)$ due to the dependence of the effective drift and diffusion coefficients on $P_b(y,t)$, we can only apply an approximation method to determine their solution. Having this in mind, we start from suitable probability distributions $P_a^{(0)}(x,t)$ as e.g. the solutions of the FOKKER-PLANCK equations which *neglect* pair interactions. These are given by

$$K_{ai}(x,t) := m^a_{ai}(x,t) \tag{6.66a}$$

and

$$Q_{aij}(x,t) := m^a_{aiaj}(x,t)\,. \tag{6.66b}$$

With $P_a^{(0)}(x,t)$ we evaluate the *effective* drift and diffusion coefficients for the next iteration step which gives the next approximation $P_a^{(1)}(x,t)$ etc. Hence our *iteration method* has the form

$$\begin{aligned}\frac{\partial}{\partial t}P_a^{(k+1)}(x,t) = &-\sum_i \frac{\partial}{\partial x_i}\Big[K^{(k)}_{ai}(x,t)P_a^{(k+1)}(x,t)\Big]\\ &+\frac{1}{2}\sum_{i,j}\frac{\partial}{\partial x_i}\frac{\partial}{\partial x_j}\Big[Q^{(k)}_{aij}(x,t)P_a^{(k+1)}(x,t)\Big]\end{aligned} \tag{6.67a}$$

with

$$\int d^m x\, P_a^{(k+1)}(x,t) = 1\,, \tag{6.67b}$$

$$K^{(k)}_{ai}(x,t) := m^a_{ai}(x,t) + \sum_b \int d^m y m^{ab}_{ai}(x,y,t)P_b^{(k)}(y,t)\,, \tag{6.67c}$$

and

$$Q^{(k)}_{aij}(x,t) := m^a_{aiaj}(x,t) + \sum_b \int d^m y\, m^{ab}_{aiaj}(x,y,t)P_b^{(k)}(y,t)\,. \tag{6.67d}$$

If the *sequences* of the functions $P_a^{(k)}(x,t)$ converge for $k \to \infty$, we have

$$P_a(x,t) = \lim_{k\to\infty} P_a^{(k)}(x,t)\,. \tag{6.68}$$

The *recursive method* (6.67) is especially useful for the determination of stationary solutions $P_a^0(x)$. It formally corresponds to the quantum mechanical *self-consistent field method* of HARTREE (cf. [49, 114]). An analogous recursive solution method can also be formulated for BOLTZMANN-like equations (4.22).

In the one-dimensional case ($m = 1$), the recursion formula for stationary solutions $P_a^0(x)$ of the BOLTZMANN-FOKKER-PLANCK equations is particularly simple. Here the *stationarity condition*

$$0 \stackrel{!}{=} -\frac{\partial}{\partial x}\left[K_a^{(k)}(x)P_a^{0(k+1)}(x)\right] + \frac{1}{2}\frac{\partial^2}{\partial x^2}\left[Q_a^{(k)}(x)P_a^{0(k+1)}(x)\right] \tag{6.69}$$

can, due to (6.33), be cast into the form

$$P_a^{0(k+1)}(x) = P_a^{0(k+1)}(x_0)\frac{Q_a^{(k)}(x_0)}{Q_a^{(k)}(x)}\exp\left(2\int\limits_{x_0}^{x} dx'\,\frac{K_a^{(k)}(x')}{Q_a^{(k)}(x')}\right) \tag{6.70}$$

with

$$\int dx\, P_a^{0(k+1)}(x) = 1\,. \tag{6.71}$$

Chapter 7
LANGEVIN Equations and Non-linear Dynamics

7.1 Introduction

For the modelling of stochastically behaving systems, apart from the master equation and the FOKKER-PLANCK equation, often also *stochastic differential equations* are used (cf. [84]). The most common one is the *(general)* LANGEVIN *equation*

$$\frac{dZ_I}{dt} = F_I(Z,t) + \sum_J G_{IJ}(Z,t)\xi_J(t) \tag{7.1}$$

where

$$\zeta_I(Z,t) := \sum_J G_{IJ}(Z,t)\xi_J(t) \tag{7.2}$$

describes the contribution to the change of the state Z which is caused by the *fluctuations*

$$\xi(t) := \Big(\xi_1(t), \dots, \xi_M(t)\Big)^{\text{tr}}. \tag{7.3}$$

The LANGEVIN equation is particularly suitable for the discussion of the *stability* of a system state against fluctuations. If the system state is *stable,* the fluctuations will become clearly noticeable only after a long time period (the *escape time*)—namely, when the fluctuations accidentally add up. However, the fluctuations play an *essential* role at *critical points (points of instability)* at which the system is subject to *phase transitions* (i.e. discontinuous changes of its properties). In this context, the different *routes to chaos* are particularly interesting.

If the fluctuations $\xi(t)$ are δ-correlated and GAUSSian distributed (cf. (7.23)), $F_I(\widehat{Z},t)$ describes just the *most probable state change* $d\widehat{Z}(t)/dt$ of an individual system. In addition, the description of the considered system by LANGEVIN equation (7.1) is equivalent to a delineation by the FOKKER-PLANCK equation

$$\frac{\partial}{\partial t}P(Z,t) = -\sum_I \frac{\partial}{\partial Z_I}\Big[k_I(Z,t)P(Z,t)\Big] + \frac{\varepsilon}{2}\sum_{I,J}\frac{\partial}{\partial Z_I}\frac{\partial}{\partial Z_J}\Big[q_{IJ}(Z,t)P(Z,t)\Big] \tag{7.4}$$

D. Helbing, *Quantitative Sociodynamics*, 2nd ed.,
DOI 10.1007/978-3-642-11546-2_7, © Springer-Verlag Berlin Heidelberg 2010

with the *drift coefficients*

$$k_I(Z,t) := F_I(Z,t) + \frac{1}{2}\sum_{J,K} G_{JK}(Z,t)\frac{\partial}{\partial Z_J}G_{IK}(Z,t) \tag{7.5a}$$

and the *diffusion coefficients*

$$\varepsilon q_{IJ}(Z,t) := \sum_K G_{IK}(Z,t)G_{JK}(Z,t)\,. \tag{7.5b}$$

Whereas the LANGEVIN equation reflects the fluctuation affected state changes of *single* systems, the FOKKER-PLANCK equation describes the corresponding time-dependent state *distribution* of an *ensemble* of equivalent systems.

According to (7.5a), the fluctuations yield a contribution to the drift via the term

$$\frac{1}{2}\sum_{J,K} G_{JK}(Z,t)\frac{\partial}{\partial Z_J}G_{IK}(Z,t)\,. \tag{7.6}$$

However, this contribution is usually negligible since $G_{IJ}(Z,t)$ vanishes for $\varepsilon \to 0$ according to (7.5b). This neglection will be *not* allowed only if the relative average jump distance δ (cf. (6.10)) is large. *Omitting the term (7.6), the equations*

$$\frac{d\widehat{Z}_I}{dt} = F_I(\widehat{Z},t) \overset{\delta\ll 1}{\cong} k_I(\widehat{Z},t) \tag{7.7}$$

for the most probable behaviour $\widehat{Z}(t)$ of an individual system have, in the coordinates $\widehat{X} := \widehat{Z}/\varepsilon$, the form

$$\frac{d\widehat{X}_I}{dt} \overset{\delta\ll 1}{\cong} m_I(\widehat{X},t)\,. \tag{7.8}$$

These fully agree with the approximate mean value equations (3.179). The equations for the most probable individual behaviour are especially suitable for the description of *single* or *rare* events. For the description of *frequently* recurring phenomena or of *ensembles* of systems, in contrast, one better applies the corrected mean value equations.

If we choose the matrices

$$\underline{G}(Z,t) \equiv \Big(G_{IJ}(Z,t)\Big) \tag{7.9}$$

in a *symmetrical* and *positive definite* form (cf. p. xvi), by (7.5) we can easily determine the LANGEVIN equation which is equivalent to FOKKER-PLANCK equation (7.4). For this purpose, we must only find $\underline{G} \equiv \underline{G}(Z,t)$ with

$$\underline{G}\,\underline{G} = \left(\sum_K G_{IK}G_{KJ}\right) = \left(\sum_K G_{IK}G_{JK}\right) = \varepsilon\Big(q_{IJ}\Big) \equiv \varepsilon\underline{q}\,. \tag{7.10}$$

Since the matrix $\varepsilon \underline{q} = \varepsilon \underline{q}(Z,t)$ is, as a consequence of (6.14c), also symmetrical and positive definite, it can be represented in the form

$$\varepsilon \underline{q}(Z,t) = \underline{S}^{-1}(Z,t)\underline{D}(Z,t)\underline{S}(Z,t) \tag{7.11a}$$

with a *diagonal matrix*

$$\underline{D}(Z,t) \equiv \Big(D_{IJ}(Z,t)\Big) = \Big(\lambda_I(Z,t)\,\delta_{IJ}\Big) \tag{7.11b}$$

and a *rotation matrix* $\underline{S}(Z,t)$. The matrix

$$\underline{G}(Z,t) := \underline{S}^{-1}(Z,t)\underline{D}^{1/2}(Z,t)\underline{S}(Z,t) \tag{7.12a}$$

with

$$\underline{D}^{1/2}(Z,t) \equiv \Big(D^{1/2}_{IJ}(Z,t)\Big) := \Big(\sqrt{\lambda_I(Z,t)}\,\delta_{IJ}\Big) \tag{7.12b}$$

fulfills the demanded relation (7.10).

7.2 Derivation[1]

In the following we will supply the proof (first given by STRATONOVICH [267]) that FOKKER-PLANCK equation (7.4), (7.5) is, on certain assumptions concerning the fluctuations ξ, equivalent to LANGEVIN equation (7.1). This is a special case of the general problem to find for a given *stochastic differential equation*

$$\frac{dZ_I(t)}{dt} = H_I\Big(Z(t), \xi(t), t\Big) \tag{7.13}$$

the equivalent *KRAMERS-MOYAL expansion*

$$\frac{\partial}{\partial t}P(Z,t) = -\sum_{l=1}^{\infty}\frac{(-\varepsilon)^{l-1}}{l!}\sum_{\substack{l_1,\ldots,l_M\\(l_1+\ldots+l_M=l)}}\frac{l!}{l_1!\ldots l_M!}$$
$$\times\frac{\partial^{l_1}}{\partial Z_1^{l_1}}\cdots\frac{\partial^{l_M}}{\partial Z_M^{l_M}}\Big[M_{l_1\ldots l_M}(Z)P(Z,t)\Big] \tag{7.14}$$

(cf. (6.12)). For the LANGEVIN equation we have

$$H_I\Big(Z(t),\xi(t)\Big) := F_I\Big(Z(t)\Big) + \sum_J G_{IJ}\Big(Z(t)\Big)\xi_J(t) \tag{7.15}$$

[1] This section can be omitted during a first reading.

if we, for the time being, assume that the coefficients F_I and G_{IJ} are not *explicitly* time-dependent ($F_I(Z,t) \equiv F_I(Z)$, $G_{IJ}(Z,t) \equiv G_{IJ}(Z)$). Obviously, the problem is solved if the *jump moments*

$$\begin{aligned} &M_{l_1 \dots l_M}\Big(Z(t)\Big) \\ &= \lim_{\Delta t \to 0} \frac{\varepsilon^{1-l}}{\Delta t} \Big\langle \Big(Z_1(t+\Delta t) - Z_1(t)\Big)^{l_1} \cdot \ldots \cdot \Big(Z_M(t+\Delta t) - Z_M(t)\Big)^{l_M} \Big\rangle_{t+\Delta t} \end{aligned} \tag{7.16}$$

(cf. (6.47)) have been determined in dependence on the functions $F_I(Z)$ and $G_{IJ}(Z)$. Here $Z(t)$ denotes the system state at time t.

First of all, for the determination of the jump moments we must evaluate the quantities

$$\Delta Z_I(\Delta t) := Z_I(t+\Delta t) - Z_I(t)\,. \tag{7.17}$$

These are obtained in the limit

$$\Delta Z_I(\Delta t) = \lim_{k\to\infty} \Delta Z_I^{(k)}(\Delta t) \tag{7.18}$$

of an iteration method which, due to

$$\Delta Z_I(\Delta t) = \int_t^{t+\Delta t} dt'\, H_I\Big(Z(t'), \xi(t')\Big)\,, \tag{7.19}$$

is given by

$$\Delta Z_I^{(1)}(\Delta t) = \int_t^{t+\Delta t} dt'\, H_I\Big(Z(t), \xi(t')\Big) \tag{7.20a}$$

and

$$\begin{aligned} \Delta Z_I^{(k+1)}(\Delta t) = \int_t^{t+\Delta t} dt' \sum_{l=0}^{\infty} \sum_{\substack{l_1,\dots,l_M \\ (l_1+\dots+l_M=l)}} \frac{\Big[\Delta Z_{I_1}^{(k)}(t'-t)\Big]^{l_1} \cdot \ldots \cdot \Big[\Delta Z_{I_M}^{(k)}(t'-t)\Big]^{l_M}}{l_1! \cdot \ldots \cdot l_M!} \\ \times \frac{\partial^{l_1}}{\partial Z_1^{l_1}} \cdots \frac{\partial^{l_M}}{\partial Z_M^{l_M}} H_I\Big(Z(t), \xi(t')\Big)\,. \end{aligned} \tag{7.20b}$$

Here the quantity

$$H_I\Big(Z(t'), \xi(t')\Big) = H_I\Big(Z(t) + \Delta Z(t'-t), \xi(t')\Big) \tag{7.21}$$

was represented as a TAYLOR series.

Those terms which are of order $O\big((\Delta t)^2\big)$ do *not* contribute to the jump moments (7.16). In spite of this, the relations we are interested in will be very complicated unless we make special assumptions about the fluctuations $\xi(t)$. These are characterized by distributions

$$P_n(\xi_n, t_n; \ldots; \xi_1, t_1)dt_1 \ldots dt_n \tag{7.22}$$

which describe the probabilities with which the fluctuations $\xi_1, \ldots, \xi_n$ occur during the n differential time intervals $[t_1, t_1 + dt_1], \ldots, [t_n, t_n + dt_n]$. In the following we assume *δ-correlated GAUSSian random processes* which are given by

$$P_n(\xi_n, t_n; \ldots; \xi_1, t_1) := \mathcal{N}_n \exp\left(-\frac{1}{2}\sum_{i,j=1}^{n} \xi_i^{\mathrm{tr}} \underline{\kappa}(t_i, t_j)\xi_j\right) \tag{7.23a}$$

with the normalization constants $\mathcal{N}_n$ and

$$\kappa_{IJ}(t_1, t_2) := \sigma_{IJ}\delta(t_1 - t_2)\,. \tag{7.23b}$$

Without loss of generality we can presuppose

$$\sigma_{IJ} \equiv \delta_{IJ} \tag{7.23c}$$

which can always be achieved by means of a suitable variable transformation.

The relation (7.23b) somehow corresponds to the MARKOV *assumption* of negligible *memory effects* since it implies that there is no correlation between fluctuations at different times. From (7.23) one can derive the relations

$$\Big\langle \xi_{I_1}(t_1) \cdot \ldots \cdot \xi_{I_{2k-1}}(t_{2k-1}) \Big\rangle = 0 \tag{7.24a}$$

and

$$\Big\langle \xi_{I_1}(t_1) \cdot \ldots \cdot \xi_{I_{2k}}(t_{2k}) \Big\rangle = \sum_{\substack{\text{all}\\ \text{pairs}}} \delta_{J_1 J_2}\delta(t_1' - t_2') \cdot \ldots \cdot \delta_{J_{2k-1} J_{2k}}\delta(t_{2k-1}' - t_{2k}') \tag{7.24b}$$

where the sum extends over all $(2k)!/(2^k k!)$ *different* pairs that can be built with the elements of the set $\{(I_1, t_1), \ldots, (I_{2k}, t_{2k})\}$ (cf. [241], pp. 23f.). In particular, we have

$$\Big\langle \xi_I(t) \Big\rangle = 0 \tag{7.25a}$$

and

$$\Big\langle \xi_{I_1}(t_1)\xi_{I_2}(t_2) \Big\rangle = \kappa_{I_1 I_2}(t_1, t_2) = \delta_{I_1 I_2}\delta(t_1 - t_2)\,. \tag{7.25b}$$

Now (7.20) allows, in connection with (7.23) and (7.24), the specification of the quantities $\Delta Z_I(\Delta t)$ as well as of the jump moments $M_{l_1 \ldots l_M}$. A detailed analysis shows that the relevant terms which are proportional to Δt can already be obtained by means of the second order approximations $\Delta Z_I^{(2)}(\Delta t)$. We find

$$\begin{aligned}
\left\langle \Delta Z_I(\Delta t) \right\rangle &= \left\langle \int_t^{t+\Delta t} dt' \left\{ H_I\Big(Z(t), \xi(t')\Big) \right.\right. \\
&\quad \left.\left. + \sum_J \underbrace{\left[\int_t^{t'} dt'' \, H_J\Big(Z(t), \xi(t'')\Big) \right]}_{=\Delta Z_J^{(1)}(t'-t)} \frac{\partial}{\partial Z_J} H_I\Big(Z(t), \xi(t')\Big) \right\} \right\rangle \\
&\quad + O\Big((\Delta t)^2\Big) \\
&= \left\langle \int_t^{t+\Delta t} dt' \left[F_I\Big(Z(t)\Big) + \sum_J G_{IJ}\Big(Z(t)\Big) \xi_J(t') \right] \right\rangle \\
&\quad + \sum_J \left\langle \int_t^{t+\Delta t} dt' \int_t^{t'} dt'' \left[F_J\Big(Z(t)\Big) + \sum_K G_{JK}\Big(Z(t)\Big) \xi_K(t'') \right] \right. \\
&\quad \times \left. \left[\frac{\partial F_I\Big(Z(t)\Big)}{\partial Z_J} + \sum_L \frac{\partial G_{IL}\Big(Z(t)\Big)}{\partial Z_J} \xi_L(t') \right] \right\rangle + O\Big((\Delta t)^2\Big) \\
&= \Delta t \cdot F_I\Big(Z(t)\Big) + \frac{\Delta t}{2} \sum_{J,K,L} \frac{\partial G_{IL}\Big(Z(t)\Big)}{\partial Z_J} G_{JK}\Big(Z(t)\Big) \delta_{KL} \\
&\quad + O\Big((\Delta t)^2\Big),
\end{aligned} \tag{7.26a}$$

$$\begin{aligned}
\left\langle \Delta Z_I(\Delta t) \Delta Z_J(\Delta t) \right\rangle &= \left\langle \left[\int_t^{t+\Delta t} dt' \, H_I\Big(Z(t), \xi(t')\Big) \right] \right. \\
&\quad \times \left. \left[\int_t^{t+\Delta t} dt'' \, H_J\Big(Z(t), \xi(t'')\Big) \right] \right\rangle + O\Big((\Delta t)^2\Big) \\
&= \left\langle \int_t^{t+\Delta t} dt' \int_t^{t+\Delta t} dt'' \left[F_I\Big(Z(t)\Big) \right.\right. \\
&\quad \left.\left. + \sum_K G_{IK}\Big(Z(t)\Big) \xi_K(t') \right] \left[F_J\Big(Z(t)\Big) \right.\right.
\end{aligned}$$

$$+ \sum_L G_{JL}\Big(Z(t)\Big)\xi_L(t'')\Big]\Big\rangle + O\Big((\Delta t)^2\Big)$$

$$= \Delta t \cdot \sum_{K,L} G_{IK}\Big(Z(t)\Big)G_{JL}\Big(Z(t)\Big)\delta_{KL} + O\Big((\Delta t)^2\Big),$$

(7.26b)

and

$$\Big\langle \Delta Z_{I_1}(\Delta t) \cdot \ldots \cdot \Delta Z_{I_l}(\Delta t)\Big\rangle = O\Big((\Delta t)^2\Big) \qquad \text{for} \qquad l \geq 3\,. \tag{7.26c}$$

This implies

$$M_I\Big(Z(t)\Big) = F_I\Big(Z(t)\Big) + \frac{1}{2}\sum_{J,K} \frac{\partial G_{IK}\Big(Z(t)\Big)}{\partial Z_J} G_{JK}\Big(Z(t)\Big), \tag{7.27a}$$

$$\varepsilon M_{IJ}\Big(Z(t)\Big) = \sum_K G_{IK}\Big(Z(t)\Big)G_{JK}\Big(Z(t)\Big), \tag{7.27b}$$

$$\varepsilon^{l-1} M_{I_1 \ldots I_M}\Big(Z(t)\Big) = 0 \qquad \text{for} \qquad l = l_1 + \ldots + l_M \geq 3\,. \tag{7.27c}$$

Relations (7.27) obviously agree with relations (7.5) (cf. (6.14b,c)). That is, if the fluctuations satisfy relations (7.23), LANGEVIN equation (7.1) is equivalent to FOKKER-PLANCK equation (7.4) with the special drift and diffusion coefficients given by (7.5).

For explicitly time-dependent coefficients, $F_I(Z)$ and $G_{IJ}(Z)$ have, in formulas (7.15), (7.26), and (7.27), only to be replaced by $F_I(Z,t)$ and $G_{IJ}(Z,t)$ since the TAYLOR expansion of

$$H_I\Big(Z(t'), \xi(t'), t'\Big) = H_I\Big(Z(t) + \Delta Z(t'-t), \xi(t'), t + (t'-t)\Big) \tag{7.28}$$

in $\Delta t = (t' - t)$ exclusively leads to additional terms which, after carrying out the integrations (7.19), are of order $O\big((\Delta t)^2\big)$.

7.3 Escape Time

Often we would like to know how long it takes on average until a system leaves a stable state Z_0 due to an accidental summation of fluctuations. For illustrative reasons, let us assume a one-dimensional system which obeys the LANGEVIN equation

$$\frac{dZ}{dt} = F(Z) + G\xi(t) \tag{7.29}$$

and thus can also be described by the FOKKER-PLANCK equation

$$\frac{\partial}{\partial t}P(Z,t) = -\frac{\partial}{\partial Z}\Big[k(Z)P(Z,t)\Big] + \frac{\varepsilon q}{2}\frac{\partial^2}{\partial Z^2}P(Z,t) \tag{7.30}$$

with $k(Z) = F(Z)$ and the constant diffusion coefficient $q = G^2$. (The constancy of the diffusion coefficients can always be reached by means of an appropriate transformation of the co-ordinates; cf. [241], pp. 88ff. Methods for the evaluation of the escape time in multidimensional cases are treated in [84], pp. 357ff.)
Let

$$\phi(Z) := -\frac{2}{\varepsilon q}\int_{Z_0}^{Z} dZ'\, k(Z') \tag{7.31}$$

be a *double well potential* with minima at Z_0 and Z_2 and a maximum at Z_1 (cf. Fig. 7.1). Then the *ARRHENIUS formula*

$$T \approx \frac{4\pi}{\varepsilon q\sqrt{\frac{\partial^2}{\partial Z^2}\phi(Z_0)\left|\frac{\partial^2}{\partial Z^2}\phi(Z_1)\right|}}\mathrm{e}^{2[\phi(Z_1)-\phi(Z_0)]/(\varepsilon q)} \tag{7.32}$$

can be derived for the average *escape time* T that a system needs to get over the potential barrier at Z_1 starting from Z_0 for the first time (cf. [84], pp. 139ff.). According to this formula, the escape time (which is also called *'first-passage time'*) increases more than exponentially with the system size $\Omega = 1/\varepsilon$. For systems which

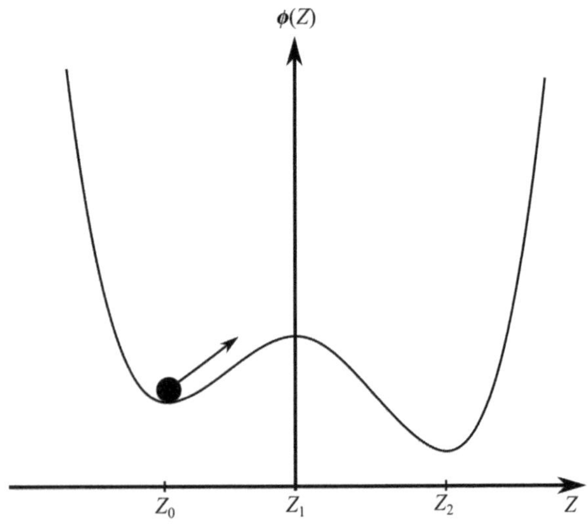

Fig. 7.1 For the depicted double well potential the states Z_0 and Z_2 are stable. Despite this, after a long time period (the *escape time*) fluctuations lead to a passing of the potential barrier at Z_1

consist of $N \gg 1$ elements the stability of state Z_0 is, due to $\Omega = 1/\varepsilon = N$, hardly endangered by fluctuations.

Whereas the transition from Z_0 to Z_1 is caused by *fluctuations,* the transition from Z_1 to Z_2 is dominated by the *drift.* Hence, for arriving in Z_2 from Z_1, a small time period is necessary in comparison with T.

7.4 Phase Transitions, LIAPUNOV Exponents, and Critical Phenomena

In order to explain *phase transitions* we best start from LANGEVIN equation (7.1) but let $F \equiv (\ldots, F_I, \ldots)^{\mathrm{tr}}$ and possibly also $\underline{G} \equiv \left(G_{IJ}\right)$ depend on a *control parameter* κ:

$$\frac{dZ}{dt} = F(Z,\kappa) + \underline{G}(Z,\kappa)\xi(t)\,. \tag{7.33}$$

κ is given by external conditions and can sometimes be arbitrarily regulated. Disregarding fluctuations $\xi(t)$ for the time being, (7.33) has the form

$$\frac{dZ}{dt} \approx F(Z,\kappa)\,. \tag{7.34}$$

After a finite *transient time* t_0 the system will in general take on a stable temporal behaviour $Z_0(\kappa, t)$. That is, for $t \geq t_0$ the solution $Z_0(\kappa, t)$ with

$$\frac{d}{dt} Z_0(\kappa, t) = F\Big(Z_0(\kappa, t), \kappa\Big) \tag{7.35}$$

will practically be located on or move on an *attractor* (e.g. a *fixed point,* a *limit cycle,* or a *torus*). Considering a system which, at the time t_0, is in a neighbouring state $Z(\kappa, t_0) \approx Z_0(\kappa, t_0)$, the deviations

$$\delta Z(\kappa, t) := Z(\kappa, t) - Z_0(\kappa, t) \tag{7.36}$$

obey the equation

$$\begin{aligned} \frac{d\,\delta Z}{dt} &= \frac{dZ}{dt} - \frac{dZ_0}{dt} \\ &= F(Z,\kappa) - F(Z_0,\kappa) \\ &= \underline{L}(\kappa, t)\,\delta Z + O\Big(\|\delta Z\|^2\Big) \end{aligned} \tag{7.37a}$$

where

$$\underline{L}(\kappa, t) \equiv \Big(L_{IJ}(\kappa, t)\Big) := \left(\frac{\partial}{\partial Z_J} F_I\Big(Z_0(\kappa, t), \kappa\Big)\right) \tag{7.37b}$$

is the JACOBIan *matrix* of the system. Since $\|\delta Z(\kappa, t_0)\| \to 0$ implies $\|\delta Z(\kappa, t)\| \to 0$, the solution of (7.37) for this limit can immediately be written as

$$\delta Z(\kappa, t) = \underline{M}(t, t_0)\, \delta Z(\kappa, t_0) \tag{7.38a}$$

with

$$\underline{M}(t, t_0) = \mathrm{T}\exp\Big(\underline{L}, t, t_0\Big) \tag{7.38b}$$

(cf. p. xxiii). Hence the deviations δZ change exponentially in time.

Now let e_I be the normalized vectors of a basis of the state space. We are interested in the temporal course of the initial deviations

$$\delta Z_I(\kappa, t_0) := \varepsilon_I(\kappa, t_0) e_I \tag{7.39}$$

in directions e_I. Let us denote their magnitude by

$$\varepsilon_I(\kappa, t) := \|\delta Z_I(\kappa, t)\| = \|\underline{M}(t, t_0)\, \delta Z_I(\kappa, t_0)\| \,. \tag{7.40}$$

A measure for the mean exponential change of the magnitude $\varepsilon_I(\kappa, t)$ of the deviations $\delta Z_I(\kappa, t)$ during the period $(t - t_0)$ of time is

$$\overline{\lambda}_I(\kappa, t) := \frac{1}{t - t_0} \ln \frac{\varepsilon_I(\kappa, t)}{\varepsilon_I(\kappa, t_0)} \tag{7.41}$$

due to

$$\varepsilon_I(\kappa, t) = \mathrm{e}^{\overline{\lambda}_I(\kappa, t)\cdot(t - t_0)} \varepsilon_I(\kappa, t_0) \,. \tag{7.42}$$

Thus a *criterion for the stability* of solution $Z_0(\kappa, t)$ with respect to small deviations $\delta Z(\kappa, t_0)$ are the *LIAPUNOV exponents* [260, 264]

$$\lambda_I(\kappa) := \lim_{t\to\infty} \lim_{\varepsilon_I(\kappa, t_0)\to 0} \overline{\lambda}_I(\kappa, t) = \lim_{t\to\infty} \lim_{\varepsilon_I(\kappa, t_0)\to 0} \left[\frac{1}{t - t_0} \ln\left(\frac{\varepsilon_I(\kappa, t)}{\varepsilon_I(\kappa, t_0)}\right)\right]. \tag{7.43}$$

It can be shown that the LIAPUNOV exponents $\{\lambda_I(\kappa)\}$ are independent of the choice of the basis $\{e_I\}$ [210].

- If all LIAPUNOV exponents $\lambda_I(\kappa)$ are negative, small deviations will be exponentially damped. Then fluctuations do not disturb the stability of the solution $Z_0(\kappa, t)$ as long as they do not accidentally sum up (cf. Sect. 7.3).
- If at least one LIAPUNOV exponent $\lambda_I(\kappa)$ is positive, a small initial deviation $\delta Z_I(\kappa, t_0)$ will grow exponentially until it usually reaches a limit due to the non-linear terms $O(\|\delta Z\|^2)$. Then a new solution $Z_1(\kappa, t)$ results. (A detailed discussion is given in [106] where it is also investigated which new patterns of motion result on which conditions.)

With respect to a variation of the control parameter κ, those values κ_i for which a LIAPUNOV exponent reaches zero and the solution $Z_i(\kappa, t)$ becomes *unstable* will be denoted as *critical points*. The phenomenon of a *discontinuous* change from one stable solution $Z_i(\kappa_i^-, t)$ to another stable solution $Z_{i+1}(\kappa_i^+, t)$ during the variation of the control parameter κ from $\kappa_i^- := \kappa_i - \delta\kappa$ to $\kappa_i^+ := \kappa_i + \delta\kappa$ is called a *'phase transition'* or, in chaos theory, a *'bifurcation'*. Phase transitions can only occur in *non-linear systems* and are very typical for those. In particular, they are denoted as *noise induced transitions* if caused by the fluctuations' contributions

$$\frac{1}{2}\sum_{J,K} \frac{\partial G_{IJ}(Z,\kappa)}{\partial Z_K} G_{JK}(Z,\kappa) \tag{7.44}$$

to the drift coefficients $k_I(Z)$ (cf. [145]).

A phase transition is preceded by *critical phenomena:* The deviation $\delta Z_I(\kappa, t_0)$ that belongs to a LIAPUNOV exponent $\lambda_I(\kappa)$ which is slightly less than zero will be very slowly damped. Hence in this connection one speaks of *critical slowing down.* Additionally, owing to the small damping, fluctuations in direction $\delta Z_I(\kappa, t)$ become strongly noticeable, i.e. the dynamics of the system is influenced by the fluctuations $\xi(t)$. These are denoted as *critical fluctuations.*

7.5 Routes to Chaos

Among the phase transitions (or bifurcations) the *'routes to chaos'* are of special interest. There are several different routes to chaos *('chaotic scenarios')* of which

- the *RUELLE-TAKENS-NEWHOUSE scenario* [197, 254] via HOPF *bifurcations* [144, 254],
- the *period doubling scenario* via *pitchfork bifurcations* [16, 70, 71, 254],
- and the *intermittency* route to chaos via a *tangent bifurcation* [16, 146, 223, 254]

are the most famous ones (cf. also [135, 183]).

The common property of all chaotic phenomena is that the slightest change of the initial state of the system can, after a certain time, lead to a completely different dynamics. Thus, within the frame of all possible *measuring accuracy,* the dynamic behaviour of a chaotic system is *not* any more *predictable* after this time *('butterfly effect').*

It already became clear in Sect. 7.4 that the LIAPUNOV exponents are the decisive criterion whether the system behaviour is chaotic or not. However, a number of other methods also allow an examination for *chaos* (cf. [135, 254, 264]):

- The measurement of the *irregularity* of the temporal course. For example, in case of *intermittency,* nearly periodic phases (so-called *intermissions*) are randomly interrupted by relative short irregular bursts (cf. Fig. 10.30).
- The analysis of *phase portraits,* i.e. of *projections* of the *trajectory* (which is the *path* of temporal evolution). If a phase portrait displays a closed curve, we will have *periodic* temporal behaviour (cf. Figs. 10.11, 10.13, and 10.24). *Chaotic* as well as *quasi-periodic behaviour* is characterized by trajectories which are not closed and do not end in a *fixed point.* In case of *quasi-periodicity* the curve closes only almost and then runs 'parallel' to one of its former curve segments (cf. Fig. 10.16). Otherwise we have *chaotic behaviour* (cf. Fig. 10.27).

- The investigation of *POINCARÉ sections (*POINCARÉ *maps),* which, instead of projections of the trajectory, depict its *intersection points* with a certain plane through the state space (cf. [222, 254]). *Chaotic* behaviour manifests itself by a cluster of infinitely many points whereas *periodic* behaviour is indicated by a finite number of points.
- The evaluation of the *power spectrum* (cf. Sect. 7.5.2). For chaotic behaviour the maxima of the spectrum are *dense* (cf. Fig. 10.27).
- The calculation of the LIAPUNOV *dimension* $D_{KY} \equiv D_{KY}(\kappa)$ which *KAPLAN* and *YORKE* conjectured to be a measure for the *dimension* of the attractor. If $\lambda'_i(\kappa)$ denote the LIAPUNOV exponents $\lambda_I(\kappa)$ in a decreasing order (i.e. $\lambda'_1 \geq \ldots \geq \lambda'_i \geq \ldots \geq \lambda'_M$) and if n is the largest integer number with

$$\sum_{i=1}^{n} \lambda'_i(\kappa) \geq 0\,, \tag{7.45}$$

the LIAPUNOV dimension is defined by [135, 152, 254]

$$D_{KY}(\kappa) := n + \sum_{i=1}^{n} \frac{\lambda'_i(\kappa)}{|\lambda'_{n+1}(\kappa)|}\,. \tag{7.46}$$

For all cases of a *fractional* LIAPUNOV dimension we have chaotic behaviour. The same applies to the *HAUSDORFF dimension* [99, 118, 180, 254] and the *correlation dimension* (cf. [254], pp. 127ff.) which are more reliable and more usual measures for the dimension of an attractor.
- The consideration of the *autocorrelation function*

$$\overline{C}(\Delta t) := \lim_{T\to\infty} \frac{1}{T} \int_{t_0}^{t_0+T} dt\, y(t) y(t+\Delta t) \tag{7.47}$$

which measures the *average correlation* of a time signal $y(t)$ with itself at a later time $t + \Delta t$. For chaotic behaviour $\overline{C}(\Delta t)$ generally decreases exponentially with growing Δt since the temporal evolution of $y(t + \Delta t)$ will become more and more unpredictable on the basis of $y(t)$ (cf. [254]).

In the following, LIAPUNOV exponents and power spectra will be discussed in more detail.

7.5.1 *RUELLE-TAKENS-NEWHOUSE Scenario and LIAPUNOV Exponents*

The definition of the LIAPUNOV exponents $\lambda_I(\kappa)$ was already given in Sect. 7.4. They are a measure for the average temporal exponential change of deviations into various directions in the course of time (cf. (7.43)). If the sum of the LIAPUNOV exponents is negative, a small volume ΔV will shrink to zero:

$$\lim_{t\to\infty} \Delta V(t) = 0\,. \tag{7.48}$$

Then neighbouring trajectories converge to one and the same *attractor* and the system is called a *dissipative* system.

Above all, we distinguish the following important cases:

(1) If all LIAPUNOV exponents are negative, there is $D_{KY} = 0$ and the trajectory of the system moves towards a *stable fixed point.*
(2) If one of the LIAPUNOV exponents is zero and the other ones are negative, we have $D_{KY} = 1$ and the trajectory approaches a *limit cycle* (cf. Figs. 10.18 and 10.24). A deviation $\delta Z_I(\kappa, t_0)$ tangential to the limit cycle stays *undeformed,* then, whereas deviations $\delta Z_J(\kappa, t_0)$ in all other directions are *compressed.*
(3) If two LIAPUNOV exponents are zero and the others are negative, there is $D_{KY} = 2$ and the trajectory converges to a path which runs on a *torus* (cf. Fig. 10.13). Deviations in one of the two rotation directions remain undeformed whereas deviations in all other directions are compressed.
(4) If at least one LIAPUNOV exponent is positive, one is zero, and the others are negative, we have a *strange attractor* (cf. Fig. 10.27). In directions which correspond to positive LIAPUNOV exponents a *stretching* of the deviations occurs, i.e. neighbouring trajectories move off from each other. A *compression*, i.e. an approach of neighbouring trajectories, however, takes place in those directions which correspond to negative LIAPUNOV exponents. The alternation between compression and stretching processes leads to a *folding* which explains the sensitive dependence on the initial conditions and causes a *tattering* of the volume element ΔV (that is responsible for the strange shape of the attractor).

    For the LIAPUNOV dimension of a strange attractor we find $D_{KY} > 2$. In addition, D_{KY} is generally no integer but a *fractional* number. Hence we speak of a *fractal dimension* [17, 180, 183].

We are confronted with a RUELLE-TAKENS-NEWHOUSE *scenario* if a system goes in succession through the situations (1)–(4) along with a variation of the control parameter κ.

7.5.2 Period Doubling Scenario and Power Spectra

The FOURIER *spectrum* of a function $y(t_0 + t)$ is defined by the function

$$\widetilde{f}(\omega) := \int_{-\infty}^{+\infty} dt\, \mathrm{e}^{\mathrm{i}\omega t}\, y(t_0 + t)\,. \tag{7.49}$$

Roughly speaking, the FOURIER *transformation* (7.49) decomposes a time-dependent function $y(t_0 + t)$ into sinusoidal contributions of different *frequencies* ω and *amplitudes* $\widetilde{f}(\omega)$. (If evaluating the FOURIER spectrum of a time-dependent function which starts at time t_0 by means of a computer, we integrate only from t_0 up to a very large value of T. The mistake connected with this appears predominantly for large values of ω which are not of particular interest.)

Often, however, the application of the logarithm

$$\log f(\omega) \tag{7.50}$$

of the *power spectrum*

$$f(\omega) := \left|\widetilde{f}(\omega)\right|^2 . \tag{7.51}$$

is preferred to the FOURIER spectrum. One interesting property of $f(\omega)$ can be discovered by means of the *convolution theorem* (cf. p. xviii):

$$\left|\widetilde{f}(\omega)\right|^2 = \int_{-\infty}^{+\infty} dt\, \mathrm{e}^{\mathrm{i}\omega t} \lim_{T\to\infty} \frac{1}{T} \int_{t_0}^{t_0+T} dt'\, y(t')y(t'+t)\,. \tag{7.52}$$

This relation agrees with the WIENER-KHINTCHINE *theorem* [158, 296]

$$f(\omega) = \int_{-\infty}^{+\infty} dt\, \mathrm{e}^{\mathrm{i}\omega t}\overline{C}(t) \tag{7.53}$$

according to which the power spectrum is the FOURIER transform of the *autocorrelation function* $\overline{C}(t)$ (cf. [84], pp. 15ff., and [264], pp. 113f.).

When considering the power spectrum $f(\omega)$ of a function $y(t) = y(t + T_0)$ with *period* T_0, we find especially large amplitudes at the frequencies

$$\omega(n) := n \cdot \omega_0 \qquad \text{with} \qquad \omega_0 := \frac{2\pi}{T_0} \qquad (n \in \{0, 1, 2, \ldots\}). \tag{7.54}$$

Thus a function with period $T_i := i\, T_0$ has its spectral maxima at the frequencies

$$\omega(n) = n\omega_i = \frac{n}{i}\omega_0 \qquad \text{with} \qquad \omega_i := \frac{2\pi}{T_i} \qquad (n \in \{0, 1, 2, \ldots\})\,. \tag{7.55}$$

If, along with a change of the control parameter κ, a system is subject to a *period doubling* (i.e. $T_{i+1} \equiv T_{i+1}(\kappa_{i+1}) = 2 \cdot T_i \equiv 2 \cdot T_i(\kappa_i)$), the power spectrum displays maxima at the frequencies

$$\omega_{i+1}(n) = \frac{n}{2}\omega_i \qquad (n \in \{0, 1, 2, \ldots\})\,. \tag{7.56}$$

That is, *additional* maxima occur in the *middle* of those maxima which already existed before the period doubling. This will become especially clear in Figs. 10.24, 10.25, 10.26 and 10.27. In the limit of infinitely many period doublings we find a transition to an *aperiodic* or *chaotic* behaviour. Then the maxima of the power spectrum are *dense,* i.e. the power spectrum possesses an *irregular* appearance (cf. Fig. 10.27).

The *period doubling route* to chaos described above often occurs in systems which are periodically forced with a frequency ω_0.

Part II
Quantitative Models of Social Processes

Overview

In Part II of this book the stochastic methods from Part I will be made applicable to the modelling of social processes. It will turn out that the differences with respect to physical models do not primarily concern *methodology* but rather *terminology* and *system specification.* This is the topic of Chap. 8.

Since the characteristic properties of a system are reflected by the *transition rates,* we will discuss in Chap. 9 how to choose these for social processes in the most appropriate way. An approach which bases on the concept of *utility maximization* is found to be particularly plausible. It takes into account the *effort* of behavioural changes and the *incompleteness of information* during a decision.

Regarding the interaction rates, it is possible to classify three different *kinds* of direct *pair interactions—imitative processes, avoidance processes,* and *compromising processes.* The resulting BOLTZMANN-*like model for behavioural changes* of individuals includes, as special cases, a number of equations which have important applications in the social sciences. Worth mentioning are the *logistic equation* for limited growth processes, the *gravity model* for migration phenomena, and the *game theoretical approaches* for the description of the cooperation or competition among individuals.

In Chap. 10 the model for behavioural changes will be specified with an example concerning *opinion formation.* It shows that, apart from other cases, *oscillatory* or *chaotic* opinion changes are possible. In addition, we will discuss how *spatial* opinion spreading by *diffusion* or *telecommunication* can be taken into account.

A reformulation of the model from Chap. 9 by means of BOLTZMANN-FOKKER-PLANCK equations and LANGEVIN equations respectively will be carried out in Chap. 9. This allows a mathematical specification of LEWIN's idea that the behaviour of individuals is guided by a *'social field'.*

Chapter 12 is dedicated to the *game dynamical equations* which result for a special case of imitative processes. Thus immediately a *stochastic formulation* of game theory becomes possible. With an example for the competition of equivalent strategies we will explain the *self-organization of behavioural conventions.*

The topic of Chap. 13 is the procedure of determining the parameters of the developed models from *empirical data.* It will show up that even in the case of *incomplete data* the quantities which are of interest can be estimated. As a concrete example, the migration data of removals between the federal states of West Germany will be examined. In this connection also the procedure of successive model reductions will be illustrated.

Of particular interest for model applications is the decomposition of the time-dependent utility functions with regard to *explanatory variables.* As examples we will discuss approaches for the behaviour of purchasers and voters. The evaluation of empirical data can afterwards be continued by means of *prognosis techniques.*

Chapter 8
Problems and Terminology

8.1 Terms

This section explains the meaning of the terms that were introduced in Part I for social processes.

8.1.1 System and Subsystems

In connection with the social sciences we will use the term *'system'* to refer to a *social system*. Its constituent *elements (units, subsystems)* are *individuals* which undergo social interactions with each other.

8.1.2 State

We will assume that individuals can be characterized by their *properties.*[1] Let the *state* x combine those properties of an individual which are important for the description of his/her *behaviour* in a given situation and which are subject to noticeable temporal changes. For example, in connection with opinion formation concerning a certain affair, x can combine the opinion i and location r of an individual:

$$x := (i, r)^{\text{tr}} . \tag{8.1}$$

8.1.3 Subpopulation

Those properties of individuals which remain practically constant in time during the considered behavioural changes but which influence these, are a convenient criterion for the subdivision of the examined total population into several, internally

[1] In this context, with *properties* we do not only mean the character traits but *all* quantities which are required for a complete description of an individual α.

DOI 10.1007/978-3-642-11546-2_8,

homogeneous *subpopulations* a. If the total population consists of N individuals and subpopulation a of N_a individuals, we obviously have

$$\sum_a N_a = N \, . \tag{8.2}$$

Per definition, with regard to an investigated affair, individuals of the same subpopulation have a *similar* behaviour pattern ('character') whereas individuals of different subpopulations *differ* in their behaviour pattern. A *behaviour pattern* is characterized by the *individual transition rates* $\widehat{w}_a$, $\widehat{w}_{ab}$. Thus, for individuals of one and the same subpopulation, we do not assume similar *behaviours* but only similar *probabilities* for equal behaviours.

Normally, by subpopulations a we distinguish different *types of behaviour* a, but in some cases also the residence location r or other quantities can be decisive for the subdivision into subpopulations. The behavioural types a can mostly be characterized by a few personal *circumstances* or *socio-economic criteria* (e.g. age, religious affiliation, education, and level of income) the relevance of which depends on the considered affair. This is the case since individuals show a tendency to *increase* the *consistency* between their attitudes, behaviours, and social environment (cf. Fig. 8.1 and [37, 74, 119, 120, 213]) as a consequence of their attempts to maximize their *utility* $U_a(x,t)$ by the respective behaviour x. $U_a(x,t)$ consists of two parts:

$$U_a(x,t) = B_a(x,t) + S(x,t) \tag{8.3}$$

(cf. [77], p. 16ff.). $B_a(x,t)$ stands for the *attitude* towards behaviour x or for the *satisfaction* which an individual of subpopulation a expects from behaviour x. $S(x,t)$ is the '*social reinforcement*', i.e. the 'reward' or the 'punishment' which an individual must expect as a consequence of behaviour x. Thus an individual has three different possibilities to maximize his/her utility:

- The individual can decide on a behaviour x' with $B_a(x',t) > B_a(x,t)$ instead of on behaviour x.
- If, due to *social pressure* $-S(x,t)$, an individual shows a behaviour x' which does not agree with his/her attitudes, this will change the valuation $B_a(x',t)$ in the sense of a *revaluation* of behaviour x'. This phenomenon is known from psychology as *dissonance reduction* [32, 74]. (By the way, an attitude change does not occur in the case of a sufficiently high social reward for a behaviour x' that is in disagreement with his/her attitudes!)
- The individual can also look for a social environment $S(x,t)$ which has a positive attitude towards his/her behaviour x: 'Birds of a feather flock together.'

All three possibilities cause the consistency between behavioural type and personal circumstances to increase.

A possible means for the empirical determination of suitable subpopulations is, for example, the *cluster analysis* (cf. [149], pp. 532ff.). The number of subpopulations which should be distinguished depends on the desired precision of the

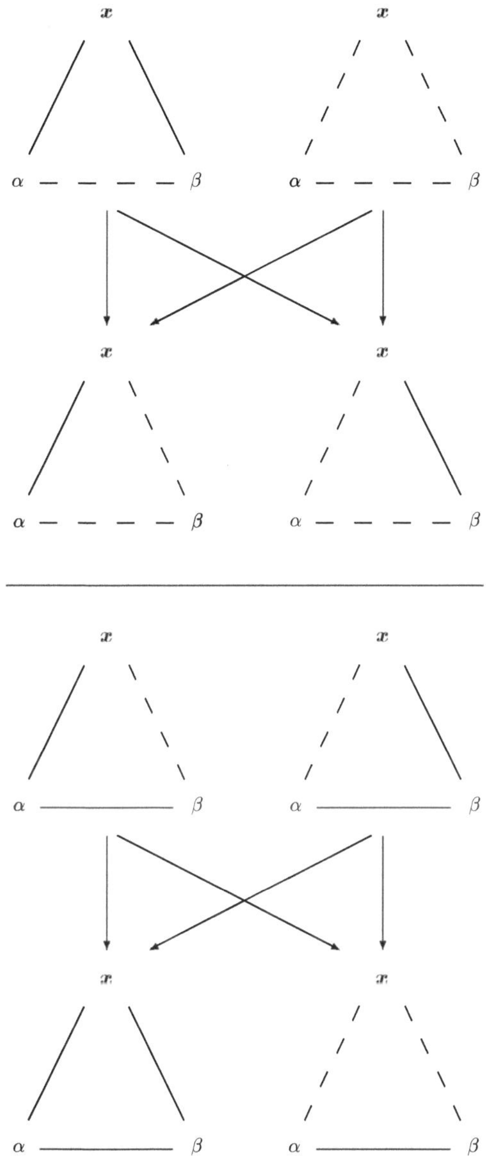

Fig. 8.1 Stability of cognitive structures according to *balance theory* [119, 120]: Depicted are the relations (attitudes) of two persons α and β towards behaviour x and the relation of the persons among each other. Positive relations (e.g. likings) are represented by *solid lines* (—) and negative relations (e.g. dislikes) by *dashed lines* (– –). Only structures with an odd number of positive relations are stable *(consistent)*. Unstable structures are denoted as *unbalanced*. If a structure is unstable, it will, in the course of time, turn into a stable one. In this connection, the most probable transitions are those which require the least changes of relations *(economy principle)* [35, 243]. These transitions are symbolized by arrows as far as they result in a change of the attitude towards behaviour x. Another possibility would be a change of the relation between persons α and β

description. However, it is not reasonable to use a lot of subpopulations because, in such a way, the resulting model will become very complicated and susceptible to accidental disturbances.

8.1.4 Socioconfiguration

The *socioconfiguration* denotes the vector

$$n := (\ldots, n_x^a, \ldots)^{\text{tr}} \tag{8.4}$$

which describes the behavioural distribution within the whole considered social system. n_x^a means the number of individuals with behaviour x within subpopulation a. Within the *total* population the number n_x of individuals with behaviour x is given by

$$n_x = \sum_a n_x^a . \tag{8.5}$$

Moreover, since the N_a individuals of subpopulation a are distributed over all possible behaviours x, we have

$$\sum_x n_x^a = N_a . \tag{8.6}$$

8.1.5 Interaction

In the context of the social sciences, with *interactions* we mean *social interactions.* These are characterized by an *exchange of information.* The decisive interactions are those which are related to a behavioural change. For example, in the case of opinion formation interactions are given by conversations or discussions. Depending on the kind of opinion changes initialized by these we distinguish various kinds of interactions—imitative processes, avoidance processes, and compromising processes.

Of special interest are *non-linear interactions* for which effects are not simply proportional to causes. So the behaviour of the social system cannot simply be understood as a sum of the behaviours of the single individuals. This is due to the fact that an individual often exerts an influence on the social system which, in turn, acts upon the individual *(feedback effect).* One possible consequence of this is the *emergence* of *self-organization phenomena.*

8.2 Problems with Modelling Social Processes

8.2.1 Complexity

The topic of *social sciences* are individuals and their *behaviours* as well as the *social structures* which result from these (groups, organizations, societies, etc.). From the point of view of natural sciences the behaviour of an individual is decisively determined by the processes in his/her *brain*. As a consequence, to model the behaviour in an exact way we seem to require an exact description of the processes within the brain. The functioning of the latter essentially bases on about 10^{12} *neurons* (nerve cells) and about 10^{15} *synapses* (connections). Consequently, the detailed dynamics of the brain is extremely complicated.[2] However, the numerous variables of the brain are *non-linearly coupled* so that, according to HAKEN's *'slaving principle'* (cf. [106], pp. 187ff.), most of the fast variables can be *adiabatically eliminated.* Thus we expect that the behavioural dynamics is determined by relatively few slowly changing variables.

In order to motivate this *phenomenologically* we proceed as follows: The behavioural changes of an individual α are, on the one hand, given by *(environmental) influences* $Q_{\alpha j}$ and, on the other hand, by variables $I_{\alpha k}$ which reflect the *internal processes* of his/her *organism,* especially of his/her *brain.* Some of the variables $I_{\alpha k}$ are influenced by variables $Q_{\alpha j}$ via *perception.* It will prove convenient to subdivide the variables $\{I_{\alpha k}\}$ into variables $\{A_{\alpha l}\}$ which change very *slowly,* variables $\{X_{\alpha m}\}$ which describe the *(macroscopic) behaviour,* and the remaining variables $\{O_{\alpha n}\}$:

$$I_\alpha \equiv (A_\alpha, X_\alpha, O_\alpha)^{\text{tr}} . \tag{8.7}$$

Then the exact description of behavioural changes is given by the transition rates

$$w_\alpha(Q'_\alpha, X'_\alpha, O'_\alpha | Q_\alpha, X_\alpha, O_\alpha; t) \equiv w(Q'_\alpha, A'_\alpha, X'_\alpha, O'_\alpha | Q_\alpha, A_\alpha, X_\alpha, O_\alpha; t) \tag{8.8}$$

which describe the change of the variables Q_α, A_α, X_α, and O_α into the variables Q'_α, A'_α, X'_α, and O'_α. The dependence on the *quasi-constant variables* $A'_\alpha = A_\alpha$ appears only in the dependence of the transition rates on α.

The number of internal variables $O_{\alpha n}$ which are not at all of interest (as far as the description of *social* processes is concerned) is extraordinarily large. However, the situation simplifies since *an organism follows the necessity to react to a change of the external situation* Q_α *fast enough by changing his/her behaviour* X_α

[2] Already in electroencephalograms (EEGs) one can detect *chaotic* time signals [14, 15].

in an appropriate way. Therefore, in many cases behavioural changes are already determined by the external situation so that they can be described by transition rates of the form

$$w_\alpha(Q'_\alpha, X'_\alpha | Q_\alpha, X_\alpha; t) \,. \tag{8.9}$$

That is, the knowledge of the internal variables O_α is not necessary. The reason for this is that the *(microscopic)* internal variables O_α, which are to manage an adequate behavioural change, must vary much faster than the *(macroscopic) situative variables* Q_α and the (macroscopic) behavioural variables X_α. Therefore, due to their larger changing rate (i.e. the faster *time scale* on which they change), variables O_α can be *adiabatically eliminated* (cf. Sect. 3.2.3). This implies an enormous *data reduction* and *information compression* for the model.

In other words: Normally the internal variables O_α are *'slaved'* (determined) by the slow situative variables Q_α in order to guarantee an appropriate *response (reaction)* X_α to an external *stimulus* Q_α (cf. Fig. 8.2). Nevertheless, the reaction X_α to a stimulus Q_α is not necessarily *unique (deterministic)* but can be one of several possible behavioural alternatives.

Clearly speaking, for the modelling of behavioural changes it is not absolutely necessary to comprehend all the different *thought processes (trains of thought)* which lead to these in detail, provided that the thought processes are significantly faster than those behavioural changes which result from them. So behavioural changes can be understood as reactions occuring with certain probabilities to the perceptions (stimuli) that take place.

The *stimulus-response scheme* is a result of long-term *evolution processes* and short-term *learning processes*. Thus it is not guaranteed in new or special situations for which an appropriate reaction scheme first has to be *learned*, e.g. by *trial and error*. It also fails in the case of very fast changes of the situative variables. Furthermore, the elimination of internal variables O_α is not admissible in the case of behavioural changes which are caused internally, e.g. by hunger or other physical needs.

In the following we will mainly consider cases in which the situative variables are given by the behaviours X_β of other individuals β. Then there is

$$Q_\alpha = (X_1, \dots, X_{\alpha-1}, X_{\alpha+1}, \dots, X_N)^{\text{tr}} \tag{8.10}$$

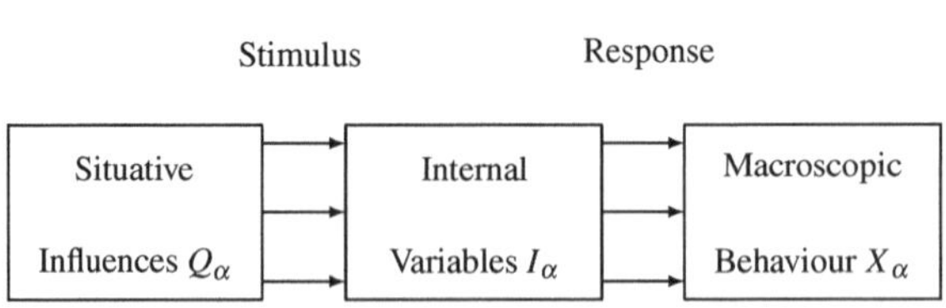

Fig. 8.2 *Stimulus-response scheme* of individual behaviour.

and

$$w_\alpha(Q'_\alpha, X'_\alpha | Q_\alpha, X_\alpha; t) = w_\alpha(X'_1, \dots, X'_\alpha, \dots, X'_N | X_1, \dots, X_\alpha, \dots, X_N; t) \,. \tag{8.11}$$

8.2.2 Individuality

The *individuality* of human beings results from their complexity which causes that there are no two persons anywhere who behave exactly in the same way. That is, for the number $n_X(Q)$ of individuals which are in situation Q and show behaviour X we have

$$n_X(Q) \approx 1 \,. \tag{8.12}$$

However, a *verifiable* or *falsifiable* model for behavioural changes presupposes the *reproducibility* of the considered behavioural changes. This can be achieved by restriction to clearly delimited and *simple situations.* Then it is not important how the single individuals behave in other situations and *areas of life.* The modeled kinds of behaviour may even be influenced by other behavioural areas as long as these influences have, *on average* of all individuals of one subpopulation, no *systematic* effect.

Simple situations can be characterized by a few situative variables $q_j \in \{Q_{\alpha j'}\}$ and they concern only a few behavioural variables $x_k \in \{X_{\alpha k'}\}$. So it is usually sufficient to distinguish a few *behavioural types* a. Consequently, simple situations evoke behavioural changes which can be described by transition rates of the form

$$w_a(q', x' | q, x; t) \,. \tag{8.13}$$

Then for the number $n_x^a(q)$ of individuals who belong to behavioural type a and show behaviour x in situation q we have

$$n_x^a(q) \gg 1 \,. \tag{8.14}$$

Simple situations are, for example, situations in which the individuals decide on

- taking a certain position concerning a given affair,
- purchasing a product of a given kind of good, or
- voting a certain party.

8.2.3 Stochasticity and Disturbances

The necessity of a finally *quantum mechanical treatment* of the microscopic processes of an organism and the quasi-adiabatic elimination of fast variables O_n have the consequence that behavioural changes are subject to *fluctuations* which stem from the dynamics of the (microscopic) internal variables O_n (cf. Sect. 3.2.3). *Because of these internally caused fluctuations for behavioural changes it is only transition probabilities that are given, i.e. a different individual behaviour can result even in seemingly identical situations q. Hence individual behaviour is unpredictable in detail.*

However, it is possible to develop *stochastic models,* i.e. statements about the *probability* of certain behaviours. We can *check* the probability statements if there exists a sufficient number n_x^a of individuals who possess the same behavioural type a, show initially the same behaviour x, and experience the same situation q. That is, we have a similar situation as in quantum mechanics where also only probability statements can be made. In quantum mechanics such probability statements can be checked by an *ensemble* of systems which possess the same interaction type, are initially in the same state (by *preparation*), and undergo the same kind of interaction.

Further reasons for the necessity of a stochastic treatment of social processes are

- unknown *situative influences (disturbances)* $Q_{j'} \notin \{q_j\}$ (cf. Sect. 3.2.2) and
- small individual variations of the behaviour pattern within one and the same behavioural type a. (The way of proceeding described in Sect. 4.3 is only approximately possible, then.)

The difficulties with disturbances are related to the *complexity* of individuals.

8.2.4 Decisions and Freedom of Decision-Making

From a *psychological* point of view the fluctuations of the internal variables show up in the *decisions* of an individual which are decisive for his/her behaviour.

According to FEGER's empirically founded model [68, 69] decisions are *conflict situations* which will occur if there can be chosen between several each other excluding *behavioural alternatives* $x \in \{x_1, \dots, x_S\}$. At the beginning of such a conflict one values its *importance.* This is decisive for the time spent on decision-making. Afterwards, an *accidental sequence of consequence anticipations* starts, i.e. a sequence

$$i_0, i_1, i_2, \dots \tag{8.15}$$

of imagined consequences i_n of the possible decisions each of which favours a certain alternative $x_n = f(i_n)$. In other words: The decision-maker becomes *conscious* of the *consequences* which the choice of the single available alternatives would bring about. The arguments considered on this occasion are of a more or less accidental order and let the decision-maker alternate between the different possible choices.

A final dccision will bc madc if k succcssivc conscqucncc anticipations (i.c. k arguments)

$$i_{n+1}, \ldots, i_{n+k} \qquad \text{with} \qquad f(i_{n+1}) = \ldots = f(i_{n+k}) = x \tag{8.16}$$

are in favour of one and the same alternative x. Here the number k is the larger the more important the decision is.

A comparison of the consequences of the different alternatives continues even *after* a decision was made. This is felt as *doubts* about the rightness of the decision made.

The *preferences* for the single alternatives x affect the decision process in such a way that they determine the probabilities with which consequence anticipations i_n occur in favour of them. If the differences between the preferences of the considered alternatives are small (so that the choice of the best alternative and thus the decision is 'difficult') the sequence of consequence anticipations up to the final decision will be particularly long (since k is chosen especially large, then). A broad spectrum of alternatives has a similar consequences (because in this case the probability that successive arguments are in favour of one and the same alternative is small).

The just described relations can be also formulated in a mathematical way. Let $U_\alpha(x|y;t)$ be the *preference* for alternative x for an individual α who has finally made decision y. Let $\widehat{p}_\alpha(x|y;t) \geq 0$ be the probability for the occurence of consequence anticipation i which, due to $f(i) = x$, is in favour of alternative x. It is plausible that this probability increases with the preference $U_\alpha(x|y;t)$. Therefore and in order to guarantee the non-negativity of $\widehat{p}_\alpha(x|y;t)$ we will assume

$$\widehat{p}_\alpha(x|y;t) := \frac{\mathrm{e}^{U_\alpha(x|y;t)}}{\sum\limits_{x'} \mathrm{e}^{U_\alpha(x'|y;t)}} = \mathrm{e}^{\widehat{U}_\alpha(x|y;t)} \tag{8.17a}$$

with the *scaled preferences*

$$\widehat{U}_\alpha(x|y;t) := U_\alpha(x|y;t) - \ln \sum_{x'} \mathrm{e}^{U_\alpha(x'|y;t)} . \tag{8.17b}$$

Then the probability $p'_\alpha(x|y;t)$ that the next k arguments are all in favour of alternative x is

$$p'_\alpha(x|y;t) = \Big[\widehat{p}_\alpha(x|y;t)\Big]^k = \mathrm{e}^{k\widehat{U}_\alpha(x|y;t)} . \tag{8.18}$$

As a consequence, the probability $p_\alpha(x|y;t)$ of a decision on alternative x is given by

$$p_\alpha(x|y;t) := \frac{\mathrm{e}^{k\widehat{U}_\alpha(x|y;t)}}{\sum_{x'} \mathrm{e}^{k\widehat{U}_\alpha(x'|y;t)}} = \mathrm{e}^{U'_\alpha(x|y;t)} \tag{8.19a}$$

provided that the final decision is actually made. In the following the quantity

$$U'_\alpha(x|y;t) := k\widehat{U}_\alpha(x|y;t) - \ln \sum_{x'} \mathrm{e}^{k\widehat{U}_\alpha(x'|y;t)} \tag{8.19b}$$

will be denoted as the *attractiveness* of decision x for an individual α who chose behaviour y before. According to (8.19a), the probability $p_\alpha(\widehat{x}|y;t)$ of a decision on the alternative $\widehat{x}$ with the largest preference $U_\alpha(\widehat{x}|y;t)$ will increase with k.

Interestingly enough, (8.17a) leads, in contrast to other conceivable approaches, to the desired property that the probability distribution $\widehat{p}_\alpha(x|y;t)$ of consequence anticipations and the probability distribution $p_\alpha(x|y;t)$ of decisions, independently of k, possess the same functional form, namely an exponential dependence. Moreover, from

$$\begin{aligned} U'_\alpha(x_1|y;t) - U'_\alpha(x_2|y;t) &= k\Big[\widehat{U}_\alpha(x_1|y;t) - \widehat{U}_\alpha(x_2|y;t)\Big] \\ &= k\Big[U_\alpha(x_1|y;t) - U_\alpha(x_2|y;t)\Big] \end{aligned} \tag{8.20}$$

follows the important relation that the attractiveness difference $U'_\alpha(x_1|y;t) - U'_\alpha(x_2|y;t)$ between two alternatives x_1 and x_2 is k times as large as the difference $U_\alpha(x_1|y;t) - U_\alpha(x_2|y;t)$ of their preferences. The larger the number k of arguments which successively must be in favour of the same alternative x for a final decision on x the more evident become the differences between the preferences of similarly valued alternatives. Thus an increase of k (which implies a prolongation of the time spent on the decision) can compensate for a small difference between the preferences for different alternatives.

Of special interest in connection with decision conflicts is also the quantity that is denoted as the *freedom of decision-making*. According to psychological investigations the (subjectively felt) freedom of decision-making increases with the *uncertainty* with respect to the final result of a decision [115, 265]. That is, the freedom of decision-making is an *entropy-like* quantity. It is the larger the larger the number of alternatives is and the more equivalent they are (with regard to their preferences). A restriction of the freedom of decision-making gives rise to *reactance* (i.e. to a kind of a *defiant reaction*): Either the limitation of freedom will be evaded if possible, or *resistance* is formed [30–32, 299].

8.2.5 Experimental Problems

Experiments in social sciences are especially difficult for various reasons:

- Too many internal variables play a role most of which are unknown.
- It is not easy to find comparable test persons for experiments.
- Many *disturbing influences* make clear and reproducible results difficult.

These problems and the possibilities to avoid them were already discussed in Sects. 8.2.1, 8.2.2 and 8.2.3. Furthermore, among other things, there are the following hindrances:

- Experiments with many test persons require much effort and high expenditures.
- In some cases experiments must be ruled out for *ethical reasons.*

However, these problems do not affect so much the models of this book for which we can, for example, use empirical data from *opinion polls* or *election* and *market analyses.*

8.2.6 Measurement of Behaviours

A further difficulty in social sciences is the *quantitative measurement* of behaviours. However, this is quite simple for the quantities which are needed for the description of *pedestrian behaviour* (cf. [123–125, 129] and [132], pp. 232ff.). These quantities are the locations, velocities, densities, routes, and demands ('shopping lists') of pedestrians.

Already more difficult is the quantitative *measurement of opinions.* For this task the meanwhile very disseminated method of *semantic differentials* was developed by OSGOOD [211, 212]. A semantic differential characterizes an opinion i by the associated *ratings* $\widehat{d}^l(i) \in [1;7]$ on L seven-point *rating scales* l that are defined by polar terms (adjectives) (cf. Fig. 8.3).

For two semantic differentials

$$\left(\widehat{d}^1(i),\ldots,\widehat{d}^l(i),\ldots,\widehat{d}^L(i)\right) \quad \text{and} \quad \left(\widehat{d}^1(j),\ldots,\widehat{d}^l(j),\ldots,\widehat{d}^L(j)\right) \tag{8.21}$$

a measure of *distance*

$$d(i,j) := \sqrt{\sum_{l=1}^{L}\left(\widehat{d}^l(i) - \widehat{d}^l(j)\right)^2} \tag{8.22}$$

can be defined which is a measure for the *dissimilarity* of opinions i and j (cf. [212], pp. 85ff.). *Factor analyses* (cf. [149], pp. 401ff.) show that most opinions can already be well characterized by using co-ordinates of a three-dimensional so-called *semantic space* where the co-ordinate axes can be interpreted as

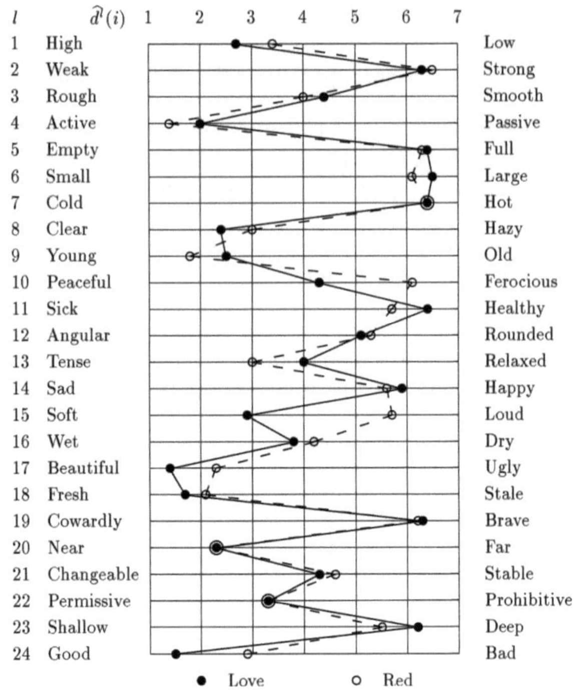

Fig. 8.3 Semantic differentials for two similar terms (according to [142], pp. 35ff.; cf. also [141])

evaluation, activity, and *potency.* The dimension of evaluation corresponds to polar terms like good/bad, kind/cruel, or beautiful/ugly, the dimension of potency to strong/weak or hard/soft, and the dimension of activity to fast/slow or active/passive (cf. [212], pp. 36ff.).

A related method for the analysis of opinions will be introduced in Sect. 13.6.

8.3 Summary

In the last sections it was shown how the stochastic methods that were discussed in Part I can be made applicable to a quantitative description of social processes. A comparison of the essential aspects of stochastic models in physics and the social sciences is given in Table 8.1.

Whereas up to now we only provided a general *modelling strategy* and a suitable *terminological frame*, in the following chapters concrete models for various social processes will be developed. In this connection we will consciously avoid a *'physicalistic' way of proceeding* which would transfer existing physical models without critical inspection to the description of social systems by simply reinterpreting the

Table 8.1 Specification and comparison of the essential terms used in stochastic models in physics and the social sciences

Aspect	Physics	Social sciences
Total system	Gas, fluid, etc.	Social system
Subsystems	Particles	Individuals
State	Location, velocity, energy level, etc.	Behaviour, opinion, strategy, etc.
Type of subsystem	Sort of particle, kind of interaction	Subpopulation, behavioural type
Vector of occupation numbers	Configuration	Socioconfiguration
Transition	State change	Behavioural change
Interactions	Attractive and repulsive forces	Imitative processes, Avoidance processes, Compromising processes
Experimental verification	Restriction to a few effects, preparation of an ensemble of identical particles	Restriction to simple situations selection of various individuals of the same behavioural type
Stochasticity/ uncertainty	Fluctuations	'Freedom of decision-making'

variables. Instead, models will be presented each of which is geared to the characteristics of the particular social system and takes into account its peculiarities in an adequate manner.

Chapter 9
Decision Theoretical Specification of the Transition Rates

9.1 Introduction

The abstract mathematical formulation developed in Part I of this book obtains a physical, chemical, biological, or sociological meaning only after the transition rates have been concretely determined. Approaches for the transition rates can be found by *plausibility considerations* but, in the end, they must be *empirically* checked (cf. Chap. 13). Especially, in connection with social sciences the decision theoretical approach

$$w_a(x'|x;t) := \nu_a(t) R_a(x'|x;t) \qquad (x' \neq x) \tag{9.1a}$$

with

$$R_a(x'|x;t) := \frac{\mathrm{e}^{U_a(x',t)-U_a(x,t)}}{D_a(x',x;t)} \tag{9.1b}$$

yields good results (cf. [289] and [294], p. 19). Here

- $\nu_a(t)$ is a rate which describes the *flexibility* of individuals of subpopulation a with respect to behavioural changes,
- $R_a(x'|x;t)$ is the *readiness* of individuals of behavioural type a for a change from behaviour x to x' during a time interval $1/\nu_a(t)$,
- the *distance* $D_a(x',x;t) = D_a(x,x';t) > 0$ is a measure for the *effort*

$$S_a(x',x;t) := \ln D_a(x',x;t) \tag{9.2}$$

  of a behavioural change from x to x',
- $U_a(x,t)$ is the *utility* of behaviour x for an individual of subpopulation a.

The larger the flexibility $\nu_a(t)$, the smaller the effort $S_a(x',x;t)$, or the larger the utility increase $U_a(x',t) - U_a(x,t)$ for a behavioural change from behaviour x to behaviour x' the larger is the corresponding transition rate $w_a(x'|x;t)$.

We will investigate approach (9.1) from various perspectives, now. In particular, in Sect. 9.2.1 we will find that (9.1) takes into account the uncertainty regarding the determination of the *(scaled) conditional utility*

D. Helbing, *Quantitative Sociodynamics*, 2nd ed.,
DOI 10.1007/978-3-642-11546-2_9,

$$U_a(x'|x,t) := U_a(x',t) - U_a(x,t) - S_a(x',x;t) \tag{9.3}$$

of a behavioural change (i.e. the *incompleteness of information* during a decision).

Finally, we will specify the transition rates $w_{ab}(x',y'|x,y;t)$ for pair interactions. These can be decomposed into several contributions due to different *kinds of interactions*, namely into imitative processes, avoidance processes, and compromising processes.

9.2 Derivation

Transition rates $w_a(x'|x;t)$ can always be written in the form

$$w_a(x'|x;t) =: \frac{1}{\Delta t} p_a(x'|x;t) \qquad (x' \neq x) \tag{9.4a}$$

with

$$p_a(x'|x;t) \geq 0\,, \qquad p_a(x|x;t) := 1 - \sum_{x'(\neq x)} p_a(x'|x;t) \tag{9.4b}$$

(cf. (3.22b)). Here Δt means a short time unit and $p_a(x'|x;t)$ the probability to carry out a behavioural change from x to x' during this time unit. Δt should be so small that we have

$$\sum_{x'(\neq x)} p_a(x'|x;t) \ll 1 \qquad \text{or} \qquad p_a(x|x;t) \approx 1 \tag{9.5}$$

so that there will not be *several* transitions within a time interval Δt.

Due to the non-negativity of $p_a(x'|x;t)$ the unique representation

$$p_a(x'|x;t) = \mathrm{e}^{U'_a(x'|x;t)} \qquad \text{with} \qquad U'_a(x'|x;t) := \ln p_a(x'|x;t) \tag{9.6}$$

is *always* possible. The exponential representation (9.6) has only implications with respect to the *interpretation* of the functions $U'_a(x'|x;t)$. Assumptions concerning the form of $p_a(x'|x;t)$ are introduced *later* along with the specification of $U'_a(x'|x;t)$ (cf. Sect. 9.2.4). In the following we will see that (9.6) can be motivated in several ways. One derivation was already provided in Sect. 8.2.4 in connection with the mathematical formulation of FEGER's decision model.

9.2.1 The Multinomial Logit Model

Let

$$U^*_a(x'|x;t) := \widehat{U}_a(x'|x;t) + \varepsilon_a(x'|x;t) \tag{9.7}$$

be the *conditional utility* which an individual of subpopulation a expects from a *behavioural change* from x to x'. Due to the *incompleteness of information* the conditional utility $U_a^*(x'|x;t)$ consists of a *known part* $\widehat{U}_a(x'|x;t)$ and an *unknown part* $\varepsilon_a(x'|x;t)$ which can only be estimated. For example, if (after having purchased product x) one decides on the purchase of a product x' the price will, as a known quantity, contribute to $\widehat{U}_a(x'|x;t)$ whereas the durability can only be estimated and will be taken into account by $\varepsilon_a(x'|x;t)$. $\varepsilon_a(x'|x;t)$ varies from individual $\alpha \in a$ to individual $\beta \in a$ and from decision to decision.

It is plausible to assume that the decision-maker will choose just that behavioural change

$$(x'|x) := x' \longleftarrow x \tag{9.8}$$

from behaviour x to behaviour x' for which the conditional utility $U_a^*(x'|x;t)$ is *maximal,* i.e. for which

$$U_a^*(x'|x;t) > U_a^*(y|x;t) \tag{9.9}$$

holds in comparison with all other possible behavioural changes $(y|x)$. Thus the probability to choose $(x'|x)$ among all possible behavioural changes $(y|x)$ is given by

$$\begin{aligned} &p_a(x'|x;t) \\ &:= \text{Prob}\Big(\Big\{\widehat{U}_a(x'|x;t) + \varepsilon_a(x'|x;t) > \widehat{U}_a(y|x;t) + \varepsilon_a(y|x;t) \quad \forall\, y \neq x'\Big\}\Big)\,. \end{aligned} \tag{9.10}$$

If the randomly varying part $\varepsilon_a(x'|x;t)$ is independent of x' and distributed according to the *extreme value distribution*

$$P_a(\varepsilon_a|x,t) := \frac{1}{h_a(x,t)} \mathrm{e}^{-[c_a(x,t)+\varepsilon_a/h_a(x,t)]} \exp\left\{-\mathrm{e}^{-[c_a(x,t)+\varepsilon_a/h_a(x,t)]}\right\} \tag{9.11}$$

(which is also denoted as the WEIBULL or GNEDENKO distribution) one obtains

$$p_a(x'|x;t) = \frac{\mathrm{e}^{\widehat{U}_a(x'|x;t)/h_a(x,t)}}{\sum\limits_y \mathrm{e}^{\widehat{U}_a(y|x;t)/h_a(x,t)}}\,. \tag{9.12}$$

(9.12) is a generalization of the *multinomial logit model* which, in its original version, does not take into account any dependence on the present behaviour x (cf. [53, 209]). $h_a(x,t)$ is a measure for the *uncertainty of information* during the decision (and will normally be independent of x). With the *affine transformation*

$$U_a'(x'|x;t) := \frac{\widehat{U}_a(x'|x;t)}{h_a(x,t)} - \ln \sum_y \mathrm{e}^{\widehat{U}_a(y|x;t)/h_a(x,t)} \tag{9.13}$$

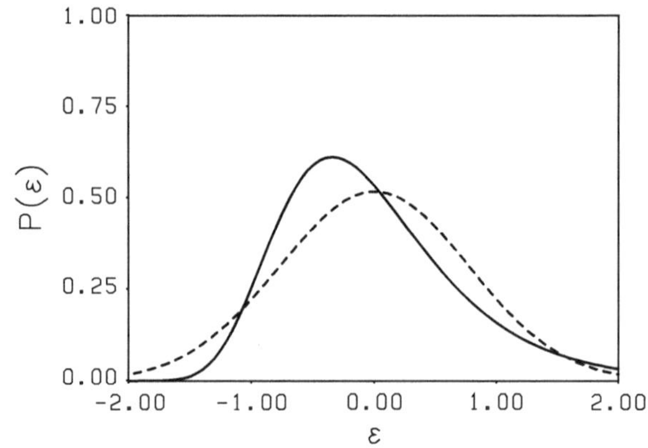

Fig. 9.1 Comparison of the *extreme value distribution* for $h(x,t) \equiv 0,6$ (—) with the *normal distribution* having the same mean value and variance (– –)

we arrive at the desired representation

$$p_a(x'|x;t) = \mathrm{e}^{U'_a(x'|x;t)} . \tag{9.14}$$

Obviously, it is just the *scaled conditional utility* $U'_a(x'|x;t)$ which is the decisive quantity for behavioural changes. With the terminology of Sect. 8.2.4, $U'_a(x'|x;t)$ can also be called the *attractiveness* of behaviour x' for an individual of subpopulation a who has chosen behaviour x before.

The application of the *extreme value distribution* is particularly reasonable since the maximum of two extreme value distributed variables is also extreme value distributed. From this point of view, it is the *'natural distribution'* for maximization problems [53]. If, instead of (9.11), we had applied the *normal distribution* (which is compared to the extreme value distribution in Fig. 9.1), probability (9.10) would have not been explicitly calculable.

9.2.2 Entropy Maximization

The distribution (9.12) resembles the *canonical distribution*

$$P(i) = \frac{\mathrm{e}^{-E'(i)/(\mathrm{k}T')}}{\sum_j \mathrm{e}^{-E'(j)/(\mathrm{k}T')}} \tag{9.15}$$

from *thermodynamics* [236, 245] where k denotes the BOLTZMANN *constant,* T' the *absolute temperature,* and $E'(i)$ the *energy* of state i. Equation (9.15) can be derived by maximizing the *(information) entropy*

$$S := -\sum_i P(i) \log_2 P(i) \quad \text{or} \quad S' := -\sum_i P(i) \ln P(i) = S \ln 2 \tag{9.16}$$

on the *constraints*

$$\sum_i P(i) \stackrel{!}{=} 1 \quad \text{(normalization)}, \tag{9.17a}$$

$$\sum_i E'(i)P(i) \stackrel{!}{=} E' \quad \text{(conservation of total energy } E'\text{)} \tag{9.17b}$$

(cf. [107], pp. 53ff.).

A similar derivation also exists for (9.12). It already became clear that, due to $U_a^*(x'|x;t) = \widehat{U}_a(x'|x;t) + \varepsilon_a(x'|x;t)$, the maximization of the *known conditional utility* $\widehat{U}_a(x'|x;t)$ does not necessarily lead to the *best* decision. Rather it is reasonable to occasionally vary behavioural changes in order to determine the *actual conditional utility* $U_a^*(x'|x;t)$ of other behavioural changes more precisely and, as a consequence, to be able to really *optimize* the behaviour. Hence the individual will undergo a certain *risk* but choose the probabilities $p_a(x'|x;t)$ of behavioural changes in such a way that the *expected* conditional utility reaches a certain percentage p of the *maximal* known conditional utility:

$$\sum_y \widehat{U}_a(y|x;t)p_a(y|x;t) \stackrel{!}{=} p \cdot \max_y \widehat{U}_a(y|x;t) =: h'_a(x,t) \tag{9.18}$$

Since there exist many of such distributions, we will select the one which is *'most likely'*, i.e. the distribution which maximizes the *information entropy*

$$-\sum_y p_a(y|x;t) \ln p_a(y|x;t) \tag{9.19}$$

on the conditions (9.4b) and (9.18). Instead, however, we can also introduce the LAGRANGE *multipliers* $\lambda_a^x(t)$ and $\mu_a^x(t)$ and maximize the expression

$$\begin{aligned} &- \sum_y p_a(y|x;t) \ln p_a(y|x;t) \\ &- \mu_a^x(t) \left(\sum_y \widehat{U}_a(y|x;t) p_a(y|x;t) - h'_a(x,t) \right) \\ &- \left(\lambda_a^x(t) - 1 \right) \left(\sum_y p_a(y|x;t) - 1 \right) \end{aligned} \tag{9.20}$$

without considering the constraints. For this purpose we must derive (9.20) with respect to $p_a(x'|x;t)$ and set the resulting term to zero:

$$-\ln p_a(x'|x;t) - 1 - \mu_a^x(t)\widehat{U}_a(x'|x;t) - \left(\lambda_a^x(t) - 1\right) \stackrel{!}{=} 0\,. \tag{9.21}$$

The solution of this equation is

$$p_a(x'|x;t) = \mathrm{e}^{-\lambda_a^x(t) - \mu_a^x(t)\widehat{U}_a(x'|x;t)} . \tag{9.22}$$

Now the LAGRANGE multiplier $\lambda_a^x(t)$ must be determined from the normalization condition. This leads to

$$p_a(x'|x;t) = \frac{\mathrm{e}^{-\mu_a^x(t)\widehat{U}_a(x'|x;t)}}{Z_a(x,t)} \tag{9.23a}$$

with the so-called *state sum (partition function)*

$$Z_a(x,t) := \sum_y \mathrm{e}^{-\mu_a^x(t)\widehat{U}_a(y|x;t)} = \mathrm{e}^{\lambda_a^x(t)} . \tag{9.23b}$$

Finally, the LAGRANGE multiplier $\mu_a^x(t)$ should be chosen in such a way that constraint (9.18) is fulfilled. This constraint is equivalent to the more familiar formulation

$$-\frac{\partial \ln Z_a}{\partial \mu_a^x(t)} = -\frac{1}{Z_a}\frac{\partial Z_a}{\partial \mu_a^x(t)} \stackrel{!}{=} h_a'(x,t) . \tag{9.24}$$

Denoting the value which fulfills (9.18) and (9.24) by $\mu_a'(x,t)$ and setting

$$h_a(x,t) := \frac{1}{\mu_a'(x,t)} , \tag{9.25}$$

(9.23) again takes on the form (9.12). The further argumentation continues like in Sect. 9.2.1.

9.2.3 FECHNER'S Law

Up to now the derivations of the probabilities of behavioural changes started from the *subjective conditional utilities* $\widehat{U}_a(x'|x;t)$. In this section we will assume that $V_a(x'|x;t) \geq 0$ reflects the *objective advantage* of a behavioural change from x to x'. Then the probabilities $p_a(x'|x;t)$ of behavioural changes $(x'|x)$ are expected to be proportional to the objective advantages $V_a(x'|x;t)$:

$$p_a(x'|x;t) = \frac{V_a(x'|x;t)}{\sum_y V_a(y|x;t)} \tag{9.26}$$

(*law of relative effect* [228, 229, 251]). According to FECHNER's *law* [67], however, the *subjective perception* $\widehat{U}_a(x'|x;t)$ of an objective quantity $V_a(x'|x;t)$ *scales logarithmically:*

$$\widehat{U}_a(x'|x;t) = h_a(x,t)\ln V_a(x'|x;t) + c_a(x,t) \tag{9.27}$$

(cf. also [21]). Inserting

$$V_a(x'|x;t) = \mathrm{e}^{[\widehat{U}_a(x'|x;t)-c_a(x,t)]/h_a(x,t)} \tag{9.28}$$

into (9.26) and using (9.13), we again obtain (9.6) (cf. [174]).

9.2.4 Utility and Distance Function

In the following the transition probability

$$p_a(x'|x;t) = \mathrm{e}^{U'_a(x'|x;t)} \tag{9.29}$$

will be specified by a concrete approach for $U'_a(x'|x;t)$ (cf. (9.35)). With this aim in view we decompose the scaled conditional utility $U'_a(x'|x;t)$ into an *antisymmetrical part*

$$A'_a(x'|x;t) := \frac{U'_a(x'|x;t) - U'_a(x|x';t)}{2} = -A'_a(x|x';t) \tag{9.30a}$$

and a *symmetrical part*

$$S'_a(x',x;t) := -\frac{U'_a(x'|x;t) + U'_a(x|x';t)}{2} = S'_a(x,x';t)\,. \tag{9.30b}$$

Then we can write

$$U'_a(x'|x;t) = A'_a(x'|x;t) - S'_a(x',x;t) \tag{9.31}$$

which is still a *general* relation. Both contributions $A'_a(x'|x;t)$ and $S'_a(x',x';t)$ have a clear interpretation:

- $S'_a(x',x;t)$ describes those contributions to transition probability $p_a(x'|x;t)$ which are independent of whether the behavioural change occurs from x to x' or vice versa, from x' to x. Thus $S'_a(x',x;t)$ can be understood as a measure for the *(subjective) effort* which accompanies the behavioural change. The quantity

$$D'_a(x',x;t) := \mathrm{e}^{S'_a(x',x;t)} = D'_a(x,x';t) \geq 0 \tag{9.32}$$

can be interpreted as a kind of *distance (dissimilarity)* between two behaviours and is a measure for the *complications* of a behavioural change or for the *incompatibility* of two behaviours. Depending on the situation it reflects the time needed, travelling expenses, removal costs, or transaction costs.

The larger the distance $D'_a(x', x; t)$ or the effort $S'_a(x', x; t)$ is the smaller the corresponding transition probability $p_a(x'|x; t)$.

- $A'_a(x'|x; t)$ describes those contributions to transition probability $p_a(x'|x; t)$ which depend on the *direction*: If $A'_a(x'|x; t)$ is large for a behavioural change from x to x', the value $A'_a(x|x'; t) = -A'_a(x'|x; t)$ for the *opposite* behavioural change from x' to x will be small. So $A'_a(x'|x; t)$ takes into account those factors which are in favour of a certain behaviour x' in comparison with another behaviour x. One often chooses the special approach

$$A'_a(x'|x; t) := U'_a(x', t) - U'_a(x, t) \tag{9.33}$$

where $U'_a(x, t)$ denotes the *utility* of behaviour x. Here the *advantages* of a behaviour yield a positive contribution to the utility whereas its *disadvantages* yield a negative one.

The larger the *utility difference*

$$\Delta_{x'x} U'_a := U'_a(x'; t) - U'_a(x, t) \tag{9.34}$$

is the higher the probability $p_a(x'|x; t)$ of a behavioural change.

In summary, the scaled conditional utility $U'_a(x'|x; t)$ for a behavioural change from x to x' is given by the utility difference $\Delta_{x'x} U'_a$ of the potential behaviour x' compared to the current behaviour x, reduced by the effort $S'_a(x', x; t)$ of the corresponding behavioural change. The resulting transition rates read

$$w_a(x'|x; t) = \frac{1}{\Delta t} p_a(x'|x; t) = \frac{1}{\Delta t} \mathrm{e}^{U'_a(x'|x;t)} \tag{9.35a}$$

with

$$p_a(x'|x; t) = \mathrm{e}^{U'_a(x',t) - U'_a(x,t) - S'_a(x',x;t)} = \frac{\mathrm{e}^{U'_a(x',t) - U'_a(x,t)}}{D'_a(x', x; t)} . \tag{9.35b}$$

However, we are not at all interested in the *'staying rate'* $w_a(x|x; t)$ because it does not have any influence on the solution of the master equation (cf. (3.23)). Thus we often prefer a representation of the form

$$w_a(x'|x; t) = \nu_a(t) \frac{\mathrm{e}^{U_a(x',t) - U_a(x,t)}}{D_a(x', x; t)} \qquad (x' \neq x) \tag{9.36a}$$

which is connected with (9.35) via the relations

$$\nu_a(t) := \frac{1}{D_a(t)\Delta t}\,, \tag{9.36b}$$

$$D_a(x',x;t) := \frac{D'_a(x',x;t)}{D_a(t)}\,, \tag{9.36c}$$

and

$$U_a(x,t) := U'_a(x,t) - C_a(t)\,. \tag{9.36d}$$

In order to define the distance $D_a(x',x;t)$ and the utility $U_a(x,t)$ in a *unique* way, we demand, for example,

$$\frac{1}{S}\sum_x U_a(x,t) \stackrel{!}{=} 0 \tag{9.37a}$$

and

$$\frac{1}{S(S-1)}\sum_{x\neq x'} D_a(x',x;t) \stackrel{!}{=} 1 \tag{9.37b}$$

where S means the number of states $x \in \gamma_a \equiv \{x_1,\dots,x_S\}$. Of course, S *must be replaced by* $\widehat{S}_a$ *if the state spaces* γ_a *consist of different numbers* $\widehat{S}_a$ *of states.* (9.37a) shifts the mean value

$$C_a(t) := \frac{1}{S}\sum_x U'_a(x,t) \tag{9.38a}$$

of the *(scaled) utilities* $U_a(x,t)$ to *zero* whereas (9.37b), by

$$D_a(t) := \frac{1}{S(S-1)}\sum_{x\neq x'} D'_a(x',x;t)\,, \tag{9.38b}$$

defines a *'unit of distance'* $D_a(t)$ in which *(scaled) distances* $D_a(x',x;t)$ are measured. $\nu_a(t)$ is a *flexibility parameter* that reflects the frequency of behavioural changes per time unit. It is the larger the smaller the unit of distance $D_a(t)$ is.

Since the scaling condition (9.37b) replaces the normalization condition (9.4b), approach (9.36) can manage without staying rates $w_a(x|x;t)$. Via relations (9.36) and (9.38) the flexibility parameter $\nu_a(t)$, distances $D_a(x',x;t)$, and utility functions $U_a(x,t)$ can be determined from empirical data $w^{\mathrm{e}}_a(x'|x;t)$ of transition rates (cf. Chap. 13).

Finally, it is worth mentioning that approach (9.36) also describes transition rates that correspond to the ARRHENIUS formula (7.32): Applying the notations from Sect. 7.3 we find

$$U'(Z) = -\frac{2}{\varepsilon q}\phi(Z) \tag{9.39a}$$

and

$$D'(Z_1, Z_0) = \frac{4\pi}{\varepsilon q\sqrt{\left|\frac{\partial^2}{\partial Z^2}\phi(Z_1)\right| \cdot \left|\frac{\partial^2}{\partial Z^2}\phi(Z_0)\right|}}\,. \tag{9.39b}$$

9.3 Pair Interaction Rates

Up to now only the transition rates for spontaneous behavioural changes have been treated. In this section we will specify the transition rates $w_{ab}(x', y'|x, y; t)$ which are related to pair interactions. A particularly simple approach would be

$$w_{ab}(x', y'|x, y; t) := \nu_{ab}(t)\, p_{ab}(x', y'|x, y; t) \tag{9.40}$$

with

$$p_{ab}(x', y'|x, y; t) := p_{ab}(x'|x; t)\, p_{ba}(y'|y; t)\,. \tag{9.41}$$

Here

- $\nu_{ab}(t) = N_b \widehat{\nu}_{ab}(t)$ is the *meeting rate* between an individual of subpopulation a and individuals of subpopulation b (which, by the way, reflects the *social network*; cf. [29, 36]),
- $p_{ab}(x', y'|x, y; t)$ means the probability that, due to an interaction of two individuals of subpopulations a and b, the former one changes the behaviour from x to x' and the latter one from y to y',
- $p_{ab}(x'|x; t)$ describes the probability that an individual of subpopulation a exchanges the behaviour x for x' due to an interaction with an individual of subpopulation b.

Especially, we have

$$\sum_{x'} \sum_{y'} p_{ab}(x', y'|x, y; t) = 1 \tag{9.42a}$$

and

$$\sum_{x'} p_{ab}(x'|x; t) = 1\,. \tag{9.42b}$$

Thus the effective *transition rates*

$$w^a(x'|x;t) = w_a(x'|x;t) + \sum_b \sum_y \sum_{y'} w_{ab}(x',y'|x,y;t) P_b(y,t) \tag{9.43a}$$

of the BOLTZMANN-like equations

$$\frac{d}{dt} P_a(x,t) = \sum_{x'} \Big[w^a(x|x';t) P_a(x',t) - w^a(x'|x;t) P_a(x,t) \Big] \tag{9.43b}$$

(cf. (4.22)) take on the form

$$w^a(x'|x;t) = w_a(x'|x;t) + \sum_b \nu_{ab}(t) p_{ab}(x'|x;t) \tag{9.44}$$

by applying (9.40), (9.41). That is, the effective transition rates lose their dependence on the behavioural distribution $P_b(y,t)$ so that the effects of pair interactions can no longer be mathematically distinguished from the effects of spontaneous transitions. (This implies that they can no longer be separately determined from empirical transition rates $w^a_{\mathrm{e}}(x'|x;t)$; cf. Chap. 13.) The reason for this dilemma is *factorization assumption* (9.41) which supposes that interacting individuals carry out their behavioural changes in a *statistically independent* way.

Therefore, we are looking for a more sophisticated approach than (9.41). Let

$$x', y' \longleftarrow x, y \tag{9.45}$$

represent pair interactions during which the interacting individuals change their behaviours from x and y to x' and y' respectively. Then we can *classify* these according to the following scheme:

$$\left.\begin{array}{l} x,x \longleftarrow x,x \\ x,y \longleftarrow x,y \end{array}\right\} (0) \tag{9.46a}$$

$$\left.\begin{array}{l} x,x \longleftarrow x,y \ (x \neq y) \\ y,y \longleftarrow x,y \ (x \neq y) \end{array}\right\} (1) \tag{9.46b}$$

$$\left.\begin{array}{l} x,y' \longleftarrow x,x \ (y' \neq x) \\ x',y \longleftarrow y,y \ (x' \neq y) \\ x',y' \longleftarrow x,x \ (x' \neq x, y' \neq x) \end{array}\right\} (2) \tag{9.46c}$$

$$\left.\begin{array}{l} x,y' \longleftarrow x,y \ (x \neq y, y' \neq y, y' \neq x) \\ x',y \longleftarrow x,y \ (x \neq y, x' \neq x, x' \neq y) \\ x',y' \longleftarrow x,y \ (x \neq y, x' \neq x, y' \neq y, x' \neq y, y' \neq x) \end{array}\right\} (3) \tag{9.46d}$$

$$\left.\begin{array}{l} y,x \longleftarrow x,y \ (x \neq y) \\ x',x \longleftarrow x,y \ (x \neq y, x' \neq x, x' \neq y) \\ y,y' \longleftarrow x,y \ (x \neq y, y' \neq y, y' \neq x) \end{array}\right\} (4) \tag{9.46e}$$

The classification (9.46) of transitions (9.45) is complete because there exist $2^2 = 4$ transitions of the kind

$$x', y' \longleftarrow x, x \tag{9.47a}$$

and $3^2 = 9$ transitions of the kind

$$x', y' \longleftarrow x, y\,. \tag{9.47b}$$

In the first case x' and y' can be chosen equal or unequal to x, in the second case equal to x, equal to y, or unequal to x and y.

By means of curly brackets five *kinds* $k \in \{0, \ldots, 4\}$ *of pair interactions* were distinguished in (9.46) which can obviously be interpreted as follows:

(0) By (0) situations are summarized in which two each other meeting individuals do not change their behaviour. Since they are not connected with any temporal change of the probabilities $P_a(x, t)$ or of the socioconfiguration n, they will not be treated any further.

(1) Interactions (1) describe *imitative processes* or *persuasion processes*, i.e. the tendency of an individual to take over the behaviour of another individual (cf. [77], pp. 451ff.).

(2) Interactions (2) reflect *avoidance processes* where an individual gives up his/her behaviour in favour of another behaviour if meeting an individual with the same behaviour. Such *aversive reactions* show up, for example, in *defiance* or *snobbish behaviour.*

(3) Interactions (3) delineate *compromising processes* (in a *broader* sense) where an individual changes his/her behaviour for a new behaviour if meeting an individual with another behaviour. One can find such behavioural changes in situations where the own behaviour cannot be maintained when confronted with another behaviour which is, however, also not convincing.
In the *narrow* sense, with a compromising process we mean the special case

$$x', x' \longleftarrow x, y \quad (x \neq y, x' \neq x, x' \neq y) \tag{9.48}$$

of

$$x', y' \longleftarrow x, y \quad (x \neq y, x' \neq x, y' \neq y, x' \neq y, y' \neq x) \tag{9.49}$$

where an individual only changes the behaviour in favour of another one if he/she can agree on this with his/her interaction partner. Thus we will call (9.48) an *agreement process.*

(4) Interactions (4), like (1), describe persuasion processes. The difference, however, is that one of the individuals changes the behaviour though he/she could persuade the other one of his/her previous behaviour. Such processes occur only very rarely, i.e. they are very improbable. Consequently, they will be excluded from our further considerations.

Instead of (9.41), we can now set up the following more sophisticated approach:

$$p_{ab}(x', y'|x, y; t) := \begin{cases} p^1_{ab}(x'|x; t) \cdot 1 = p^1_{ab}(x'|x; t) & \text{if } x' = y \text{ and } y' = y \\ 1 \cdot p^1_{ba}(y'|y; t) = p^1_{ba}(y'|y; t) & \text{if } y' = x \text{ and } x' = x \\ p^4_{ab}(x'|x; t) \cdot 0 = 0 & \text{if } x' = y \text{ and } y' \neq y \\ 0 \cdot p^4_{ba}(y'|y; t) = 0 & \text{if } y' = x \text{ and } x' \neq x \\ p^k_{ab}(x'|x; t) p^k_{ba}(y'|y; t) & \text{otherwise.} \end{cases} \tag{9.50}$$

With this approach we give up the statistical independence of formulation (9.41) because the transition probabilities depend on the kind k of the pair interaction. This is a function of the behaviours shown by two individuals at the beginning and the end of their interaction:

$$k \equiv k(x', y'; x, y) . \tag{9.51}$$

Apart from this we assumed for an individual who persuades another individual that he/she will keep the behaviour with probability 1 (confer first and second line). That is, he/she changes the behaviour with probability 0 (confer third and fourth line).

From (9.46) and (9.50) we obtain the following transition rates for imitative processes, avoidance processes, and compromising processes:

$$\begin{aligned} w^1_{ab}(x', y'|x, y; t) = {} & \nu_{ab}(t) p^1_{ba}(x|y; t) \delta_{xx'} \delta_{xy'} (1 - \delta_{xy}) \\ & + \nu_{ab}(t) p^1_{ab}(y|x; t) \delta_{yy'} \delta_{yx'} (1 - \delta_{xy}) , \end{aligned} \tag{9.52a}$$

$$\begin{aligned} w^2_{ab}(x', y'|x, y; t) = {} & \nu_{ab}(t) p^2_{ab}(x|x; t) p^2_{ba}(y'|x; t) \delta_{xy} \delta_{xx'} (1 - \delta_{xy'}) \\ & + \nu_{ab}(t) p^2_{ab}(x'|y; t) p^2_{ba}(y|y; t) \delta_{xy} \delta_{yy'} (1 - \delta_{yx'}) \\ & + \nu_{ab}(t) p^2_{ab}(x'|x; t) p^2_{ba}(y'|x, t) \delta_{xy} (1 - \delta_{xx'}) \\ & \times (1 - \delta_{xy'}) , \end{aligned} \tag{9.52b}$$

$$\begin{aligned} & w^3_{ab}(x', y'|x, y; t) \\ & = \nu_{ab}(t) p^3_{ab}(x|x; t) p^3_{ba}(y'|y; t) \delta_{xx'} (1 - \delta_{xy})(1 - \delta_{yy'})(1 - \delta_{xy'}) \\ & + \nu_{ab}(t) p^3_{ab}(x'|x, t) p^3_{ba}(y|y; t) \delta_{yy'} (1 - \delta_{xy})(1 - \delta_{xx'})(1 - \delta_{yx'}) \\ & + \nu_{ab}(t) p^3_{ab}(x'|x; t) p^3_{ba}(y'|y, t)(1 - \delta_{xy})(1 - \delta_{xx'})(1 - \delta_{yy'}) \\ & \times (1 - \delta_{yx'}) . \end{aligned} \tag{9.52c}$$

The factors δ_{xy} stem from constraints of the kind $x = y$ within classification (9.46), the factors $(1 - \delta_{xy})$, in contrast, from constraints of the type $x \neq y$.

Inserting the total transition rate

$$w_{ab}(x', y'|x, y; t) := \sum_k w^k_{ab}(x', y'|x, y; t) \tag{9.53}$$

into the general BOLTZMANN-like equations

$$\begin{aligned}\frac{d}{dt}P_a(x,t) = &\sum_{x'}\Big[w_a(x|x';t)P_a(x',t) - w_a(x'|x;t)P_a(x,t)\Big] \\ &+\sum_b\sum_{x'}\sum_y\sum_{y'} w_{ab}(x,y'|x',y;t)P_b(y,t)P_a(x',t) \\ &-\sum_b\sum_{x'}\sum_y\sum_{y'} w_{ab}(x',y'|x,y;t)P_b(y,t)P_a(x,t)\end{aligned} \tag{9.54}$$

(cf. (9.43)) and using (9.52), (9.42) we find the explicit formula

$$\begin{aligned}\frac{d}{dt}P_a(x,t) = &\sum_{x'}\Big[w_a(x|x';t)P_a(x',t) - w_a(x'|x;t)P_a(x,t)\Big] \\ &+\sum_b\sum_{x'(\neq x)} \nu_{ab}(t)\Big[p^1_{ab}(x|x';t)P_b(x,t)P_a(x',t) \\ &- p^1_{ab}(x'|x;t)P_b(x',t)P_a(x,t)\Big] \\ &+\sum_b\sum_{x'(\neq x)} \nu_{ab}(t)\Big[p^2_{ab}(x|x';t)P_b(x',t)P_a(x',t) \\ &- p^2_{ab}(x'|x;t)P_b(x,t)P_a(x,t)\Big] \\ &+\sum_b\sum_{x'(\neq x)} \nu_{ab}(t)\Big[p^3_{ab}(x|x';t)\sum_{y(\neq x',x)} P_b(y,t)P_a(x',t) \\ &- p^3_{ab}(x'|x;t)\sum_{y(\neq x,x')} P_b(y,t)P_a(x,t)\Big].\end{aligned} \tag{9.55}$$

Despite the differences between formulations (9.50) and (9.41) it turns out that for the special case

$$p^k_{ab}(x'|x;t) \equiv p_{ab}(x'|x;t) \tag{9.56}$$

(9.55) again leads to the BOLTZMANN-like equations (9.43b) with the special transition rates (9.44). These differences appear, however, in the master equation (5.50), the covariance equations (5.22), and even in the corrected mean value equations (5.47).

In the following, for $p^k_{ab}(x'|x;t)$ with $x' \neq x$, we will assume a relation of the kind

$$p^k_{ab}(x'|x;t) := f^k_{ab}(t) R^a(x'|x;t) . \tag{9.57a}$$

As a consequence, for the *staying probability* $p^k_{ab}(x|x;t)$ this implies

$$p^k_{ab}(x|x;t) = 1 - f^k_{ab}(t) \sum_{x'(\neq x)} R^a(x'|x;t) . \tag{9.57b}$$

Here

- $f^k_{ab}(t)$ is a measure for the *frequency of interactions of kind k* for individuals of subpopulation a in interactions with individuals of subpopulation b,
- $R^a(x'|x;t)$ is a measure for the *readiness* of an individual of subpopulation a to change from behaviour x to x' in interactions with other individuals.

In order to arrive at (9.57) we choose for $p^k_{ab}(x'|x;t)$ an approach analogous to that of Sect. 9.2.4:

$$p^k_{ab}(x'|x;t) := \frac{1}{D^k_{ab}(t)} \frac{\mathrm{e}^{U^k_{ab}(x',t) - U^k_{ab}(x,t)}}{D^k_{ab}(x',x;t)} \qquad (x' \neq x) \tag{9.58a}$$

with

$$D^k_{ab}(x,x';t) = D^k_{ab}(x',x;t) =: \mathrm{e}^{S^k_{ab}(x',x;t)} \geq 0 . \tag{9.58b}$$

The quantities $U^k_{ab}(x,t)$ have also the meaning of *(scaled) utility functions* and the quantities $D^k_{ab}(x',x;t)$ the interpretation of *(scaled) distance functions,* however, this time for behavioural changes which are induced by interactions of kind k with individuals of subpopulation b. For reasons of simplicity, in the following we will assume that for the scaled utility and the scaled distance it is neither of importance to which subpopulation b an interaction partner belongs nor which kind k of interaction occurs:

$$U^k_{ab}(x,t) := U^a(x,t) , \tag{9.59a}$$

$$D^k_{ab}(x',x;t) := D^a(x',x;t) . \tag{9.59b}$$

Thus the scaled utility and distance functions depend only on the *acting* individual. This assumption is not always justified but it can be dropped if required. In any case, assumption (9.59) still admits that the utility functions $U^a(x,t)$ and distance functions $D^a(x',x;t)$ which describe behavioural changes due to *pair interactions* differ from the utility functions $U_a(x,t)$ and distance functions $D_a(x',x;t)$ which concern *spontaneous* behavioural changes. The dependence of the transition probabilities $p^k_{ab}(x'|x;t)$ on the subpopulation b to which an interaction partner belongs and on the kind k of interaction is described by the *'units of distance'* $D^k_{ab}(t)$. These determine, by

$$f^k_{ab}(t) := \frac{1}{D^k_{ab}(t)}, \tag{9.60}$$

the frequencies with which individuals of subpopulation a show behavioural changes of kind k in interactions with individuals of subpopulation b. For reasons of uniqueness, we again introduce constraints of the form

$$\frac{1}{S}\sum_x U^a(x,t) \stackrel{!}{=} 0 \tag{9.61a}$$

and

$$\frac{1}{S(S-1)}\sum_{x \neq x'} D^a(x',x;t) \stackrel{!}{=} 1\,. \tag{9.61b}$$

Now, with the definition

$$R^a(x'|x;t) := \frac{\mathrm{e}^{U^a(x',t)-U^a(x,t)}}{D^a(x',x;t)} \qquad (x' \neq x)\,, \tag{9.62a}$$

we obtain the desired relation (9.57) which simplifies the BOLTZMANN-like equations (9.55) considerably. After introducing the abbreviations

$$\nu^k_{ab}(t) := \nu_{ab}(t) f^k_{ab}(t) \tag{9.62b}$$

and

$$\begin{aligned} \nu^a(x',x;t) := \sum_b \Big[&\Big(\nu^1_{ab}(t) - \nu^3_{ab}(t)\Big) P_b(x',t) \\ &+ \Big(\nu^2_{ab}(t) - \nu^3_{ab}(t)\Big) P_b(x,t) + \nu^3_{ab}(t)\Big]\,, \end{aligned} \tag{9.62c}$$

Eqs. (9.55) take on the clear form

$$\frac{d}{dt}P_a(x,t) = \sum_{x'}\Big[w^a(x|x';t)P_a(x',t) - w^a(x'|x;t)P_a(x,t)\Big] \tag{9.62d}$$

(cf. (9.43)) with the effective transition rates

$$w^a(x'|x;t) = w_a(x'|x;t) + \nu^a(x',x;t)R^a(x'|x;t) \qquad (x' \neq x)\,. \tag{9.62e}$$

Here

- $\nu^a(x', x; t)$ are rates which are a measure for the total frequency of interactions of an individual of subpopulation a with other individuals,
- $\nu^k_{ab}(t)$ is the *rate of interactions of kind k* between an individual of subpopulation a and individuals of subpopulation b.

By assumptions (9.50), (9.57), and (9.62a) (which can be dropped if required) the general BOLTZMANN-like model (9.43) was considerably simplified and specified. Nevertheless, the resulting model is still able to describe the essential characteristics of imitative, avoidance, and compromising processes (cf. Chap. 10). Moreover, it contains a number of established models as special cases.

9.3.1 Special Applications in the Social Sciences

In the following, in order to illustrate the versatility and explanatory power of the BOLTZMANN-like equations (9.62), we will discuss some special cases which are well-founded and widely disseminated in the social, regional and economic sciences. Many of these special cases are empirically validated so that probably also the BOLTZMANN-like model in its *general* form (9.62) will be proved.

- The *logistic model* [217, 281] describes *limited growth processes.* Examples of applications are the growth of towns as well as the *diffusion* of information or innovations [18, 110, 187, 206] (cf. also p. 2).

  In order to derive the logistic model let us distinguish two behaviours $x \in \{1, 2\}$ and one subpopulation a only. Then there is $A = 1$ and $P_1(1, t) = 1 - P_1(2, t)$. $x = 2$ could have the meaning to apply a certain strategy (e.g. the newest technology) whereas $x = 1$ would then mean not to do so. If only imitative processes

$$2, 2 \longleftarrow 1, 2 \tag{9.63}$$

  and spontaneous *decay processes*

$$1 \longleftarrow 2 \tag{9.64}$$

  occur, we finally find the *logistic equation*

$$\begin{aligned} \frac{d}{dt} P_1(2, t) &= -w_1(1|2; t) P_1(2, t) + \nu^1_{11}(t) R^1(2|1; t) \Big[1 - P_1(2, t) \Big] P_1(2, t) \\ &= A(t) P_1(2, t) \Big[B(t) - P_1(2, t) \Big] \end{aligned} \tag{9.65}$$

  (cf. (4.66), (10.67)) with

$$A(t) := \nu^1_{11}(t) R^1(2|1; t) \tag{9.66}$$

and

$$B(t) := 1 - \frac{w_1(1|2;t)}{\nu_{11}^1(t)R^1(2|1;t)} . \tag{9.67}$$

- The *gravity model* delineates *exchange processes* between different locations x. It is used to describe e.g. good exchange, migration, or traffic flows between various towns [235, 305] (cf. also p. 3).

  The gravity model results for $w_a(x'|x;t) \equiv 0$, $\nu_{ab}^1(t) \equiv 1$, $\nu_{ab}^2(t) \equiv 0 \equiv \nu_{ab}^3(t)$, and $A = 1$. Then, for the *flow of exchange* $w^1(x'|x;t)P_1(x,t)$ from x to x', we get

$$w^1(x'|x;t)P_1(x,t) = \nu_{11}^1(t)e^{U^1(x',t)-U^1(x,t)} \frac{P_1(x',t)P_1(x,t)}{D^1(x',x;t)} . \tag{9.68}$$

  This is, on the one hand, proportional to the probability $P_1(x,t)$ which is a measure for the number of individuals who are at place x and might be interested in an exchange with another place. On the other hand, it is proportional to the probability $P_1(x',t)$ which is a measure for the number of potential exchange partners at place x'. Apart from this, the exchange rate is reciprocally proportional to the *transaction costs* $D^1(x,x';t)$, i.e. to the *costs* of exchange. $D^1(x,x';t)$ is often written as a function of the *metrical* distance $\|x - x'\|$ between x and x': $D^1(x,x';t) \equiv D^1(\|x - x'\|)$.

  In the original version of the gravity model the term $\exp[U^1(x',t) - U^1(x,t)]$ does not appear which corresponds to the case $U^1(x,t) \equiv const.$ If this case does not apply, it will be reasonable to denote (9.68) as the *generalized gravity model.* For the latter we obtain the *dynamical* equation

$$\frac{d}{dt}P(x,t) = \nu(t)\sum_{x'} \left[e^{U(x,t)-U(x',t)} - e^{U(x',t)-U(x,t)} \right] \frac{P(x,t)P(x',t)}{D(x,x';t)} \tag{9.69}$$

  where, for reasons of clarity, all subscripts and superscripts were omitted.

- The *game dynamical equations* [137, 139, 271, 303] model the *competition* or *cooperation* of individuals. They belong to the most powerful mathematical models in the social and economic sciences. The game dynamical equations also play a key role in the *theory of evolution* where they are known as the *mutation-selection equations* or EIGEN-FISHER *equations* [60, 78]. Their derivation and detailed discussion is the topic of Chap. 12 (cf. also pp. 3ff.).

9.4 Properties of the Utility Approach

9.4.1 Stationary Distribution

Neglecting pair interactions ($w_{ab} \equiv 0$) for the time being and inserting (9.1) into (9.43) we have, for time-independent transition rates, the master equation

$$\frac{d}{dt}P_a(x,t) = \sum_{x'} \left[\nu_a \frac{e^{U_a(x)-U_a(x')}}{D_a(x,x')} P_a(x',t) - \nu_a \frac{e^{U_a(x')-U_a(x)}}{D_a(x',x)} P_a(x,t) \right] . \quad (9.70)$$

Due to $D_a(x,x') = D_a(x',x)$ its *stationary solution* is obviously

$$P_a^0(x) = \frac{e^{2U_a(x)}}{\sum_{x'} e^{2U_a(x')}} \quad (9.71)$$

and fulfills condition (3.118) of *detailed balance.* As a solution of the corresponding *configurational master equation* (5.50) we find, according to (5.3),

$$P(n,t) = \prod_{a=1}^{A} P^a(n^a,t) \quad (9.72a)$$

with

$$P^a(n^a,t) \equiv P^a(n^a_{x_1},\dots,n^a_{x_S};t) := \frac{N_a!}{\left(\sum_{j=1}^{S} e^{2U_a(x_j)}\right)^{N_a}} \prod_{i=1}^{S} \frac{\left[e^{2U_a(x_i)}\right]^{n^a_{x_i}}}{n^a_{x_i}!} . \quad (9.72b)$$

In the first moment it seems that a similar solution as for (9.70) can also be found for imitative processes: The corresponding BOLTZMANN-like equations

$$\begin{aligned}\frac{d}{dt}P_a(x,t) = \sum_{x'}\sum_{b} \Bigg[& \nu^1_{ab} \frac{e^{U^a(x)-U^a(x')}}{D^a(x,x')} P_b(x,t) P_a(x',t) \\ & - \nu^1_{ab} \frac{e^{U^a(x')-U^a(x)}}{D^a(x',x)} P_b(x',t) P_a(x,t) \Bigg]\end{aligned} \quad (9.73)$$

let us expect the stationary solution $P_a^0(x)$ with

$$\mathrm{e}^{U^a(x)-U^a(x')} P_a^0(x') \sum_b \nu_{ab}^1 P_b^0(x) = \mathrm{e}^{U^a(x')-U^a(x)} P_a^0(x) \sum_b \nu_{ab}^1 P_b^0(x') \quad \forall a, x . \tag{9.74}$$

Indeed, the approach

$$P_a^0(x) := \mathcal{N}_a \, \mathrm{e}^{2U^a(x)} \sum_b \nu_{ab}^1 P_b^0(x) \tag{9.75}$$

with the normalization constant $\mathcal{N}_a$ satisfies condition (9.74). However, at the same time it is a *system of linear equations* which can be solved only on certain conditions (cf. [20], pp. 120ff.) which are usually not fulfilled.

9.4.2 Contributions to the Utility Function

The utility function can be often split up into several contributions of different meaning (cf. also Sect. 13.6):

$$U_a(x,t) = \sum_l c_a^l U_a^l(x,t) . \tag{9.76}$$

Here the functions $U_a^l(x,t)$ are denoted as *explanatory variables.* If $U_a^l(x,t)$ reflect the contributions of objective *advantages* $V_a^l(x,t) \geq 0$, due to FECHNER's law they will enter in the utility function *logarithmically* (cf. Sect. 9.2.3):

$$U_a^l(x,t) = h_a^l \ln V_a^l(x,t) + \ln C_a^l = \ln \left[C_a^l \left(V_a^l(x,t) \right)^{h_a^l} \right] . \tag{9.77}$$

Then (9.76) can be represented in the COBB-DOUGLAS *form* [55, 199]

$$\mathrm{e}^{U_a(x,t)} = C_a \prod_l \left(V_a^l(x,t) \right)^{h_a^l} \tag{9.78}$$

in which the quantities $V_a^l(x,t)$ enter via *power laws.*

Chapter 10
Opinion Formation Models

10.1 Introduction

Since the expression of an opinion represents a behaviour, the change of (expressed) opinions is a suitable example for behavioural changes, particularly because, by means of *opinion polls,* a direct comparison with empirical data is possible (cf. Chap. 13). The laws of opinion formation have already long been the topic of social research. For example, OSGOOD and TANNENBAUM with their *congruity principle* [213], HEIDER with his *balance theory* [119, 120], and FESTINGER with his *dissonance theory* [74] (cf. Fig. 8.1) developed formal models which deal with the stability of attitudes and with the probability of certain attitude changes. An overview of the state of related research is given in [77, 220].

In contrast to the above mentioned models, in this chapter we will treat models which describe the temporal change of the proportion $P(i, t)$ of individuals who show a certain behaviour i. Such models are of big interest for a *quantitative understanding* of opinion formation processes and for the *prognosis* of *trends* in *public opinion* (e.g. of tendencies concerning the attitudes of voters).

COLEMAN's *diffusion model* was one of the first mathematical models of this kind. It distinguished two types of opinion changes—opinion changes by *direct* interactions and opinion changes which are induced by *external* influences. Stochastic models for the *diffusion of rumours and news* were presented by BARTHOLOMEW [18]. In the literature one can find still many more *diffusion models* [50, 98, 110, 157, 178, 247].

The most advanced model stems from WEIDLICH and HAAG [285, 286, 289, 292, 294]. It was taken up by several authors in one or another modification [65, 257, 274]. WEIDLICH and HAAG's model bases on a stochastic formulation with the *master equation* and assumes that, independently of each other, individuals exchange their opinions for other ones. This happens with a certain transition rate (transition probability per time unit) which depends on the frequencies of the different expressed opinions. In other words, the approach of WEIDLICH and HAAG assumes *indirect interactions* between individuals. These can best be imagined to be induced by the *mass media* (i.e. TV, radio, newspapers, or magazines) (cf. Fig. 10.1).

D. Helbing, *Quantitative Sociodynamics*, 2nd ed.,

DOI 10.1007/978-3-642-11546-2_10,

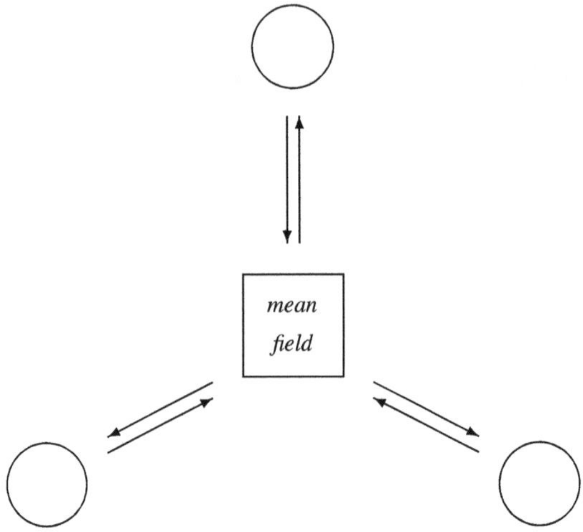

Fig. 10.1 Schematic representation of the $2N$ indirect interactions between $N = 3$ individuals. Indirect interactions are induced by a *mean field* which describes the effects of *public opinion* and the *media*

However, opinion changes are also possible by means of *direct interactions* which play a particularly important role in conversations or discussions (cf. Fig. 10.2). In this connection we will restrict ourselves to direct *pair interactions* since they normally yield the most significant contribution.

Their mathematical description bases on the BOLTZMANN-*like equations* (cf. Sect. 9.3) which also include the effects of *spontaneous* behavioural changes.

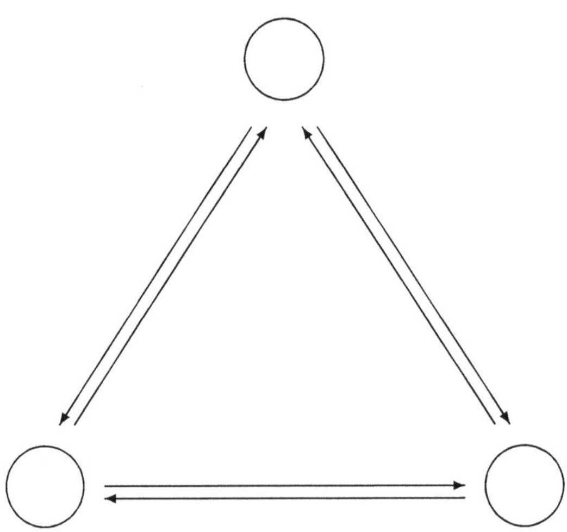

Fig. 10.2 Schematic representation of the $N!$ direct interactions between $N = 3$ individuals. Direct interactions dominate in conversations or discussions

According to Sect. 9.3, three *kinds* of direct pair interactions can be distinguished—*imitative processes (persuasion processes), avoidance processes,* and *compromising processes.* All three kinds of interactions can, in the course of time, lead to stationary proportions of opinions. However, there can also occur more complex patterns of opinion changes as, for example, *oscillatory* or *chaotic* opinion changes.

Some representative cases will be illustrated by means of computer simulations. By dividing the total population into different *subpopulations* several *behavioural types a* are taken into account. The interaction frequency between the individuals of different subpopulations is given by their *likings.*

Of course, the models for direct and indirect interactions can be integrated in a general model (cf. Fig. 10.3). Moreover, a treatment of the *spatial diffusion* of opinions is also possible.

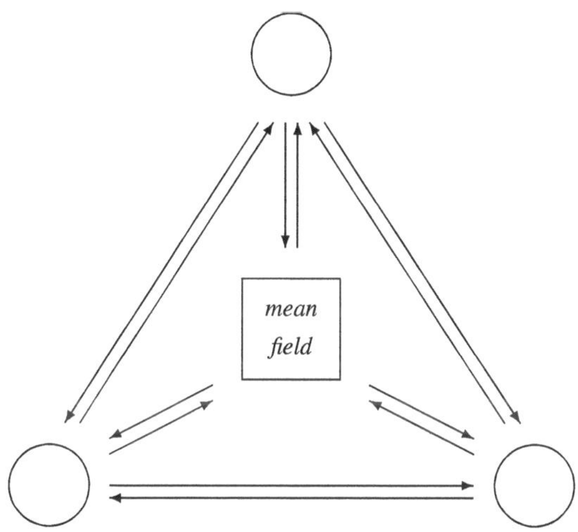

Fig. 10.3 Direct and indirect interactions of individuals can be described together by one general mathematical model

10.2 Indirect Interactions

In the following we will briefly introduce the WEIDLICH-HAAG *model* for *spontaneous* behavioural changes due to *indirect interactions* (cf. [289, 292] and p. 10). In its most general form it was developed for *migration processes (removal processes)* [288, 294] but it can be also applied to the *development of towns* [194, 293, 295] or to *opinion formation processes* [285, 286, 290, 292].

The behavioural model of WEIDLICH and HAAG results for $\widehat{w}_{ab} \equiv 0$ and bases on the master equation

$$\frac{d}{dt}P(n,t) = \sum_{n'} \Big[w(n|n';t)P(n',t) - w(n'|n;t)P(n,t) \Big] \qquad (10.1)$$

with the configurational transition rates

$$w(n'|n;t) := \begin{cases} w_a(j|i;n;t)n_i^a & \text{if } n' = n_{ji}^a \\ 0 & \text{otherwise} \end{cases} \tag{10.2a}$$

(cf. (5.50)). Here we have utilized $x \equiv i$ since, by the states x, we only distinguish opinions $i \in \{1, \ldots, S\}$.

One speaks of *indirect* interactions due to the dependence of the *individual* transition rates

$$w_a(j|i;n;t) := \nu_a(j,i;t)\mathrm{e}^{U_a(j,n;t)-U_a(i,n;t)} \tag{10.2b}$$

and the utilities $U_a(i,n;t)$ on the socioconfiguration n. The corresponding approximate mean value equations (5.16) read

$$\frac{d\langle n_i^a\rangle}{dt} = \sum_j \nu_a(j,i;t)\left[\mathrm{e}^{U_a(i,\langle n\rangle)-U_a(j,\langle n\rangle)}\langle n_j^a\rangle - \mathrm{e}^{U_a(j,\langle n\rangle)-U_a(i,\langle n\rangle)}\langle n_i^a\rangle\right]. \tag{10.3}$$

Since these are *non-linear* they are expected to display a complex temporal evolution (cf. Sects. 7.4 and 7.5).

For (10.2b) the WEIDLICH-HAAG model usually assumes the relations

$$\nu_a(i,j;t) \equiv \nu_a(j,i;t) \tag{10.4a}$$

and

$$U_a(i,n;t) := \delta_a(i) + \sum_b \widehat{\kappa}_{ab}\frac{n_i^b}{N_b} - \sum_{b,c}\widehat{\sigma}_{abc}\frac{n_i^b}{N_b}\frac{n_i^c}{N_c}. \tag{10.4b}$$

$\nu_a(j,i;t)$ denotes the *flexibility* or *mobility* of subpopulation a with respect to a transition from i to j. $\delta_a(i)$ is the *inclination*[1] of individuals of subpopulation a towards opinion i. $\widehat{\kappa}_{ab}$ are *coupling parameters* which describe the influence of the behavioural distribution within subpopulation b on the behavioural changes of subpopulation a. Sometimes $\widehat{\kappa}_{ab}$ can be also interpreted as *agglomeration parameters* (cf. [293]). For $\widehat{\kappa}_{ab} > 0$ the quantity $\widehat{\kappa}_{ab}$ reflects the *social pressure* which is exerted by the behavioural majority. Finally, the coefficients $\widehat{\sigma}_{abc}$ have the meaning of *saturation parameters. In the following we will assume the common case*

$$\nu_a(j,i;t) \equiv \nu_a(t) \equiv \nu \tag{10.5}$$

[1] In the work of WEIDLICH and HAAG this quantity is denoted as *preference*. However, in order to distinguish $\delta_a(x,t)$ from $U_\alpha(x|y;t)$ the author decided to introduce another term.

and

$$\widehat{\sigma}_{abc} \equiv 0 \tag{10.6}$$

if nothing else is mentioned.

10.2.1 A Period Doubling Route to Chaos

An interesting case is found for $A = 2$ subpopulations and $S = 3$ different opinions $i \in \{1, 2, 3\}$. Namely, for

$$\delta_a(i) \equiv 0\,, \qquad \left(\widehat{\kappa}_{ab}\right) := \begin{pmatrix} 1 & -1.5 \\ \kappa & 1 \end{pmatrix}, \tag{10.7}$$

along with a variation of the *control parameter* $\kappa \in [0, 1.5]$, we traverse a *period doubling scenario* which leads to *chaos* (cf. [237]). A more detailed description of this scenario is given in Sect. 7.5.2.

10.2.2 A Ruelle-Takens-Newhouse Route to Chaos

Now let us again assume $S = 3$ different possible opinions i but $A = 3$ subpopulations. Then, for

$$\delta_a(i) \equiv 0\,, \qquad \left(\widehat{\kappa}_{ab}\right) := \begin{pmatrix} 1.7 & 1.5 & -1.5 \\ -1.5 & 1.7 & 1.5 \\ -\kappa & -1.5 & 1.7 \end{pmatrix}, \tag{10.8}$$

we pass through the RUELLE-TAKENS-NEWHOUSE scenario along with a variation of the control parameter $\kappa \in [-1.5, 1.5]$ (cf. Sect. 7.5.1). For increasing values of κ as attractor we subsequently find a *limit cycle,* a *torus,* and a *strange attractor* (cf. [103]).

10.3 Direct Pair Interactions

In conversations or discussions opinion changes occur due to *direct interactions* (cf. [77], pp. 451ff., and [220]). The following considerations are restricted to *pair interactions (dyadic interactions)* since they normally play the most significant role. Even in groups the most frequent interactions are alternating pair interactions—not always but in many cases. In situations where simultaneous interactions between more than two persons are essential, a generalization is necessary. This is possible with the methods presented in Sect. 5.4.

Pair interactions are described by the configurational master equation (10.1) with the special transition rates

$$w(n'|n;t) := \begin{cases} \widehat{w}_{ab}(i',j'|i,j;t) n_i^a n_j^b & \text{if } n' = n^{ab\,ab}_{i'j'i\,j} \\ 0 & \text{otherwise} \end{cases} \quad (10.9)$$

(cf. Sect. 5.2.2). The approximate mean value equations of this master equation are, according to (9.54), the Boltzmann *equations*

$$\begin{aligned} \frac{d}{dt} P_a(i,t) = & \sum_b \sum_{i'} \sum_j \sum_{j'} w_{ab}(i,j|i',j';t) P_a(i',t) P_b(j',t) \\ & + \sum_b \sum_{i'} \sum_j \sum_{j'} w_{ab}(i',j'|i,j;t) P_a(i,t) P_b(j,t) \end{aligned} \quad (10.10)$$

($w_{ab} \equiv N_b \widehat{w}_{ab}$). Because of

$$P_a(i,t) \geq 0 \quad \text{and} \quad \sum_i P_a(i,t) = 1 \,, \quad (10.11)$$

the quantity $P_a(i,t) = \langle n_i^a \rangle_t / N_a$ can be interpreted as the *proportion (fraction, share)* of individuals of subpopulation a with opinion i or as the *probability* with which a randomly selected individual of subpopulation a holds opinion i. Among the *total population* the proportion $P(i,t)$ of individuals with opinion i is given by

$$P(i,t) := \frac{\langle n_i \rangle}{N} = \sum_a \frac{\langle n_i^a \rangle}{N} = \sum_a \frac{N_a}{N} \frac{\langle n_i^a \rangle}{N_a} = \sum_a \frac{N_a}{N} P_a(i,t) \quad (10.12)$$

where N denotes the size of the total population and N_a the number of individuals of behavioural type a.

10.3.1 Kinds of Pair Interactions

As derived in Sect. 9.3, three kinds of direct pair interactions can be distinguished:

1. *Imitative processes (persuasion processes)* which describe the tendency to take over the opinion of another individual.
2. *Avoidance processes* which reflect an aversive behaviour where an individual changes his/her opinion when meeting another individual with the same opinion (defiance behaviour, snob effect).
3. *Compromising processes* which describe the readiness of an individual to accept a new opinion if the old opinion cannot be maintained when confronted with the opinion of another individual whose opinion does not persuade either.

These kinds of pair interactions can be described by transition rates (9.52) together with (9.57). The corresponding BOLTZMANN equations read

$$\frac{d}{dt}P_a(i,t) = \sum_{i'}\Big[w^a(i|i';t)P_a(i',t) - w^a(i'|i;t)P_a(i,t)\Big] \tag{10.13a}$$

with the *effective transition rates*

$$w^a(i'|i;t) := \nu^a(i',i;t)R^a(i'|i;t) \qquad (i' \neq i) \tag{10.13b}$$

and

$$\nu^a(i',i;t) := \sum_b \Big[\Big(\nu^1_{ab}(t) - \nu^3_{ab}(t)\Big)P_b(i',t) + \Big(\nu^2_{ab}(t) - \nu^3_{ab}(t)\Big)P_b(i,t) + \nu^3_{ab}(t)\Big]. \tag{10.13c}$$

- $\nu^1_{ab}(t)P_b(i',t)$ describes persuasion processes by individuals of subpopulation b with opinion i'.
- $\nu^2_{ab}(t)P_b(i,t)$ reflects avoidance processes with respect to individuals of subpopulation b holding opinion i.
- $\nu^3_{ab}(t)[1 - P_b(i',t) - P_b(i,t)]$ delineates compromising processes during the remaining interactions with individuals of subpopulation b.

In the following *computer simulations* we choose for the *readiness* $R^a(i'|i;t)$ to exchange opinion i for opinion i' due to a pair interaction the approach that was motivated in Sect. 9.3:

$$R^a(i'|i;t) = \frac{\mathrm{e}^{U^a(i',t)-U^a(i,t)}}{D^a(i',i;t)} \tag{10.14}$$

with $D^a(i',i;t) \equiv 1$. In addition, we assume the case of $S = 3$ different opinions $i \in \{1,2,3\}$. The latter can be understood as a *prototype* of opinion formation situations because the opinions concerning a given affair can be classified according to the schemes 'pro', 'contra', 'indifferent', or 'positive', 'negative', 'neutral'. Furthermore, two subpopulations $a \in \{1,2\}$ are distinguished which

(a) either have the *same* preferences ($U^1(1,t) = U^2(1,t) = c$, $U^1(2,t) = U^2(2,t) = 0$, $U^1(3,t) = U^2(3,t) = -c$)
(b) or *different* preferences ($U^1(1,t) = U^2(2,t) = c$, $U^1(2,t) = U^2(1,t) = 0$, $U^1(3,t) = U^2(3,t) = -c$).

In the following figures the opinion with the largest preference ($U^a(i,t) = c$) is always represented by a solid line (—), the opinion with medium preference ($U^a(i,t) = 0$) by a dashed line (– –), and the opinion with the lowest preference ($U^a(i,t) = -c$) by a dotted line ($\cdots$).

10.3.1.1 Persuasion Processes

($\nu^1_{ab}(t) \equiv 1$, $\nu^2_{ab}(t) \equiv 0 \equiv \nu^3_{ab}(t)$, $c = 0.5$):

If both subpopulations prefer the *same* opinion, then, as expected, only the opinion with the largest preference ($i = 1$) will survive (cf. Fig. 10.4). However, if the subpopulations prefer *different* opinions (one of them $i = 1$, the other one $i = 2$), a certain proportion of individuals will become persuaded of that opinion which is preferred by the other subpopulation (cf. Fig. 10.5).

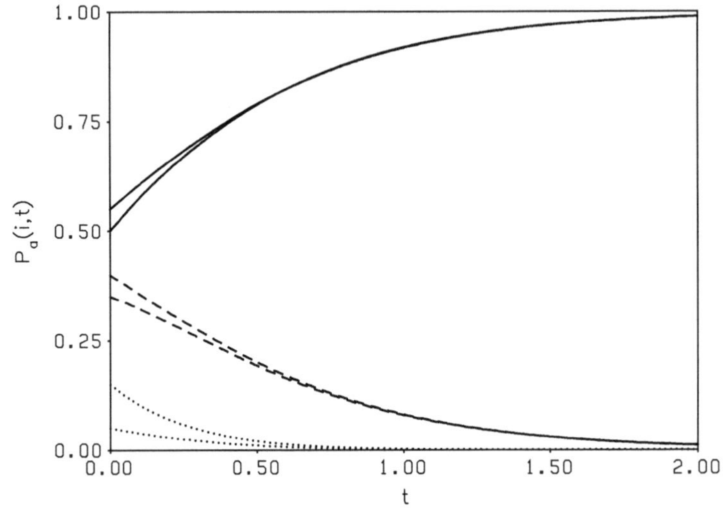

Fig. 10.4 Persuasion processes for *equal* preferences: The preferred opinion (—) wins out

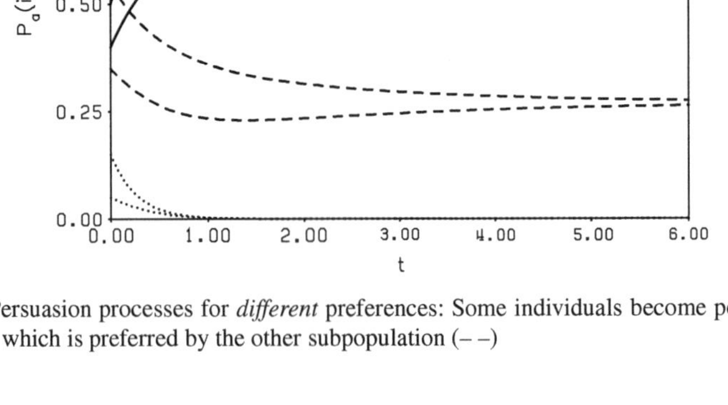

Fig. 10.5 Persuasion processes for *different* preferences: Some individuals become persuaded of the opinion which is preferred by the other subpopulation (– –)

10.3.1.2 Avoidance Processes

$(\nu^2_{ab}(t) \equiv 1,\, \nu^1_{ab}(t) \equiv 0 \equiv \nu^3_{ab}(t),\, c = 1)$:

If both subpopulations prefer the *same* opinion, avoidance processes in favour of less preferred opinions will frequently occur (cf. Fig. 10.6). However, the preferred opinion will gain a significantly *higher* proportion within a subpopulation if the other subpopulation favours a *different* opinion (cf. Fig. 10.7). Avoidance processes are only rarely necessary, then.

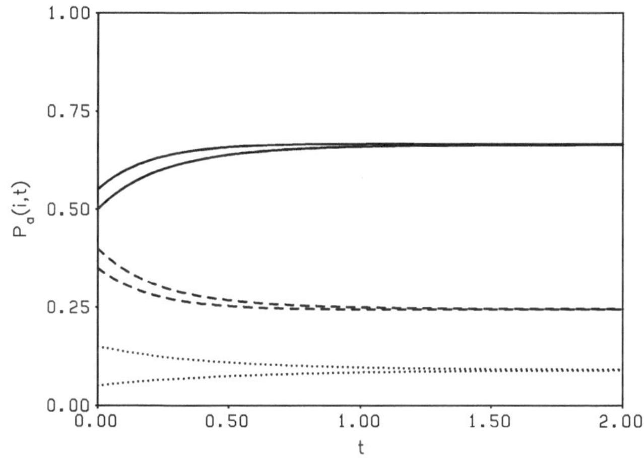

Fig. 10.6 Avoidance processes for *equal* preferences have the consequence that the favoured opinion is held only by a relatively small proportion of individuals

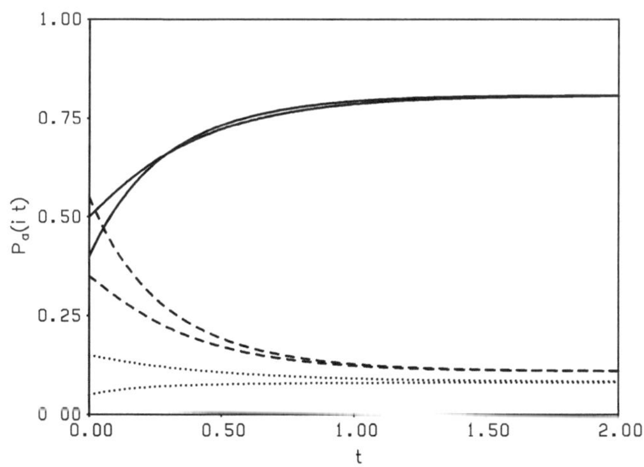

Fig. 10.7 In the case of *different* preferences avoidance processes occur rarely and their effect is relatively weak

10.3.1.3 Compromising and Persuasion Processes

$(\nu^3_{ab}(t) \equiv 1 \equiv \nu^1_{ab}(t), \nu^2_{ab}(t) \equiv 0, c = 0.5)$:

The preferred opinion will, as expected, completely win out if both subpopulations *agree* upon it (cf. Fig. 10.8). If, however, the subpopulations prefer *different* opinions, the compromising processes will become relevant: Even an opinion which is not preferred by any of the subpopulations ($i = 3$) will be chosen by a certain proportion of individuals $(\cdots)$, then (cf. Fig. 10.9).

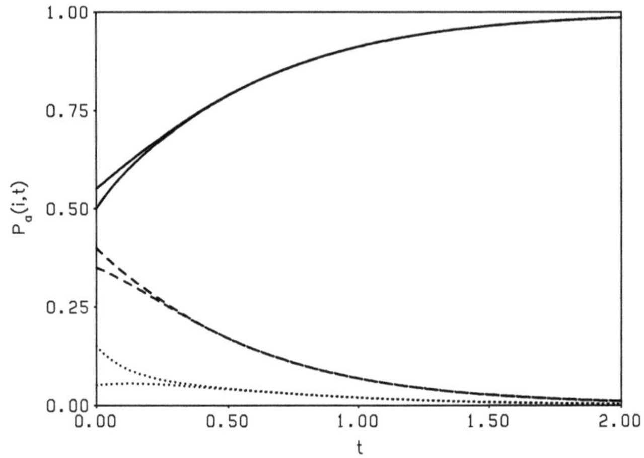

Fig. 10.8 Compromising and persuasion processes for *equal* preferences: The readiness for compromises has only a small relevance (cf. Fig. 10.4)

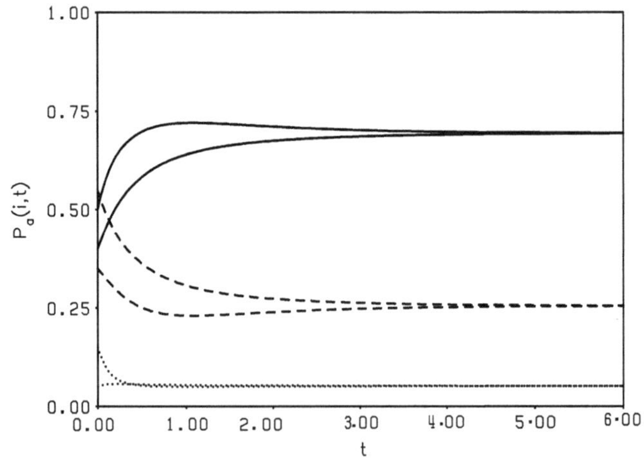

Fig. 10.9 Compromising processes and persuasion processes for *different* preferences: A certain proportion of individuals decides in favour of a compromise $(\cdots)$

10.3.2 Oscillations

Opinion changes due to persuasion processes can also occur in an *oscillatory* way. In order to prove this let us, first, consider the case in which the subpopulations do not influence each other ($\nu^1_{ab}(t) := \nu^1_{aa}\delta_{ab}$). The corresponding BOLTZMANN equations are obviously

$$\frac{d}{dt} P_a(i,t) = \nu^1_{aa} P_a(i,t) \sum_j E^a_{ij} P_a(j,t) \tag{10.15a}$$

with the *antisymmetrical matrix*

$$\underline{E}^a \equiv \left(E^a_{ij}\right) := \left(R^a(i|j) - R^a(j|i)\right) = -\left(E^a_{ji}\right). \tag{10.15b}$$

Equations (10.15) have at least the S *stationary solutions* $P^l_a(i)$ with

$$P^l_a(i) := \delta_{il} = \begin{cases} 1 \text{ if } i = l \\ 0 \text{ if } i \neq l \end{cases} \qquad l \in \{1, \dots, S\}. \tag{10.16}$$

We will now carry out a *linear stability analysis,* i.e. a *first* order TAYLOR expansion of Eqs. (10.15) around each stationary solution $P^l_a(i)$: The terms of quadratic and higher orders in

$$\delta P^l_a(i,t) := P_a(i,t) - P^l_a(i) \tag{10.17a}$$

can be neglected in comparison with the linear terms if the *deviations* $\delta P^l_a(i,t)$ from the stationary solution $P^l_a(i)$ are sufficiently small which implies

$$\delta P^l_a(i,t) \ll 1\,. \tag{10.17b}$$

This procedure leads to the equations

$$\frac{d}{dt}\delta P^l_a(i,t) = \nu^1_{aa} \cdot \begin{cases} -E^a_{li}\,\delta P^l_a(i,t) & \text{if } i \neq l \\ \sum\limits_j E^a_{lj}\,\delta P^l_a(j,t) & \text{if } i = l \end{cases} \tag{10.18}$$

where $\delta P^l_a(i,t) > 0$ for $i \neq l$ and $\delta P^l_a(l,t) = -\sum\limits_{l(\neq l)} \delta P^l_a(i,t) < 0$.

Consequently, the stationary solution $P^l_a(i)$ is stable on the condition

$$E^a_{li} > 0\,, \quad \text{i.e.} \quad R^a(l|i) > R^a(i|l) \qquad \forall\, i(\neq l)\,. \tag{10.19}$$

Otherwise it is unstable. Due to $E^a_{li} = -E^a_{il}$ maximally *one* of the solutions $P^l_a(i)$ can be stable. If such a stable solution exists, only one opinion can survive because of $P^l_a(i) = \delta_{il}$.

If *none* of the solutions $P^l_a(i)$ is stable (which, by the way, cannot happen for the approach (10.14)), a further stationary solution $P^0_a(i)$ might exist which is defined by

$$\sum_j E^a_{ij} P^0_a(j) \stackrel{!}{=} 0 \quad \forall i \tag{10.20}$$

and (10.11). With (10.17), a linear stability analysis leads to

$$\frac{d}{dt}\delta P^0_a(i,t) = \nu^1_{aa} P^0_a(i) \sum_j E^a_{ij}\, \delta P^0_a(j,t) = \sum_j M^a_{ij}\, \delta P^0_a(j,t)\,. \tag{10.21}$$

Since the matrix $\underline{E}^a$ is antisymmetrical, $\underline{E}^a$ as well as the matrix

$$\underline{M}^a \equiv \left(M^a_{ij}\right) := \left(\nu^1_{aa} P^0_a(i) E^a_{ij}\right) \tag{10.22}$$

possess only *imaginary eigenvalues* (cf. [182]). As a consequence, $\delta P^0_a(i,t)$ and $P_a(i,t)$ show an *oscillatory* behaviour. Owing to the *invariant of motion*

$$\prod_i P_a(i,t)^{P^0_a(i)} = const. \tag{10.23}$$

(cf. [182] and Sect. 10.3.4) and the normalization condition $\sum_i P_a(i,t) = 1$, the *trajectory* $(P_a(1,t), \dots, P_a(S,t))^{\text{tr}}$ moves on a $(S-2)$-dimensional *hypersurface.*

In order to illustrate this, let us consider the particular case with $S = 3$ opinions and

$$\left(R^a(i|j)\right) := \begin{pmatrix} 0 & 0 & 1 \\ 1 & 0 & 0 \\ 0 & 1 & 0 \end{pmatrix}, \qquad \text{i.e.} \qquad \left(E^a_{ij}\right) = \begin{pmatrix} 0 & -1 & 1 \\ 1 & 0 & -1 \\ -1 & 1 & 0 \end{pmatrix}. \tag{10.24}$$

Then the stationary solutions $P^l_a(i)$ ($l \in \{1, \dots, S\}$) are unstable. $P^0_a(i) \equiv 1/S$ is a further stationary solution with purely imaginary eigenvalues

$$\lambda_{a,j} := -\mathrm{i}\omega_{a,j} := \mathrm{i}\frac{2\nu^1_{aa}}{S} \sin\left(\frac{2\pi}{S}\right). \tag{10.25}$$

Equation (10.15) has the special form

$$\frac{d}{dt}P_a(i,t) = \nu^1_{aa} P_a(i,t)\Big[P_a(i-1,t) - P_a(i+1,t)\Big] = \lambda_a(i,t) P_a(i,t) \tag{10.26a}$$

with $i \equiv i \bmod S$ (cf. p. xx) where

$$\lambda_a(i,t) := \nu_{aa}^1 \Big[P_a(i-1,t) - P_a(i+1,t) \Big] \tag{10.26b}$$

can be interpreted as a *time-dependent growth rate.*

- The proportion $P_a(i,t)$ of opinion i grows as long as the proportion $P_a(i-1,t)$ of opinion $(i-1)$ is larger than the proportion $P_a(i+1,t)$ of opinion $(i+1)$, i.e. $\lambda_a(i,t) > 0$.
- The increase of the proportion $P_a(i,t)$ of opinion i induces an increase of the proportion $P_a(i+1,t)$ of opinion $(i+1)$ by $\lambda_a(i+1,t) > 0$ and a decrease of the proportion $P_a(i-1,t)$ of opinion $(i-1)$ by $\lambda_a(i-1,t) < 0$.
- As soon as the proportion $P_a(i+1,t)$ of opinion $(i+1)$ exceeds the proportion $P_a(i-1,t)$ of opinion $(i-1)$ a decrease of the proportion $P_a(i,t)$ of opinion i begins due to $\lambda_a(i,t) < 0$.
- These phases return again and again because of the *cyclical order* of the S opinions (because of $i \equiv i \bmod S$) (cf. Fig. 10.10).

The above situation can occur if the readiness $R^a(i-1|i)$ to exchange the present opinion i for the previously held opinion $(i-1)$ is smaller than the readiness $R^a(i+1|i)$ to exchange the present for a new opinion $(i+1)$. After a number of opinion changes one possibly ends up in the original opinion and the *cycle* starts from the very beginning: So to say, one goes round in circles. Such cases are known from *fashion* or *economy.*

Due to (10.23) and (10.11), for $S = 3$ the trajectory $(P_a(1,t), \dots, P_a(S,t))$ moves on a one-dimensional 'hypersurface' (in a *cyclical* way). For its shape, from (10.11) and (10.23), we obtain the relations

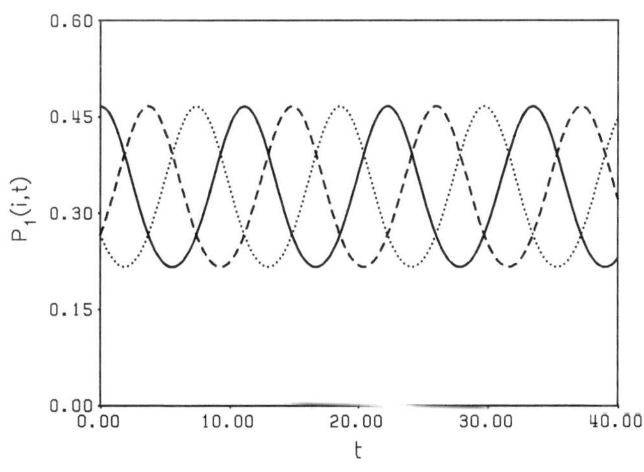

Fig. 10.10 Oscillations for persuasion processes between 3 different opinions. The different opinions win over the majority of the individuals in an alternating but determined order

$$P_a(2,t) = \frac{1-P_a(1,t)}{2} \pm \sqrt{\left[\frac{1-P_a(1,t)}{2}\right]^2 - \frac{C_a}{P_a(1,t)}} \qquad (10.27a)$$

and

$$P_a(3,t) = 1 - P_a(1,t) - P_a(2,t) \qquad (10.27b)$$

with

$$C_a := P_a(1,t_0)P_a(2,t_0)P_a(3,t_0)\,. \qquad (10.27c)$$

The corresponding cycles are depicted in Fig. 10.11 for various initial conditions.

An examination of Eq. (10.26) shows that for $S > 3$ opinions the temporal evolution becomes more and more complex with increasing number S of different opinions (cf. Figs. 10.12, 10.13 and 10.14). In particular, under certain circumstances, the *order* in which the different opinions gain the majority of individuals varies. Over short periods of time the course of $P_1(i,t)$ appears not necessarily periodic any more (cf. Fig. 10.12). However, a consideration of the *phase portrait* indicates that over long periods the temporal course is still periodic (or at least *quasi-periodic*) since the trajectory closes itself after a certain period of time (at least almost) (cf. Figs. 10.13 and 10.14).

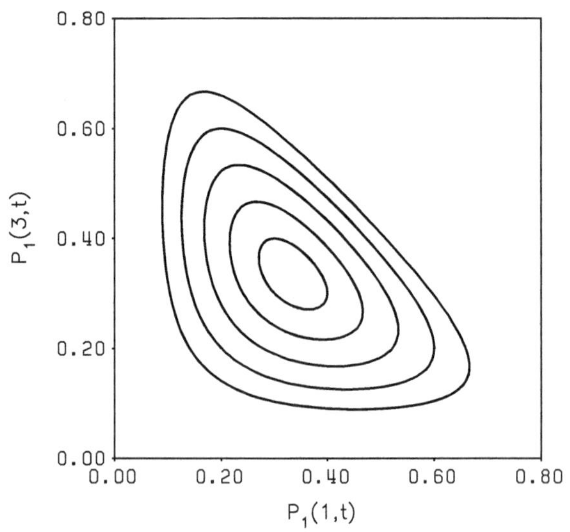

Fig. 10.11 Phase portraits of oscillations between 3 different opinions for various initial conditions

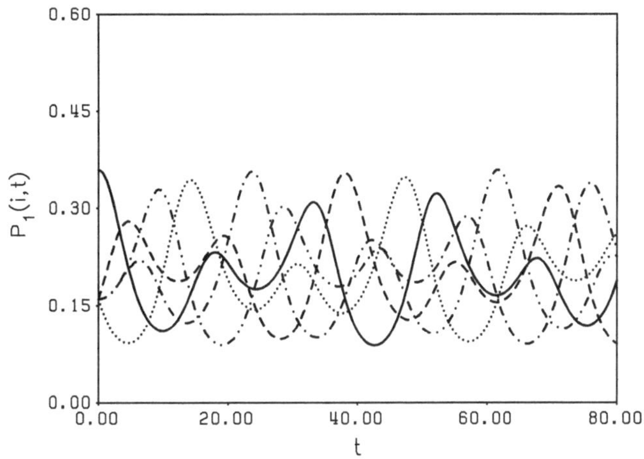

Fig. 10.12 Oscillations for persuasion processes between 5 opinions: Over a short time interval the course of $P_1(i, t)$ is not periodic. The different opinions win over the majority of individuals in a *varying* order

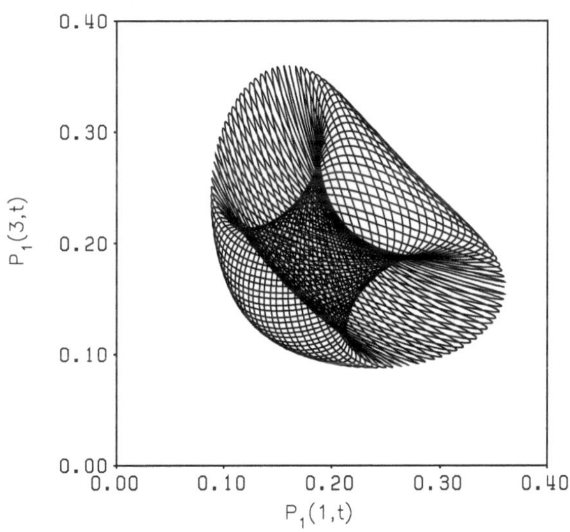

Fig. 10.13 Phase portrait of oscillations for persuasion processes between 5 opinions: The trajectory moves on a *torus*. Over a long time the trajectory closes (at least almost) so that the long-term behaviour of $P_1(i, t)$ is periodic (or at least quasi-periodic)

10.3.3 Influence of the Interaction Frequencies

In the following we will examine the influence of the *interaction frequencies* $f^k_{ab}(t)$ *of kind* k (cf. (9.57)) and of the related *interaction rates* $\nu^k_{ab}(t) := \nu_{ab}(t) f^k_{ab}(t)$ respectively. The interaction frequency $f^k_{ab}(t)$ of an individual of subpopulation a

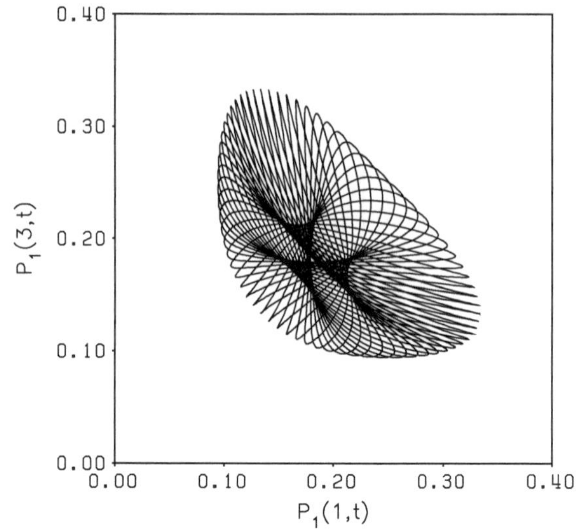

Fig. 10.14 Phase portrait of oscillations for persuasion processes between 6 opinions: The long-term behaviour is periodic since the trajectory forms a close curve

with individuals of subpopulation b certainly depends on the *liking* $l_{ab}(t)$ which individuals of subpopulation a have for individuals of subpopulation b.
For example, with $0 \le l_{ab}(t) \le 1$ and the approach

$$f^1_{ab}(t) := f^1_a(t) l_{ab}(t) , \tag{10.28a}$$

$$f^2_{ab}(t) := f^2_a(t)\Big(1 - l_{ab}(t)\Big) , \tag{10.28b}$$

$$f^3_{ab}(t) := f^3_a(t) l_{ab}(t) , \tag{10.28c}$$

we could model the fact that the frequency of imitative processes and compromising processes increases with the liking whereas the frequency of avoidance processes lessens when the liking grows. Here $f^1_a(t)$ is a measure for the *overall frequency* of imitative processes within subpopulation a, $f^2_a(t)$ denotes the overall frequency of avoidance processes, and $f^3_a(t)$ that of compromising processes.

Figures 10.15, 10.16, 10.17, 10.18, 10.19 and 10.20 illustrate the time-dependent solution of Eqs. (10.13) for two subpopulations ($a = 1$: —, $a = 2$: – –),

$$\Big(R^a(i|j;t)\Big) := \begin{pmatrix} 0 & 0 & 1 \\ 1 & 0 & 0 \\ 0 & 1 & 0 \end{pmatrix} , \tag{10.29}$$

and various interaction rates $\nu^1_{ab}(t)$ ($f^2_a(t) \equiv 0 \equiv f^3_a(t)$).

In the case

$$\underline{\nu}^1(t) \equiv \Big(\nu^1_{ab}(t)\Big) := \begin{pmatrix} 1 & 0 \\ 0 & 0.5 \end{pmatrix} \tag{10.30}$$

we find *uncoupled* oscillations in both subpopulations with different frequencies (cf. Fig. 10.15 and 10.16).

If we have

$$\underline{\nu}^1(t) \equiv \left(\nu^1_{ab}(t)\right) := \begin{pmatrix} 1 & 1 \\ 0 & 0.5 \end{pmatrix}, \tag{10.31}$$

the behaviour of subpopulation 1 (—) will depend on the behaviour of subpopulation 2 (– –): The oscillation frequency of subpopulation 1 adapts itself to the one of subpopulation 2 (cf. Fig. 10.17 and 10.18). This phenomenon is called *frequency*

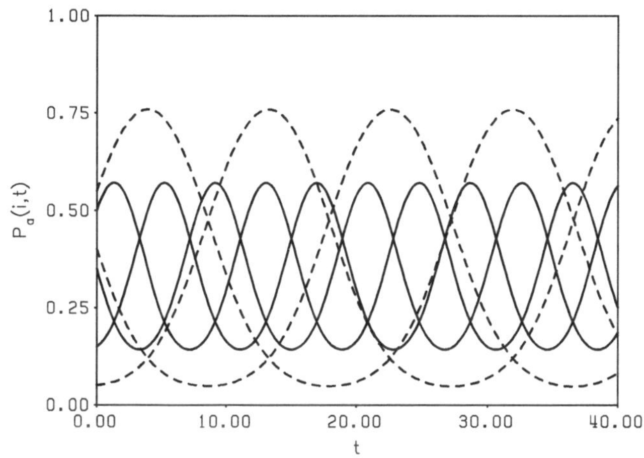

Fig. 10.15 Uncoupled oscillations of different frequencies for persuasion processes between 3 opinions in two subpopulations

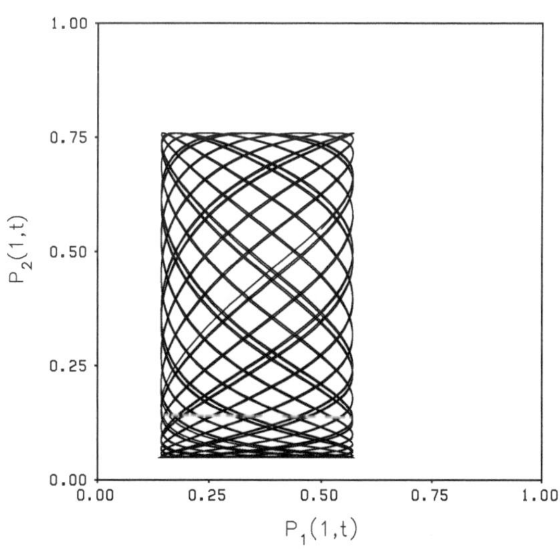

Fig. 10.16 Phase portrait of uncoupled oscillations: After some time the trajectory runs parallel to its previous segment which indicates a *quasi-periodic* temporal evolution

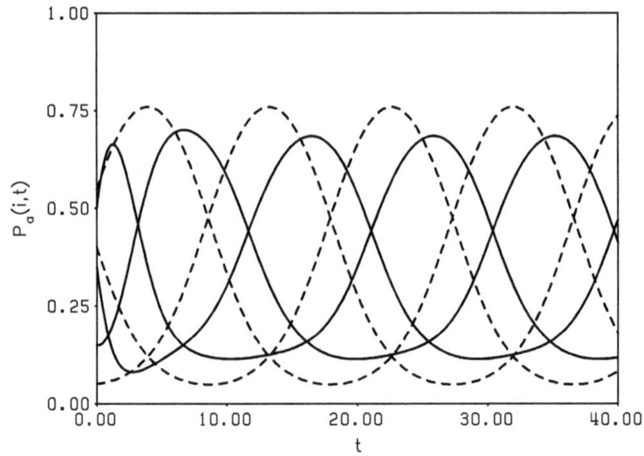

Fig. 10.17 One-sidedly coupled oscillations for persuasion processes between 3 opinions in two subpopulations are connected with an adaption of the frequency *(frequency locking)*

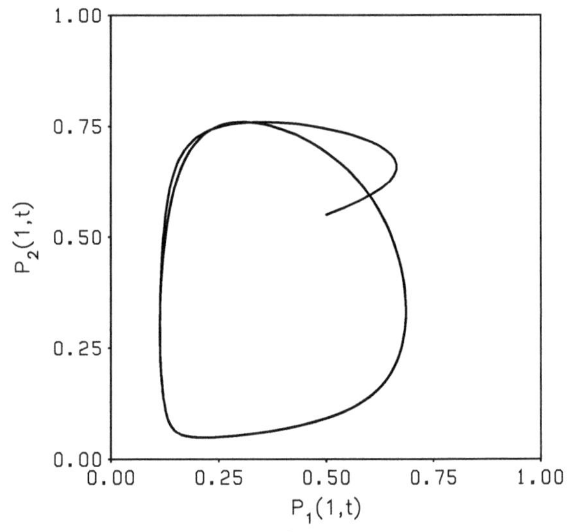

Fig. 10.18 Phase portrait of one-sidedly coupled oscillations displaying a *limit cycle* as a consequence of frequency locking

locking. Here, the individuals of subpopulation 2 can, in a certain way, be understood as *trendsetters* or *opinion leaders*.

Finally, for

$$\underline{\nu}^1(t) \equiv \left(\nu^1_{ab}(t)\right) := \begin{pmatrix} 1 & 1 \\ 0.5 & 0.5 \end{pmatrix} , \tag{10.32}$$

both subpopulations influence each other *mutually:* The oscillations damp each other (cf. Fig. 10.19 and 10.20).

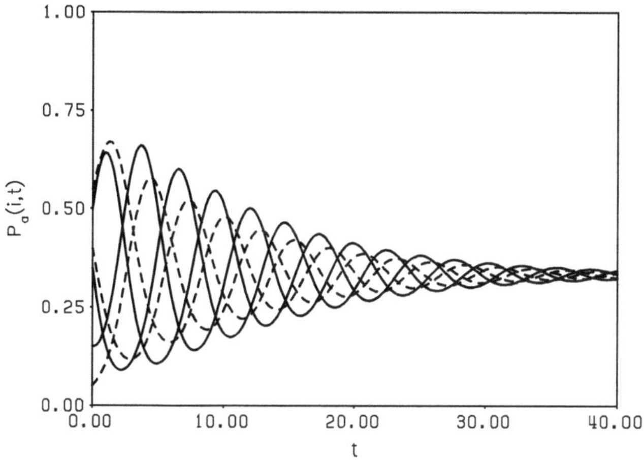

Fig. 10.19 Mutually coupled oscillations for persuasion processes between 3 opinions in two subpopulations result in a damping of the oscillations

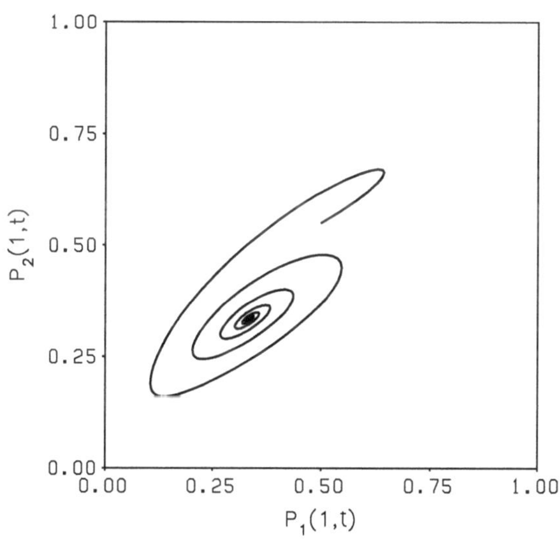

Fig. 10.20 Phase portrait of mutually coupled oscillations showing a spiral motion towards a *fixed point*

10.3.4 Period Doubling Scenarios and Chaos

According to naive everyday experience opinion formation processes can also go off in a *chaotic* way (if one thinks, for example, of *turbulent* arguments regarding controversial affairs). In this section it will turn out that the BOLTZMANN equations (10.10) for opinion formation can, indeed, display a chaotic temporal evolution. Although the concrete interaction processes and parameter values that are assumed in the following are not empirically founded, for illustrative reasons we will choose an interpretation in terms of opinion formation processes. A number of other BOLTZMANN equations showing chaos can be found in [1, 87, 258].

Let us consider the case of one single subpopulation ($A = 1$) and direct pair interactions between $S = 9$ opinions given by the following interaction schemes (cf. (9.46)):

$$\left.\begin{aligned} 1,9 &\xleftarrow{k_1} 4,9 \\ 4,9 &\xleftarrow{k_1'} 1,9 \end{aligned}\right\} (1) \tag{10.33a}$$

$$\left. 2,9 \xleftarrow{k_2} 1,9 \right\} (2) \tag{10.33b}$$

$$\left. 1,6 \xleftarrow{k_3} 4,6 \right\} (3) \tag{10.33c}$$

$$\left.\begin{aligned} 3,3 &\xleftarrow{k_4} 1,1 \\ 1,9 &\xleftarrow{k_4'} 3,9 \\ 1,3 &\xleftarrow{k_4''} 2,3 \end{aligned}\right\} (4) \tag{10.33d}$$

$$\left.\begin{aligned} 6,6 &\xleftarrow{k_5} 8,6 \\ 7,7 &\xleftarrow{k_5'} 6,7 \\ 8,8 &\xleftarrow{k_5''} 7,8 \end{aligned}\right\} (5) \tag{10.33e}$$

$$\left.\begin{aligned} 4,5 &\xleftarrow{k_6} 4,4 \\ 4,4 &\xleftarrow{k_6'} 5,4 \end{aligned}\right\} (6) \tag{10.33f}$$

Here the rates k_l represent the interaction rates $w(i', j'|i, j)$ of the interactions

$$i', j' \xleftarrow{k_l} i, j \,. \tag{10.34}$$

A social interpretation of the above interactions can be given on the basis of classification scheme (9.46):

The interactions summarized by (1) obviously describe a compromising behaviour where opinion 9 *catalyses* a change between opinions 1 and 4 without itself being subject to a change. Opinion 9 also catalyses a transition from opinion 1 to opinion 2 which is delineated by interaction (2). Besides, since opinion 9 itself is not subject to any opinion changes, it symbolizes a *mature* and *mediating* opinion. Opinion 6 has also a mediating effect, namely on the transition from opinion 4 to opinion 1 which is given by interactions (3).

Interactions (4) reflect an *interaction chain* where in the first interaction process a mutual avoidance behaviour in favour of opinion 3 occurs if opinion 1 meets opinion 1. Opinion 3, however, is replaced by opinion 1 as soon as it is confronted with the mediating opinion 9. During the time until this happens, however, opinion 3 catalyses transitions from opinion 2 to opinion 1. Opinion 2 as well as opinion 1 are characterized by a particular *readiness for compromises*.

By (5) a *cyclic* interaction chain of *persuasion processes* is described where opinion 6 mediates the change of opinion 8 to opinion 6, opinion 7 the change of opinion 6 to opinion 7, and opinion 8 the change of opinion 7 to opinion 8 (cf. Sect. 10.3.2).

Finally, by (6) an avoidance behaviour is delineated where opinion 4 is exchanged for opinion 5 if confronted with opinion 4. With a certain rate, however, also the reverse process occurs where opinion 5 is persuaded of opinion 4. Thus opinion 4 is characterized by some degree of *undecidedness*.

The temporal evolution of the proportions $P(i,t)$ of the 9 different opinions is reflected by the following BOLTZMANN equations:

$$\begin{aligned}\frac{d}{dt}P(1,t) &= k_1 P(4,t)P(9,t) - k_1' P(1,t)P(9,t) - k_2 P(1,t)P(9,t)\\ &\quad + k_3 P(4,t)P(6,t) - 2k_4[P(1,t)]^2 + k_4' P(3,t)P(9,t)\\ &\quad + k_4'' P(2,t)P(3,t)\,, \end{aligned} \tag{10.35a}$$

$$\frac{d}{dt}P(2,t) = k_2 P(1,t)P(9,t) - k_4'' P(2,t)P(3,t)\,, \tag{10.35b}$$

$$\frac{d}{dt}P(3,t) = 2k_4[P(1,t)]^2 - k_4' P(3,t)P(9,t)\,, \tag{10.35c}$$

$$\begin{aligned}\frac{d}{dt}P(4,t) &= -k_1 P(4,t)P(9,t) + k_1' P(1,t)P(9,t) - k_3 P(4,t)P(6,t)\\ &\quad - k_6[P(4,t)]^2 + k_6' P(4,t)P(5,t)\,, \end{aligned} \tag{10.35d}$$

$$\frac{d}{dt}P(5,t) = k_6[P(4,t)]^2 - k_6' P(4,t)P(5,t)\,, \tag{10.35e}$$

$$\frac{d}{dt}P(6,t) = k_5 P(6,t)P(8,t) - k_5' P(6,t)P(7,t)\,, \tag{10.35f}$$

$$\frac{d}{dt}P(7,t) = k_5' P(6,t)P(7,t) - k_5'' P(7,t)P(8,t)\,, \tag{10.35g}$$

$$\frac{d}{dt}P(8,t) = k_5'' P(7,t)P(8,t) - k_5 P(6,t)P(8,t)\,, \tag{10.35h}$$

$$\frac{d}{dt}P(9,t) = 0\,. \tag{10.35i}$$

We immediately find $P(9,t) = P(9,0)$. By means of the transformation

$$P(i,t) =: y_i(\tau)P(9,0) \times \begin{cases} \widehat{\alpha} \text{ if } i \in \{1,2,4,6,7,8\}\,, \\ \widehat{\beta} \text{ if } i = 3\,, \\ \widehat{\gamma} \text{ if } i = 5 \end{cases} \tag{10.36}$$

with

$$\tau(t) := k_1' P(9,0) t \tag{10.37}$$

and

$$\widehat{\alpha} := \sqrt{\frac{k_1' k_4'}{2k_4 k_4''}}\,, \tag{10.38a}$$

$$\widehat{\beta} := \frac{k_1'}{k_4''}\,, \tag{10.38b}$$

$$\widehat{\gamma} := \widehat{\alpha}\frac{k_6}{k_6'}\,, \tag{10.38c}$$

the BOLTZMANN equations (10.35) go over into the following *scaled equations:*

$$\begin{aligned}\frac{d}{d\tau} y_1(\tau) = {} & \widehat{A} y_4(\tau) - (\widehat{B}+1) y_1(\tau) + \kappa y_4(\tau) y_6(\tau) \\ & - \varepsilon\Big\{[y_1(\tau)]^2 - y_3(\tau)\Big\} + y_2(\tau) y_3(\tau)\,,\end{aligned} \tag{10.39a}$$

$$\frac{d}{d\tau} y_2(\tau) = \widehat{B} y_1(\tau) - y_2(\tau) y_3(\tau)\,, \tag{10.39b}$$

$$\varepsilon' \frac{d}{d\tau} y_3(\tau) = \varepsilon\Big\{[y_1(\tau)]^2 - y_3(\tau)\Big\}\,, \tag{10.39c}$$

$$\begin{aligned}\frac{d}{d\tau} y_4(\tau) = {} & -\widehat{A} y_4(\tau) + y_1(\tau) - \kappa y_4(\tau) y_6(\tau) \\ & - \widehat{D} y_4(\tau)\Big[y_4(\tau) - y_5(\tau)\Big]\,,\end{aligned} \tag{10.39d}$$

$$\widehat{E}\frac{d}{d\tau} y_5(\tau) = \widehat{D} y_4(\tau)\Big[y_4(\tau) - y_5(\tau)\Big]\,, \tag{10.39e}$$

$$\frac{d}{d\tau} y_6(\tau) = y_6(\tau)\Big[\widehat{C} y_8(\tau) - \widehat{C}' y_7(\tau)\Big]\,, \tag{10.39f}$$

$$\frac{d}{d\tau} y_7(\tau) = y_7(\tau)\Big[\widehat{C}' y_6(\tau) - \widehat{C}'' y_8(\tau)\Big]\,, \tag{10.39g}$$

$$\frac{d}{d\tau} y_8(\tau) = y_8(\tau)\Big[\widehat{C}'' y_7(\tau) - \widehat{C} y_6(\tau)\Big]\,. \tag{10.39h}$$

Here we have

$$\begin{aligned} & \widehat{A} := k_1/k_1'\,, \quad \widehat{B} := k_2/k_1'\,, \\ & \widehat{C} := \widehat{\alpha} k_5/k_1'\,, \quad \widehat{C}' := \widehat{\alpha} k_5'/k_1'\,, \quad \widehat{C}'' := \widehat{\alpha} k_5''/k_1'\,, \\ & \widehat{D} := \widehat{\alpha} k_6/k_1'\,, \quad \widehat{E} := k_6/k_6'\,, \quad \kappa := \widehat{\alpha} k_3/k_1'\,, \\ & \varepsilon := 2\widehat{\alpha} k_4/k_1'\,, \quad \varepsilon' := k_1'/(k_4''\widehat{\alpha})\,. \end{aligned} \tag{10.40}$$

As a particular example let us consider the case

$$\begin{aligned}
&\widehat{A} := 0\,, \quad \widehat{B} := 1.2\,,\\
&\widehat{C} := 0.46\,, \quad \widehat{C}' := 0.46\,, \quad \widehat{C}'' := 0.46\,,\\
&\widehat{D} := 100\,, \quad \widehat{E} := 10000\,, \quad \kappa = \mathit{variable}\,,\\
&\varepsilon := 0.01\,, \quad \varepsilon' := 0.0001\,.
\end{aligned} \tag{10.41}$$

For

$$k_1' := 1 \qquad \text{and} \qquad \widehat{\alpha} := 1 \tag{10.42}$$

this implies

$$k_1 = 0\,, \tag{10.43a}$$

$$k_2,\, k_5,\, k_5',\, k_5'' \sim 1\,, \tag{10.43b}$$

$$k_4,\, k_6' \sim 0.01\,, \tag{10.43c}$$

$$k_4',\, k_6 \sim 100\,, \tag{10.43d}$$

$$k_4'' \sim 10000\,. \tag{10.43e}$$

That is, the interactions

$$3, 3 \xleftarrow{k_4} 1, 1\,, \tag{10.44a}$$

$$4, 4 \xleftarrow{k_6'} 5, 4 \tag{10.44b}$$

are slow, the interactions

$$1, 9 \xleftarrow{k_4'} 3, 9\,, \tag{10.45a}$$

$$4, 5 \xleftarrow{k_6} 4, 4 \tag{10.45b}$$

are fast, and the interaction

$$1, 3 \xleftarrow{k_4''} 2, 3 \tag{10.46}$$

is rapid compared to the remaining interactions. Therefore, the interaction chain (4) can be formally contracted to

$$1, 1, 1 \longleftarrow 2, 1, 1 \tag{10.47}$$

which is equivalent to an interaction between three individuals. With

$$y_i(\tau) \sim 1\,, \tag{10.48a}$$

$$\widehat{\beta} \sim 0.0001\,, \tag{10.48b}$$

$$\widehat{\gamma} \sim 10000 \tag{10.48c}$$

this implies that proportion $P(3, t)$ is very small and proportion $P(5, t)$ very huge compared to the remaining proportions $P(i, t)$ $(i \notin \{3, 5\})$. As a consequence, opinion 3 represents only a *temporary stage* which is left immediately. Opinion 5 represents the opinion of the large *majority.*

The computer simulations of Eqs. (10.39) with parameters (10.41) and initial conditions

$$\begin{aligned}
y_1(0) &:= 0.6\,, & y_2(0) &:= 1.8\,, \\
y_3(0) &:= 0.36\,, & y_4(0) &:= 1\,, \\
y_5(0) &:= 1\,, & y_6(0) &:= 1.12\,, \\
y_7(0) &:= 1 + 0.12\sin(2\pi/3)\,, & y_8(0) &:= 1 + 0.12\sin(4\pi/3)
\end{aligned} \tag{10.49}$$

show several *period doubling scenarios* to chaos along with a variation of the control parameter $\kappa \in [0, 0.55]$. In Fig. 10.21 the temporal evolution of $y_1(\tau) \equiv y_1(\kappa, \tau)$ is depicted for various values of the control parameter κ. In contrast, Figs. 10.22, 10.23, 10.24, 10.25, 10.26 and 10.27 show, apart from the *power spectra* $f(\omega) \equiv f(\kappa, \omega)$ of the functions $y_1(\tau) \equiv y_1(\kappa, \tau)$, the y_1-y_2 projections of the trajectories (the phase portraits) after they converged to their attractors. (For a detailed discussion of the interpretation of these figures cf. Sect. 7.5.2).

Chaotic behaviour is characterized by the *irregularity* of the phase portraits and power spectra (cf. e.g. Fig. 10.25). If there are distinct maxima in chaotic power spectra this indicates *intermittency,* i.e. random chaotic bursts in an otherwise nearly periodic temporal evolution (cf. Fig. 10.28).

Figures 10.22, 10.23, 10.24 and 10.25 on the one hand and Figs. 10.26 and 10.27 on the other hand indicate the existence of *several* period doubling scenarios. This becomes particularly clear in Fig. 10.29 which depicts the largest LIAPUNOV exponents $\lambda_i(\kappa)$. The *critical points* κ_j where *period doubling* occurs are characterized by the fact that the largest LIAPUNOV exponent grows up to zero for decreasing values of $\kappa > \kappa_j$ and falls down again for $\kappa < \kappa_j$. Chaotic behaviour is observed if the largest LIAPUNOV exponent is greater than zero.

The above mentioned period doubling scenarios can be related to a simpler system due to the circumstance that the parameters (10.41) occuring in the scaled

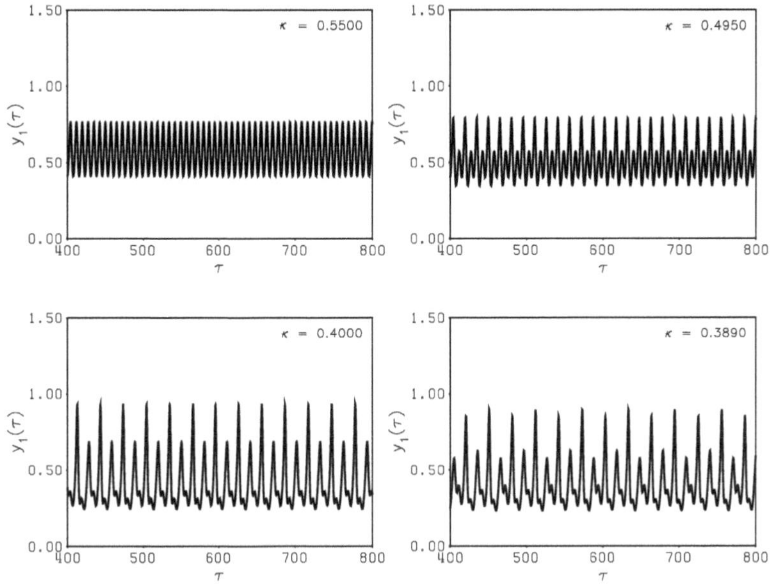

Fig. 10.21 Temporal evolution of the scaled coordinate $y_1(\tau) \equiv y_1(\kappa, \tau)$ after the transient time for various values of the control parameter κ ($\kappa = 0.55$: Period 1; $\kappa = 0.495$: Period 2; $\kappa = 0.4$: Period 4; $\kappa = 0.389$: Period 8). The *period* reflects after how many oscillations the temporal behaviour recurs

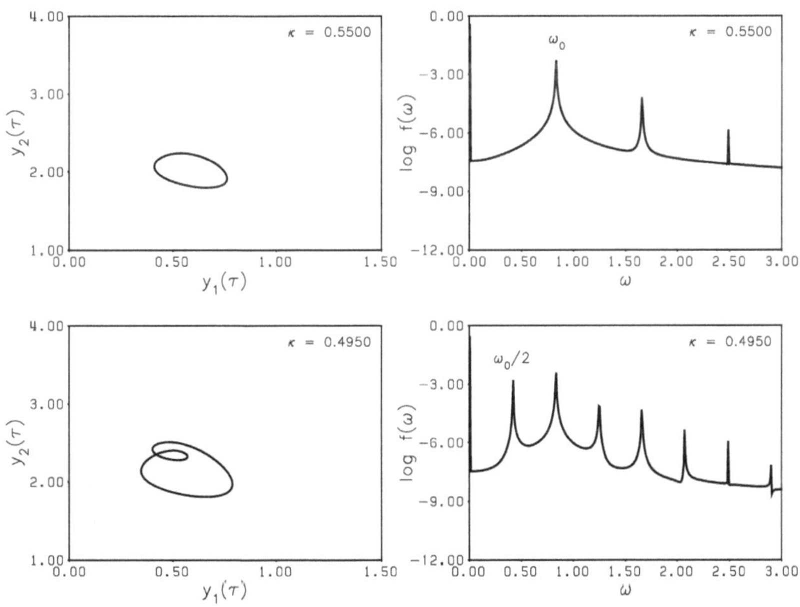

Fig. 10.22 *Phase portraits* (*below*) and corresponding *power spectra* $f(\omega)$ (*above*) for various values of the control parameter κ ($\kappa = 0.55$: Period 1; $\kappa = 0.495$: Period 2). In the phase portrait a period doubling manifests itself by an additional loop (or indentation), in the power spectrum by additional maxima in the *middle* of those maxima which were found *before* the period doubling

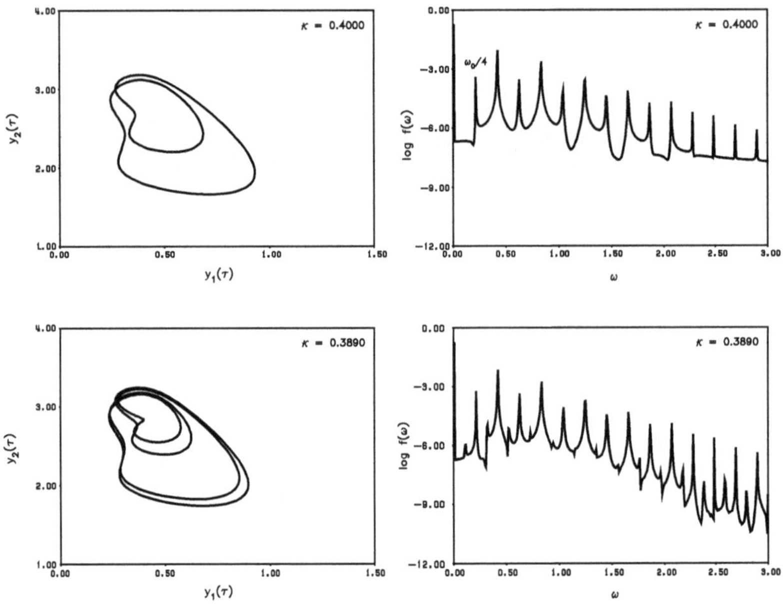

Fig. 10.23 Phase portraits (*below*) and related power spectra $f(\omega)$ (*above*) for various values of the control parameter κ ($\kappa = 0.4$: Period 4; $\kappa = 0.389$: Period 8)

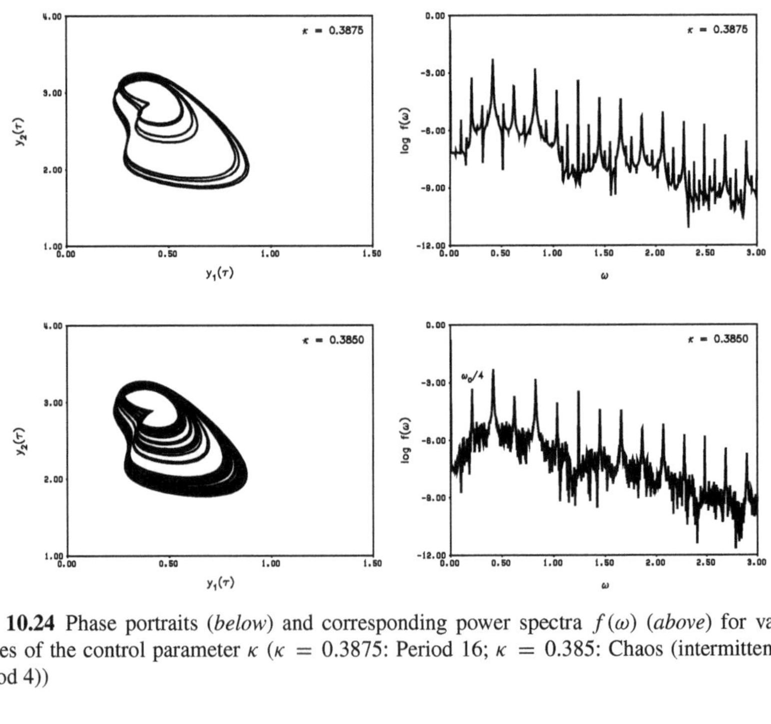

Fig. 10.24 Phase portraits (*below*) and corresponding power spectra $f(\omega)$ (*above*) for various values of the control parameter κ ($\kappa = 0.3875$: Period 16; $\kappa = 0.385$: Chaos (intermittency of period 4))

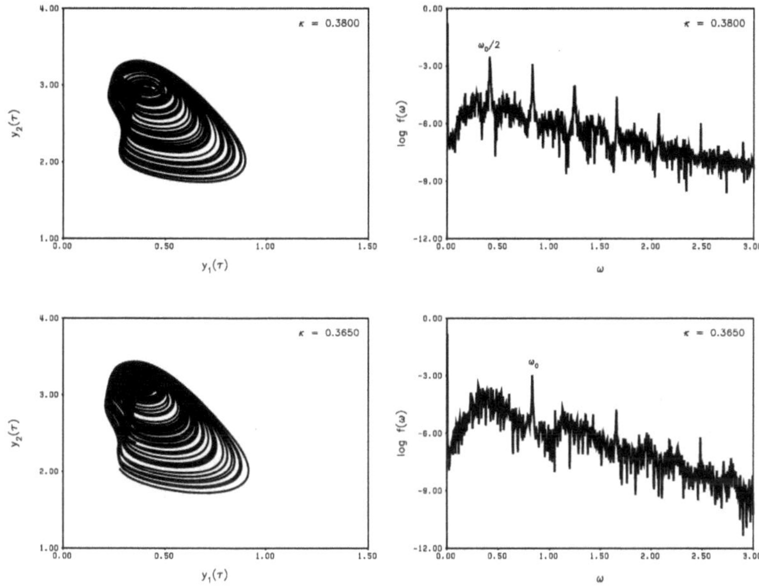

Fig. 10.25 Phase portraits (*below*) and related power spectra $f(\omega)$ (*above*) for various values of the control parameter κ ($\kappa = 0.38$: Chaos (intermittency of period 2); $\kappa = 0.365$: Chaos (intermittency of period 1)). The phase portraits have a *fractal dimension* (cf. Sect. 7.5) and are called *strange attractors*

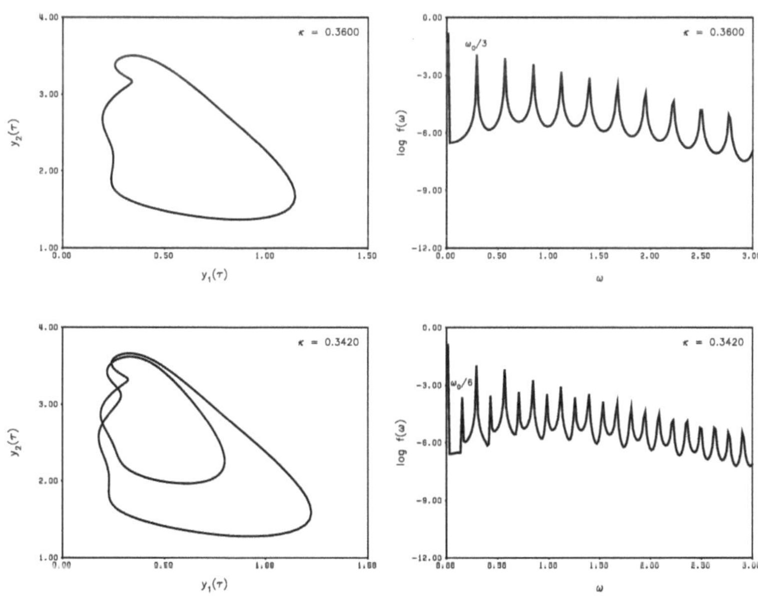

Fig. 10.26 Phase portraits (*below*) and corresponding power spectra $f(\omega)$ (*above*) for various values of the control parameter κ ($\kappa = 0.36$: Period 3; $\kappa = 0.342$: Period 6)

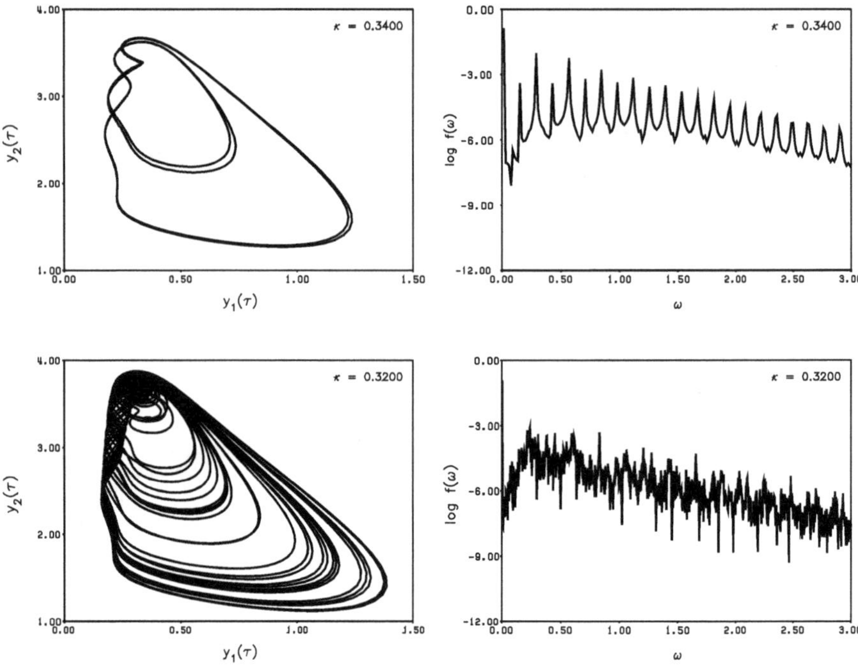

Fig. 10.27 Phase portraits (*below*) and related power spectra $f(\omega)$ (*above*) for various values of the control parameter κ ($\kappa = 0.34$: Period 12; $\kappa = 0.32$: Chaos)

equations (10.39) possess completely different *magnitudes*. As a consequence, we have

$$\left\{[y_1(\tau)]^2 - y_3(\tau)\right\} = \frac{\varepsilon'}{\varepsilon}\frac{d}{d\tau}y_3(\tau) \approx 0\,, \tag{10.50a}$$

$$y_4(\tau)\Big[y_5(\tau) - y_4(\tau)\Big] \approx \frac{1}{\widehat{D}}\frac{d}{d\tau}y_4(\tau) \approx 0\,, \tag{10.50b}$$

$$\frac{d}{d\tau}y_5(\tau) = \frac{\widehat{D}}{\widehat{E}}y_4(\tau)\Big[y_4(\tau) - y_5(\tau)\Big] \approx 0\,, \tag{10.50c}$$

i.e.

$$y_3(\tau) \approx [y_1(\tau)]^2\,, \tag{10.51a}$$

$$y_4(\tau) \approx y_5(\tau)\,, \tag{10.51b}$$

$$y_5(\tau) \approx y_5(0)\,. \tag{10.51c}$$

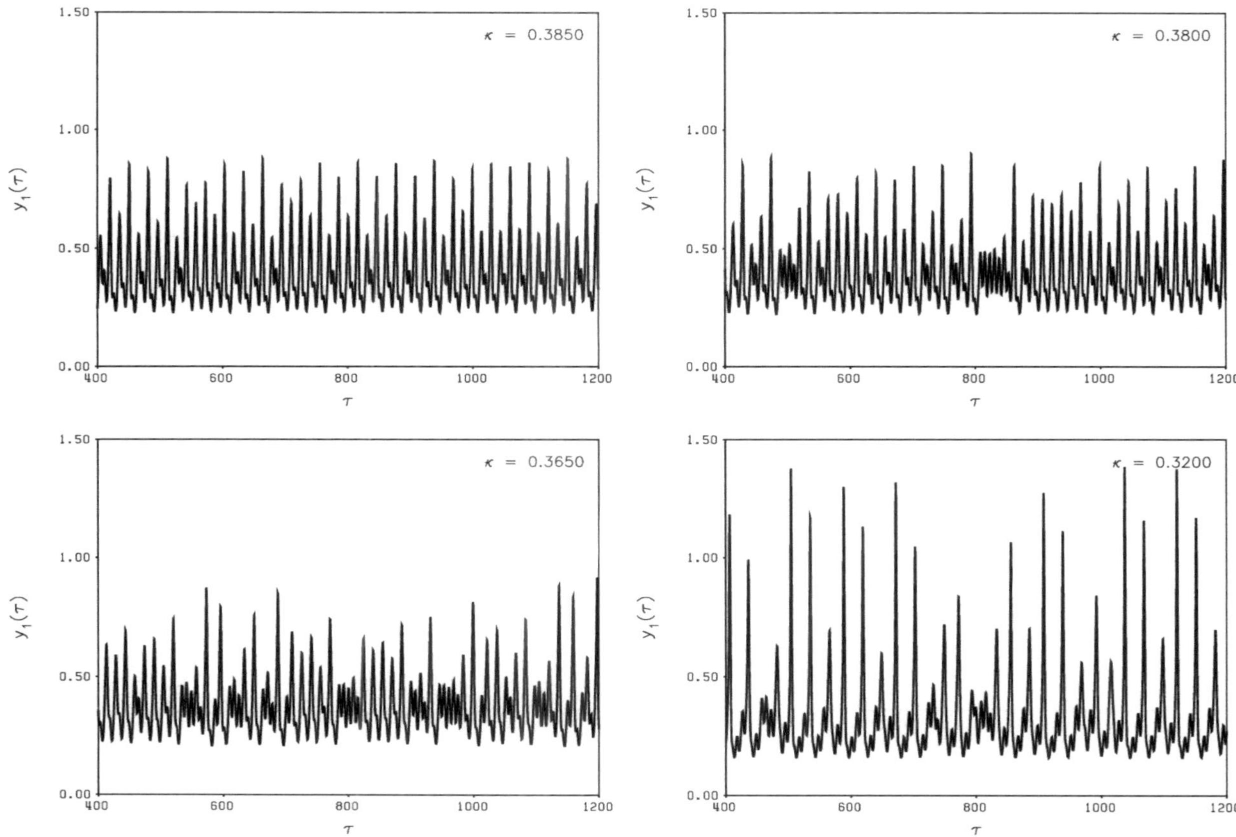

Fig. 10.28 Chaotic temporal evolution of the scaled co-ordinate $y_1(\tau)$ after the transient time for various values of the control parameter κ ($\kappa = 0.385$: Intermittency of period 4; $\kappa = 0.38$: Intermittency of period 2; $\kappa = 0.365$: Intermittency of period 1; $\kappa = 0.32$: Chaos). *Intermittency* is characterized by nearly periodic phases (so-called intermissions) that are randomly interrupted by irregular bursts

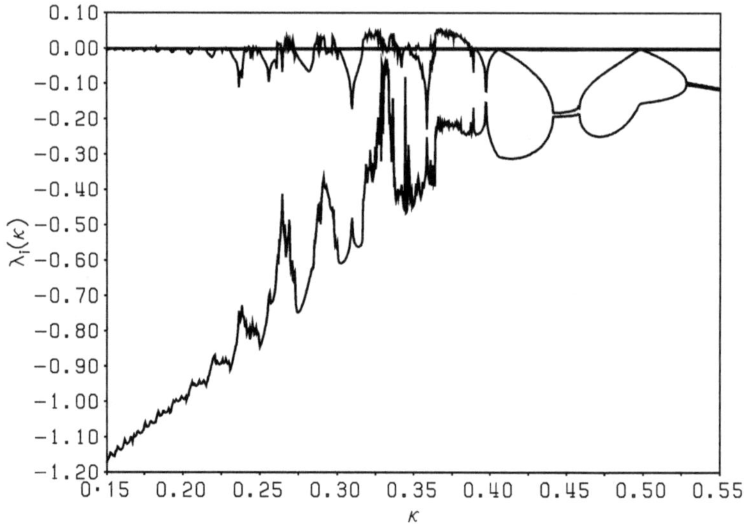

Fig. 10.29 The largest LIAPUNOV *exponents* for the considered BOLTZMANN equations: Critical points where *period doubling* occurs via a *pitchfork bifurcation* are characterized by the fact that the largest LIAPUNOV exponent just reaches zero. If it *exceeds* zero, the temporal behaviour of the system is *chaotic*

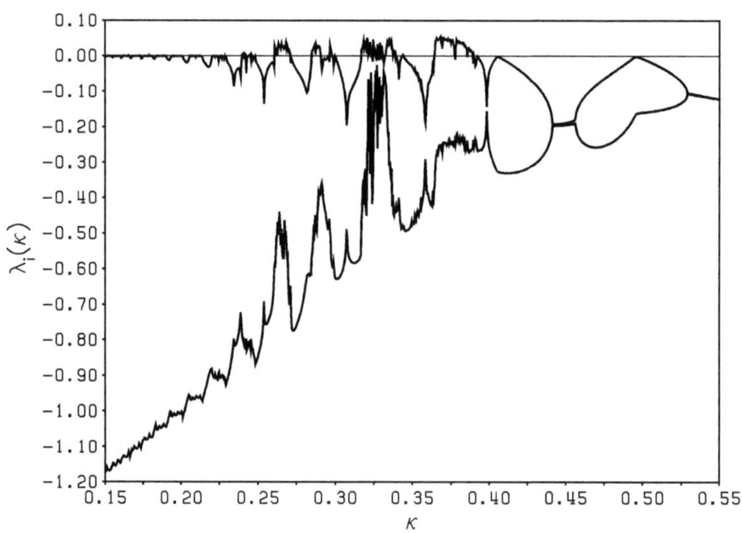

Fig. 10.30 For the reduced system (10.52), (10.39f–h) of the scaled BOLTZMANN equations the largest LIAPUNOV exponents show, in dependence of κ, a similar course as those of the exact equation system (10.39) depicted in Fig. 10.21. The *separation of time scales,* i.e. the *adiabatic elimination* and the approximation of *quasi-constance* for some variables seems, therefore, to be justified for the most part—even in the chaotic regime

Summarizing this, $y_3(\tau)$ can be *adiabatically eliminated.* In contrast, $y_4(\tau)$ and $y_5(\tau)$ are *quasi-constant.* Thus the equations for $y_1(\tau)$ and $y_2(\tau)$ simplify to

$$\frac{d}{d\tau} y_1(\tau) = \widehat{A} - (\widehat{B} + 1) y_1(\tau) + [y_1(\tau)]^2 y_2(\tau) + \kappa y_6(\tau) , \tag{10.52a}$$

$$\frac{d}{d\tau} y_2(\tau) = \widehat{B} y_1(\tau) - [y_1(\tau)]^2 y_2(\tau) \tag{10.52b}$$

where $y_6(\tau)$ displays non-linear oscillations (cf. Sect. 10.3.2). The relation of these equations with those of the *sinusoidally forced Brusselator* (cf. [112, 113, 258]) is obvious.

Figure 10.30 shows the largest LIAPUNOV exponents for the *reduced* system of Eqs. (10.52), (10.39f–h). It reveals a strong agreement with Fig. 10.29 though chaotic systems usually react in a very sensitive way on small variations. Consequently, there are also areas where considerable deviations are found, e.g. for $0.325 \le \kappa \le 0.35$.

10.4 Generalizations

The approach of the previous sections can be generalized in various aspects:

- The results in Sects. 10.2 and 10.3 obviously hold only for *local* or *spatially homogeneous* cases of opinion formation. If, apart from opinion changes, the effects of location changes are to be considered, we must represent by the state x (or y) not only the opinion i (or j) but also the location r (or s) of an individual:

$$x := (i, r), \qquad y := (j, s) . \tag{10.53}$$

- The restriction to opinion changes can be dropped because the previous discussion can be directly extended to more general behavioural changes.
- If we want to model pair interactions and indirect interactions *together*, we must use the configurational master equation with the transition rates that were introduced in Sect. 5.5. The resulting approximate mean value equations (5.16) are BOLTZMANN-*like equations* of the form (9.54) in which, however, the transition rates

$$w_a(x'|x; t) \equiv w_a(x'|x; \langle n \rangle; t) \tag{10.54a}$$

$$w_{ab}(x', y'|x, y; t) \equiv w_{ab}(x', y'|x, y; \langle n \rangle; t) \tag{10.54b}$$

depend on the mean value $\langle n \rangle$. Here transition rates (10.54b) describe pair interactions that implicitly depend on indirect interactions. Examples for the combination of spontaneous transitions and *indirect pair interactions* will be discussed in connection with the game dynamical equations (cf. Sect. 12.4).

- Up to now, due to the MARKOV *assumption* (3.17) and (3.91) respectively, we neglected *memory effects*. If necessary, however, these can be taken into account by the *generalized* BOLTZMANN-*like equations* (cf. [47]) which are derivable from the generalized master equation (3.6) in a similar way as in Chap. 4 (or Chap. 5). They read

$$\frac{d}{dt}P_a(x,t) = \int_{t_0}^{t} dt' \sum_{x'} \Big[w^a_{t-t'}(x|x';t')P_a(x',t') - w^a_{t-t'}(x'|x;t')P_a(x,t') \Big] \tag{10.55a}$$

with the effective transition rates

$$w^a_{t-t'}(x'|x;t') := w_a^{t-t'}(x'|x;t') + \sum_b \sum_y \sum_{y'} w_{ab}^{t-t'}(x',y'|x,y;t')P_b(y,t')\,. \tag{10.55b}$$

The memory effects go back partly to *short-term memory* and partly to *long-term memory*. For the effects of short-term memory the MARKOV assumption will normally be justified. The long-term memory shows up in the *increase of knowledge and experience* as well as in the *behavioural type* but it has, in most cases, probably only a subordinate influence on the dynamics of the medium-term social processes we are mainly interested in.

10.5 Spatial Spreading of Opinions

This section is dedicated to approaches for the description of *spatial* aspects of opinion formation which, up to now, were put aside. There are several processes which can lead to a spatial dissemination of opinions—on the one hand *diffusion processes* and *migration processes*, on the other hand *telecommunication processes*. Of course, both kinds of processes might also occur simultaneously.

10.5.1 Opinion Spreading by Diffusion

In the case of opinion spreading by *diffusion* we will suppose that opinions disseminate spatially by random *location changes* of the individuals. With

$$w_a(i',r'|i,r;t) := w_{a,i}(r'|r;t)\delta_{ii'} \tag{10.56a}$$

and

$$w_{ab}(i',r',j',s'|i,r,j,s;t) := w^r_{ab}(i',j'|i,j;t)\delta(r'-r)\delta(s'-s)\delta(s-r) \tag{10.56b}$$

we assume that location changes and opinion changes occur independently of each other. Moreover, the location changes happen spontaneously whereas the opinion changes stem from pair interactions at location $r = s$. The approach

$$w_{a,i}(r'|r;t) = \nu^r_{a,i}(t) p_{a,i}(r'|r;t) \tag{10.57a}$$

with

$$p_{a,i}(r'|r;t) := \frac{\mathrm{e}^{-\|r'-r\|/R_{a,i}}}{2\pi(R_{a,i})^2} \tag{10.57b}$$

guarantees that location changes arise only between *neighbouring* locations with a probability that decreases with growing distance $\|r' - r\|$ from the destination r'. Equation (9.54) for the spreading of opinions then reads

$$\begin{aligned}\frac{d}{dt} P_a(i,r;t) = \sum_b \sum_{i'} \sum_j \sum_{j'} \Big[& w^r_{ab}(i,j'|i',j;t) P_a(i',r;t) P_b(j,r;t) \\ & - w^r_{ab}(i',j'|i,j;t) P_a(i,r;t) P_b(j,r;t) \Big] \\ + \int d^2r' \Big[& w_{a,i}(r|r';t) P_a(i,r';t) - w_{a,i}(r'|r;t) P_a(i,r;t) \Big] .\end{aligned} \tag{10.58}$$

However, if $\nu^r_{a,i}(t)$ is independent of r, one often prefers the representation

$$\begin{aligned}\frac{d}{dt} P_a(i,r;t) = \sum_b \sum_{i'} \sum_j \sum_{j'} \Big[& w^r_{ab}(i,j'|i',j;t) P_a(i',r;t) P_b(j,r;t) \\ & - w^r_{ab}(i',j'|i,j;t) P_a(i,r;t) P_b(j,r;t) \Big] \\ & + D_{a,i}(t) \Delta_r P_a(i,r;t)\end{aligned} \tag{10.59a}$$

in which Δ_r denotes the LAPLACE *operator.* Eq. (10.59a) results from Eq. (10.58) by carrying out a second order TAYLOR approximation (cf. Chap. 6). Especially, there is

$$\begin{aligned} D_{a,i}(t) &= \int d^2r' \, (r' - r)^2 w_{a,i}(r'|r;t) \\ &= \int_0^{2\pi} d\varphi \int_0^{\infty} d\Delta r' \, (\Delta r')^3 \frac{\nu_{a,i}(t)}{2\pi(R_{a,i})^2} \mathrm{e}^{-\Delta r'/R_{a,i}} \\ &= 6\nu_{a,i}(t)(R_{a,i})^2 \end{aligned} \tag{10.59b}$$

with $\Delta r' := \|r' - r\|$. Equation (10.59) is known from *chemistry* as the *reaction-diffusion equation* and can produce *spatial waves* or *spirals* (cf. [153]). PUU has suggested spatial oscillations also in connection with similar models of opinion spreading [227]. However, the dissemination of opinions by diffusion applies only to a few situations, e.g. to crowded halls.

Another situation for which a *slow* diffusion by *migration* of individuals is assumed was examined by SCHWEITZER et al. [257]. They subdivided the space of the system into *discrete boxes* and chose for the *migration rates* the approach

$$w_{a,i}(r'|r;t) \equiv w_{a,i}(r'|r;n;t) := \nu_{a,i}(t)\mathrm{e}^{-c_a|i|n^a_{i,r}/N_a} . \tag{10.60}$$

According to this, the migration rate $w_{a,i}(r'|r;t)$ is proportional to the *mobility* $\nu_{a,i}(t)$ which depends on the opinion $i \in \{0, \pm 1, \pm 2, \dots\}$ of the individuals. For $c_a > 0$ the migration effect is *repulsive* and the individuals distribute all over the system. Whereas individuals with a *neutral* position ($i = 0$) tend to an *equipartition* due to $w_{a,0}(r'|r;t) = \nu_{a,0}(t)$, individuals with an extreme position (with large $|i|$) are inclined towards concentrating in small spatial areas, i.e. towards forming an *'autonomous scene'*. For large $|i|$, the *settledness* $1/w_{a,i}(r'|r;t)$ is large only if the concentration $n^a_{i,r}/N_a$ of individuals of the same opinion is high.

10.5.2 Opinion Spreading by Telecommunication

In cases

- where individuals, on the one hand, change their place but, on the other hand, always return to the same place r (their flat) or
- where individuals stay at a fixed place r but have an opinion exchange by *telephone* (or *correspondence*) with individuals at other places

we must apply another model. Since in such cases the place r can be treated as a quasi-constant, characteristic quantity of an individual, it formally plays a similar role as the behavioural type a. This implies

$$w_a(i', r'|i, r; t) := w^a_r(i'|i;t)\delta_{rr'} , \tag{10.61a}$$

$$\begin{aligned} w_{ab}(i', r', j', s'|i, r, j, s; t) &:= w^{ab}_{rs}(i', j'|i, j; t)\delta_{rr'}\delta_{ss'} \\ &= \nu^{ab}_{rs}(t)p_{ab}(i', j'|i, j; t)\delta_{rr'}\delta_{ss'} . \end{aligned} \tag{10.61b}$$

For the *distance dependence* of the interactions we could, for example, assume

$$\nu^{ab}_{rs}(t) := \nu_{ab}(t)\mathrm{e}^{-\|s-r\|/R_{ab}} \tag{10.61c}$$

because exchanges of views will become *less frequent* with growing distance $\|s - r\|$. If we define

$$P_a(i, r; t) =: P_a(r) P_r^a(i, t) \tag{10.62a}$$

with

$$P_a(r) := \frac{N_a^r}{N_a}, \tag{10.62b}$$

we have

$$\sum_i P_r^a(i, t) = 1 \tag{10.62c}$$

and Eq. (9.54) for the spatial spreading of opinions takes on the form

$$\begin{aligned} \frac{d}{dt} P_r^a(i, t) = \sum_{i'} \Big[w_r^a(i|i'; t) P_r^a(i', t) - w_r^a(i'|i; t) P_r^a(i, t) \Big] \\ + \sum_s \sum_b \sum_{i'} \sum_j \sum_{j'} \Big[w_{rs}^{ab}(i, j'|i', j; t) P_r^a(i', t) P_b(s) P_s^b(j, t) \\ - w_{rs}^{ab}(i', j'|i, j; t) P_r^a(i, t) P_b(s) P_s^b(j, t) \Big] . \end{aligned} \tag{10.63}$$

Since N_a^r denotes the number of individuals of subpopulation a who are at place r we obtain

$$\sum_r N_a^r = N_a \quad \text{and} \quad \sum_r P_a(r) = 1 . \tag{10.64}$$

Consequently, $P_a(r)$ has the meaning of the *population distribution* of subpopulation a over the places r. $P_r^a(i, t)$ is the *proportion* of individuals of subpopulation a at place r with opinion i and thus it describes the *probability* with which an accidentally selected individual of subpopulation a at location r will hold opinion i.

The simplest case of the spatial spreading of opinions is the dissemination of *news* or *rumours*. Then we have only to distinguish the *ignorance* ($i = 1$) and the *knowledge* ($i = 2$) of the concerned news (i.e. $i \in \{1, 2\}$). The spreading of news bases on pair interactions of the kind

$$2, 2 \longleftarrow 2, 1 , \tag{10.65a}$$

$$2, 2 \longleftarrow 1, 2 , \tag{10.65b}$$

i.e. on 'persuasion processes'. *Processes of forgetting* the news are modeled by spontaneous transitions:

$$1 \longleftarrow 2 . \tag{10.66}$$

Other processes will be neglected. If we, in addition, restrict ourselves to one single subpopulation ($A = 1$), the Eq. (10.63) for the spreading of news takes on the simple form

$$\frac{d}{dt}P_r(2,t) = \sum_s w_{rs}(2,2|1,2;t)P(s)\Big[1 - P_r(2,t)\Big]P_s(2,t) - w_r(1|2;t)P_r(2,t) \tag{10.67}$$

where we took into account $P_r(1,t) = 1 - P_r(2,t)$ and dropped the subscript/superscript a. In the limit $R \to 0$ (cf. (10.61c)), (10.67) has the form of the *logistic equation* (cf. (9.65)) and describes a *local* spreading of news.

For the population distribution $P(s)$ the homogeneity assumption $P(s) \equiv const.$ is normally not justified. Due to the population concentration in towns $P(s)$ varies considerably. An idealized scheme of the spatial population distribution is given by CHRISTALLER's *system of central places* which is illustrated in Fig. 10.31. Already Eq. (10.67) is no longer solvable analytically and numerically only with the effort of much computer time. Due to the spatial *self-similarity* of the system of central places, however, the methods from *renormalization theory* possibly allow the determination of approximate solutions for the spatial spreading of opinions (cf. [117, 136, 143]).

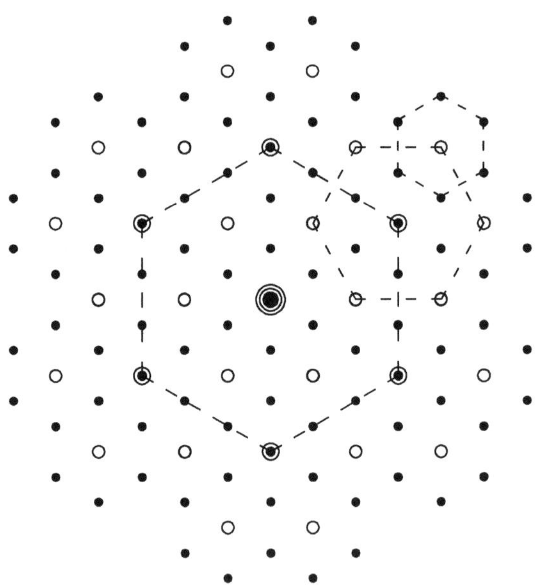

Fig. 10.31 Schematic representation of the population distribution according to **CHRISTALLER's** *system of central places* (cf. [39]): By means of circles of varying size towns with different numbers of inhabitants are represented. The population distribution has approximately a self-similar *hexagonal structure* which, however, in reality is distorted by topography

Finally, let us stress that Eq. (10.63) can, due to

$$P_b(s)w_{rs}^{ab}(i', j'|i, j; t) = N_a^r \widehat{w}_{rs}^{ab}(i', j'|i, j; t)\,, \tag{10.68}$$

with the abbreviations

$$a := (a, r), \qquad b := (b, s)\,, \tag{10.69}$$

also be written in the familiar form

$$\begin{aligned} \frac{d}{dt}P_a(i,t) = \sum_{i'} \Big[w_a(i|i';t)P_a(i',t) - w_a(i'|i;t)P_a(i,t) \Big] \\ + \sum_b \sum_{i'} \sum_j \sum_{j'} \Big[w_{ab}(i,j'|i',j;t)P_a(i',t)P_b(j,t) \\ - w_{ab}(i',j'|i,j;t)P_a(i,t)P_b(j,t) \Big] \end{aligned} \tag{10.70}$$

(cf. (9.54) with $x \equiv i$). Here there are

$$\sum_b := \sum_{b,s}\,, \tag{10.71a}$$

$$P_a(i,t) := P_r^a(i,t)\,, \tag{10.71b}$$

$$w_a(i'|i;t) := w_r^a(i'|i;t)\,, \tag{10.71c}$$

$$w_{ab}(i',j'|i,j;t) := N_a^r \widehat{w}_{rs}^{ab}(i',j'|i,j;t)\,. \tag{10.71d}$$

In contrast to the right-hand side, the subscripts on the left-hand side of (10.71) have the meaning that was defined in (10.69). The possibility of representation (10.70) bases on the circumstance that, in this section, location r is treated as a characteristic, quasi-constant quantity of an individual, i.e. in the same way as the behavioural type a.

Chapter 11
Social Fields and Social Forces

11.1 Introduction

From physics we know that nothing changes if there is no *force* at work. Conversely, the effect of a force is characterized by a change of the state of a system. Similar relations hold in connection with human behaviour. An individual will not change his/her behaviour if he/she is satisfied with his/her situation, i.e. if there is no *reason* for a change like a (physical or psychological) need, social pressure, etc. Therefore, we will try to describe the *reasons* for behavioural changes by *social forces*.

A related approach was already suggested by LEWIN [169] who developed the *'social field theory'* according to which the behaviour of an individual is determined by a *'social field'*. This reflects the influences of the *physical* and *social environment*, e.g. of *public opinion*, social *norms*, and trends. Although LEWIN *formalized* his ideas, a fully mathematical formulation has long been missing. Such a formulation will be presented in this chapter in terms of differential equations. Since the behaviour of an individual is not completely determined by the environment (cf. Sects. 8.2.3 and 8.2.4) these equations must be *stochastic* in order to take into account the individual behavioural variations.

The behaviour of individuals in a social field shows an analogy to the behaviour of electrons in an electric field (e.g. of a nucleus). In particular, individuals as well as electrons, on the one hand, influence the *effective* (social or electric) *field* which, on the other hand, affects their behaviour (cf. Sect. 6.6). However, when developing a model for the influence of a social field on behaviour, we must not be tempted by this analogy to apply the SCHRÖDINGER *equation* [250] which is not suitable for describing social processes. Instead, we have to start from the BOLTZMANN-*like equations* which were obtained for behavioural changes in Chap. 9. A reformulation of these equations by means of a second order TAYLOR approximation leads to special *diffusion equations*. Because of their relation to the BOLTZMANN equation and the FOKKER-PLANCK *equation* they are called 'BOLTZMANN-FOKKER-PLANCK *equations*' (cf. Sect. 6.6). These can be transformed into mathematically equivalent LANGEVIN *equations* (cf. Chap. 7).

According to the BOLTZMANN-FOKKER-PLANCK equations and the LANGEVIN equations the *systematic* and the *most probable* behavioural change respectively are

D. Helbing, *Quantitative Sociodynamics*, 2nd ed.,
DOI 10.1007/978-3-642-11546-2_11, © Springer-Verlag Berlin Heidelberg 2010

given by vectorial quantities $K_a(x,t)$ which can be interpreted as *social forces.* They describe concrete *motivations to act* which are evoked in an individual as a reaction to his/her environment and are the drive behind behavioural changes. Individual behavioural variations due to the *'freedom of decision-making'* are taken into account in the equations by *diffusion coefficients.*

In simple cases social forces can be expressed by *potentials* $V_a(x,t)$ which possess the above described characteristics of *social fields.* A social field can be imagined as a range of mountains in behavioural space the gradient (steepness) of which is a measure for the social force which acts at a certain location of behavioural space, i.e. on a certain behaviour.

It turns out that social forces $K_a(x,t)$ are noticeable only for those individuals whose behaviours x deviate from minima ('valleys') of the social field (from the *'norm'*). However, a behaviour which in subpopulation a is exposed to a social force can in *another* subpopulation b *correspond* to the norm.

The social force model has the advantage that it provides a very intuitive description of social processes. It is suitable for the delineation of opinion formation, purchase pattern [261], group dynamics, or pedestrian motion [123, 124, 134]. However, due to the TAYLOR approximation, it is only applicable if

1. the behavioural changes can be represented in more or less *continuous* behavioural spaces (which may differ from one subpopulation to another),
2. the behavioural changes mainly take place in *small* steps.

In cases where these conditions are not fulfilled we must use the BOLTZMANN-like equations instead.

11.2 Derivation

In the following we will assume to be confronted with *continuous* or at least *quasi-continuous behavioural spaces* γ_a. The m dimensions of these spaces correspond to various characteristic *aspects* of the considered behaviours. The BOLTZMANN-like equations (9.43) then have the form

$$\begin{aligned} \frac{d}{dt} P_a(x,t) &= \int d^m x' \Big[w^a(x|x';t) P_a(x',t) - w^a(x'|x;t) P_a(x,t) \Big] \\ &= \int d^m y \Big[w^a[y|x-y;t] P_a(x-y,t) - w^a[y|x;t] P_a(x,t) \Big] \end{aligned} \tag{11.1a}$$

with the effective transition rates

$$\begin{aligned} w^a[x'-x|x;t] &:= w^a(x'|x;t) \\ &:= w_a(x'|x;t) + \sum_b \int d^m y' \int d^m y \, w_{ab}(x',y'|x,y;t) P_b(y,t) \,. \end{aligned} \tag{11.1b}$$

From Sect. 6.6 we know that Eqs. (11.1) can, by means of a second order TAYLOR approximation, be transformed into special *diffusion equations,* namely the BOLTZMANN-FOKKER-PLANCK equations

$$\begin{aligned}\frac{\partial}{\partial t} P_a(x,t) = &-\sum_{i=1}^{m} \frac{\partial}{\partial x_i}\Big[K_{ai}(x,t)P_a(x,t)\Big] \\ &+\frac{1}{2}\sum_{i,j=1}^{m} \frac{\partial}{\partial x_i}\frac{\partial}{\partial x_j}\Big[Q_{aij}(x,t)P_a(x,t)\Big]\end{aligned} \tag{11.2a}$$

with the *effective drift coefficients*

$$K_{ai}(x,t) := \int d^m y\, y_i w^a[y|x;t] = \int d^m x'\, \Delta x'_i w^a(x'|x;t) \tag{11.2b}$$

and the *effective diffusion coefficients*

$$Q_{aij}(x,t) := \int d^m y\, y_i y_j w^a[y|x;t] = \int d^m x'\, \Delta x'_i \Delta x'_j w^a(x'|x;t) \tag{11.2c}$$

($\Delta x'_i := x'_i - x_i$). Whereas the drift coefficients $K_{ai}(x,t)$ reflect the *systematic motion* ('drift') of distribution $P_a(x,t)$, the diffusion coefficients $Q_{aij}(x,t)$ describe the *broadening* of distribution $P_a(x,t)$ by *fluctuations* which result from individual variations of the behavioural changes (cf. Sect. 6.5.1).

In the following we will apply the special BOLTZMANN-like equations with the effective transition rates

$$w^a(x'|x;t) := w_a(x'|x;t) + \nu^a(x',x;t)R^a(x'|x;t) \qquad (x' \neq x)\,, \tag{11.3a}$$

$$\begin{aligned}\nu^a(x',x;t) := \sum_b \nu_{ab}(t)\Big[&\Big(f^1_{ab}(t) - f^3_{ab}(t)\Big)P_b(x',t) \\ &+\Big(f^2_{ab}(t) - f^3_{ab}(t)\Big)P_b(x,t) + f^3_{ab}(t)\Big]\end{aligned} \tag{11.3b}$$

which were developed in Chap. 9 for the description of behavioural changes. Then the effective drift and diffusion coefficients split up into contributions by *spontaneous transitions* ($k = 0$), *imitative processes* ($k = 1$), *avoidance processes* ($k = 2$), and *compromising processes* ($k = 3$):

$$K_{ai}(x,t) := \sum_{k=0}^{3} K^k_{ai}(x,t)\,, \tag{11.4a}$$

$$Q_{aij}(x,t) := \sum_{k=0}^{3} Q^k_{aij}(x,t)\,. \tag{11.4b}$$

Here we introduced the abbreviations

$$K^0_{ai}(x,t) := \int d^m x' \, \Delta x'_i w_a(x'|x;t) \, , \tag{11.5a}$$

$$K^1_{ai}(x,t) := \sum_b \nu_{ab}(t) f^1_{ab}(t) \int d^m x' \, \Delta x'_i R^a(x'|x;t) P_b(x',t) \, , \tag{11.5b}$$

$$K^2_{ai}(x,t) := \sum_b \nu_{ab}(t) f^2_{ab}(t) \int d^m x' \, \Delta x'_i R^a(x'|x;t) P_b(x,t) \, , \tag{11.5c}$$

$$K^3_{ai}(x,t) := \sum_b \nu_{ab}(t) f^3_{ab}(t) \int d^m x' \, \Delta x'_i R^a(x'|x;t) \times \Big(1 - P_b(x',t) - P_b(x,t)\Big) \, , \tag{11.5d}$$

and

$$Q^0_{aij}(x,t) := \int d^m x' \, \Delta x'_i \Delta x'_j w_a(x'|x;t) \, , \tag{11.6a}$$

$$Q^1_{aij}(x,t) := \sum_b \nu_{ab}(t) f^1_{ab}(t) \int d^m x' \, \Delta x'_i \Delta x'_j R^a(x'|x;t) P_b(x',t) \, , \tag{11.6b}$$

$$Q^2_{aij}(x,t) := \sum_b \nu_{ab}(t) f^2_{ab}(t) \int d^m x' \, \Delta x'_i \Delta x'_j R^a(x'|x;t) P_b(x,t) \, , \tag{11.6c}$$

$$Q^3_{aij}(x,t) := \sum_b \nu_{ab}(t) f^3_{ab}(t) \int d^m x' \, \Delta x'_i \Delta x'_j R^a(x'|x;t) \times \Big(1 - P_b(x',t) - P_b(x,t)\Big) \, . \tag{11.6d}$$

In particular, $K^0_{ai}(x,t)$ and $Q^0_{aij}(x,t)$ delineate behavioural changes which are induced by the *physical environment.*

The BOLTZMANN-FOKKER-PLANCK equations (11.1) can, with the method introduced in Chap. 7, be transformed into mathematical equivalent LANGEVIN *equations.* They read

$$\frac{dx_{\alpha i}}{dt} = F_{ai}(x_\alpha,t) + \sum_{j=1}^{m} G_{aij}(x_\alpha,t)\xi_{\alpha j}(t) \tag{11.7a}$$

with

$$K_{ai}(x_\alpha,t) =: F_{ai}(x_\alpha,t) + \frac{1}{2} \sum_{j,k=1}^{m} \left[\frac{\partial}{\partial x_{\alpha j}} G_{aik}(x_\alpha,t) \right] G_{ajk}(x_\alpha,t) \tag{11.7b}$$

and

$$Q_{aij}(x_\alpha,t) =: \sum_{k=1}^{m} G_{aik}(x_\alpha,t) G_{ajk}(x_\alpha,t) \, . \tag{11.7c}$$

Here $x_\alpha(t)$ denotes the time-dependent behaviour of an individual α of subpopulation a. The vector

$$\zeta_{\alpha i}(x_\alpha, t) := \sum_{j=1}^{m} G_{aij}(x_\alpha, t)\xi_{\alpha j}(t) \tag{11.8}$$

describes the contribution to the change of behaviour $x_\alpha(t)$ which is caused by individual variations *(fluctuations)* $\xi_\alpha(t)$ of the behaviour.

If the diffusion coefficients $Q_{aij}(x_\alpha, t)$ and thus the coefficients $G_{aij}(x_\alpha, t)$ are small quantities (cf. (11.7c)) which is often the case, we have $F_{ai}(x_\alpha, t) \approx K_{ai}(x_\alpha, t)$ (cf. (11.7b)). Then Eq. (11.7a) can be written in the form

$$\frac{dx_\alpha}{dt} \approx K_a(x_\alpha, t) + \textit{individual fluctuations}\,. \tag{11.9}$$

11.3 The Social Force Model

For the following reasons it appears to be justified to denote the vectorial quantity

$$K_a(x, t) := \Big(K_{a1}(x, t), \ldots, K_{ai}(x, t), \ldots, K_{am}(x, t)\Big)^{\text{tr}} \tag{11.10}$$

as the *social force* which acts on individuals of subpopulation a:

- The various contributions $K_a^k(x, t)$ to $K_a(x, t)$ add like *vectors* as it must be demanded from forces. Especially, oppositely oriented contributions with the same absolute value cancel out each other. By the terms

$$K_a^1(x, t)\,, \quad K_a^2(x, t)\,, \quad K_a^3(x, t) \tag{11.11}$$

  which depend on the behavioural distributions $P_b(y, t)$ in other subpopulations b, the contributions due to *pair interactions* among individuals are taken into account.
- $K_a(x_\alpha, t)$ can be regarded as the drive of the temporal change dx_α/dt of the behaviour $x_\alpha(t)$ of an individual α belonging to subpopulation a. Namely, the vectorial quantity $K_a(x_\alpha, t)$ describes, according to (11.9) and (7.8), the *most probable behavioural change* and, according to (11.2) and Sect. 6.5.1, the *systematic influence* on the direction and magnitude of the behavioural change of individual α.
- If $K_a(x, t)$ vanishes this corresponds, due to

$$\nabla_x V_a(x, t) = K_a(x, t) = \mathbf{0}\,, \tag{11.12}$$

  to an *extremum* of the social field $V_a(x, t)$. Then, according to (11.9), the behavioural distribution $P_a(x, t)$ will be only subject to changes which are caused by fluctuations and described by diffusion coefficients $Q_{aij}(x, t)$.

- Nevertheless, social forces are certainly no physical forces obeying the NEWTON*ian laws of mechanics* [101, 198]. For example, in physics we have the law *'actio = reactio'* according to which a body that exerts a force on another body feels a *counterforce* of the same strength but opposite direction. In the social sciences, however, this is generally not valid. Moreover, the effect of a social force does not base on an *energy* and *momentum exchange* via so-called *virtual particles* but on an *information exchange* which evokes a behavioural change only indirectly via complicated *psychological and mental processes* (cf. Fig. 11.1). Individuals are *active systems* that produce the state changes (behavioural changes) themselves.

  Therefore, social force $K_a(x,t)$ has rather the meaning of a concrete *motivation* to carry out a certain behavioural change. It is a *reaction* to the information which an individual obtains by the *perception* of his/her physical and social environment. This reaction is the result of a *decision process* during which from various available behavioural alternatives that one is selected which comes next to the personal goals and interests, i.e. which brings about the maximal utility (cf. Sect. 8.2.4).

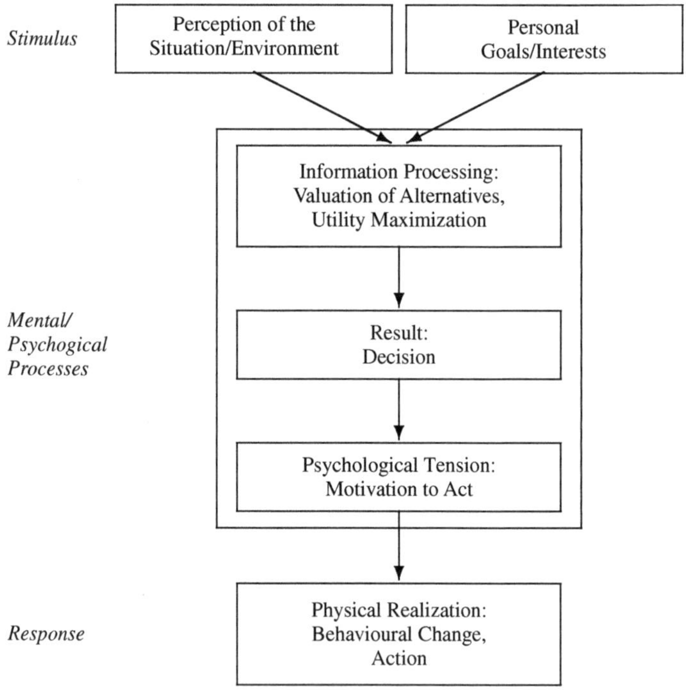

Fig. 11.1 A complete model of behavioural changes must contain three levels of description—the *perceived stimuli,* their influence on the *psychological motivation to act,* and the subsequent *physical realization.* According to the concept of the social force model, a sensory *stimulus* leads to a behavioural *reaction (response)* that depends on the personal goals and is chosen from a set of behavioural alternatives with the objective of utility maximization

- According to Chap. 9, the spontaneous transition rates for social processes often have the form

$$w_a(x'|x;t) := \nu_a(t)\frac{e^{U_a(x',t)-U_a(x,t)}}{D_a(x',x;t)}\,. \tag{11.13}$$

  Let us examine the case where there is no difference between the utilities $U_a(x,t)$ of different behaviours x which could evoke a *spontaneous* behavioural change ($U_a(x) \equiv U_a$). Furthermore, let the effort $S_a(x',x;t)$ of a behavioural change depend only on its magnitude $\|x'-x\|$:

$$D_a(x',x;t) \equiv D_a(\|x'-x\|;t)\,. \tag{11.14}$$

  As expected, in this case the contribution $K_a^0(x,t)$ to the social force $K_a(x,t)$ which is responsible for *spontaneous* behavioural changes vanishes. Systematic behavioural changes only occur due to pair interactions among individuals, then. By means of an appropriate choice of the state space γ_a (with the method of *multidimensional scaling* [161, 301]) relation (11.14) can always be achieved (at least approximately).

In simple cases it is possible to express a social force $K_a(x,t)$ by a *social field* $V_a(x,t)$. If the social force obeys the *integrability conditions*

$$\frac{\partial K_{ai}(x,t)}{\partial x_j} \stackrel{!}{=} \frac{\partial K_{aj}(x,t)}{\partial x_i} \qquad \forall\, i,j\,, \tag{11.15}$$

a time-dependent *potential*

$$V_a(x,t) := -\int^{x} dx' \cdot K_a(x',t) \tag{11.16a}$$

exists so that the social force $K_a(x,t)$ is given by its *gradient* ∇_x:

$$K_a(x,t) = -\nabla_x V_a(x,t)\,, \quad \text{i.e.} \quad K_{ai}(x,t) = -\frac{\partial}{\partial x_i}V_a(x,t)\,. \tag{11.16b}$$

Clearly spoken, due to (11.16b), we can imagine the social field $V_a(x,t)$ as a range of mountains in behavioural space, the steepness of which is a measure for the social force which acts at a certain location x of the behavioural space, i.e. on a certain behaviour x (cf. Fig. 11.2). This range of mountains varies dynamically and describes the effects of the physical and social environment including public opinion, trends, and norms on the behaviour.

If the diffusion coefficients are constant under variation of x ($Q_{aij}(x) \equiv Q_{aij}$), according to (6.29), (6.31) the *formal stationary solution* of (11.2) is given by

$$P_a^0(x) = P_a^0(x_0)\,e^{-\phi_a(x)} \tag{11.17}$$

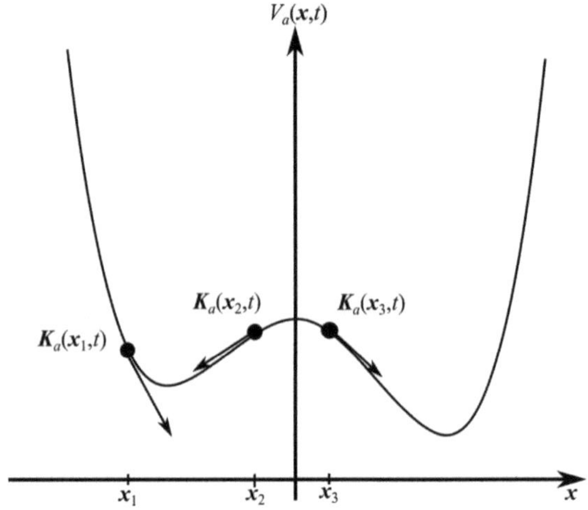

Fig. 11.2 A *social field* $V_a(x, t)$ is comparable to a range of mountains. Here a rather smooth version with one hill and two valleys is depicted for a one-dimensional behavioural space. The gradient (slope) $\nabla_x V_a(x, t)$ at a certain point $x \equiv x$ of the behavioural space is a measure for the direction and magnitude of the *social force* $K_a(x, t)$ that acts on the corresponding behaviour

with

$$\phi_a(x) := -2 \int_{x_0}^{x} dx' \cdot \underline{Q}_a^{-1} K_a(x') = 2 \int_{x_0}^{x} dx' \cdot \underline{Q}_a^{-1} \nabla_{x'} V_a(x') \,. \tag{11.18}$$

The stationary solution is then determined by the stationary social field $V_a(x)$. However, since the diffusion coefficients $\underline{Q}_a(x) \equiv \big(Q_{aij}(x)\big)$ themselves nor-mally vary with x, the relation for the stationary solution is in general considerably more complicated and can possibly not even be described by means of potential representation (11.17) (cf. Sect. 6.4.1). Relation (11.18), already, is an *implicit equation* since the stationary drift coefficients $K_{ai}(x)$ and diffusion coefficients $Q_{aij}(x)$ depend on the stationary probability distributions $P_b^0(y)$ according to relations (11.1b), (11.2b), and (11.2c). Thus for (11.18) there only exists a *recursive solution* (cf. Sect. 6.6.1).

In summary, we can state the following about the developed model:

- Behavioural changes are guided by a social force $K_a(x, t)$ which describes the *most probable individual behavioural change* or the *systematic influence* on the direction and magnitude of a behavioural change. In simple cases the social force can be expressed by a social field.
- The individual variations of behaviour are reflected by *diffusion coefficients* $Q_{aij}(x, t)$.
- Due to the *dependence of the social force on the interactions between individuals* the social field is not simply given externally. The form of the social field is rather affected by the behaviour of individuals.
- The influence of the *event history* and of *memory effects* can be taken into account by the *generalized* BOLTZMANN-FOKKER-PLANCK *equations* which are derivable

from the *generalized master equation* (3.6) in total analogy to Chap. 6. These differ from the *ordinary* BOLTZMANN-FOKKER-PLANCK equations only by an additional integration over past times $t' < t$ and read

$$\begin{aligned}\frac{\partial}{\partial t}P_a(x,t) = \int\limits_{t_0}^{t} dt' \Bigg\{ &-\sum_{i=1}^{m} \frac{\partial}{\partial x_i}\Big[K_{ai}^{t-t'}(x,t')P_a(x,t')\Big] \\ &+\frac{1}{2}\sum_{i,j=1}^{m}\frac{\partial}{\partial x_i}\frac{\partial}{\partial x_j}\Big[Q_{aij}^{t-t'}(x,t')P_a(x,t')\Big]\Bigg\}\end{aligned} \tag{11.19a}$$

with the effective drift coefficients

$$K_{ai}^{t-t'}(x,t') := \int d^m x' \; \Delta x_i' w_{t-t'}^a(x'|x;t') \,, \tag{11.19b}$$

the effective diffusion coefficients

$$Q_{aij}^{t-t'}(x,t') := \int d^m x' \; \Delta x_i' \Delta x_j' w_{t-t'}^a(x'|x;t') \,, \tag{11.19c}$$

and the effective transition rates

$$\begin{aligned} w_{t-t'}^a(x'|x;t') &:= w_a^{t-t'}(x'|x;t') \\ &+\sum_b \int d^m y' \int d^m y \, w_{ab}^{t-t'}(x',y'|x,y;t')P_b(y,t') \,. \end{aligned} \tag{11.19d}$$

11.3.1 Comparison with LEWIN's 'Social Field Theory'

The above developed model can be regarded as a mathematical formulation of LEWIN's concept of a *'social field theory'* (which certainly is no field theory in the sense of physics).

From social psychology it is known that the behaviour of an individual is given by the totality of *environmental influences* (including social influences) and his/her *personality*. Inspired by the theory of electromagnetic fields, social field theory assumes that the environmental influences can be considered as a *dynamic force field* which should be mathematically representable.

If an individual is in *equilibrium* with the dynamic force field, a temporal change of the field will induce a *psychological tension*. This finally causes a behavioural change which compensates for the psychological tension and thus restores the equilibrium.

In the following we will show up the connection of the just sketched social field theory with the social force model (cf. Table 11.1):

Table 11.1 Terminological interrelations between LEWIN's social field theory and the social force model

Social field theory	Social force model
Dynamic force field	Social field
Psychological tension	Social force (motivation to act)
Personality	Type of behaviour (subpopulation)
Equilibrium with the dynamic force field	Minimum of the social field

- According to the discussion of Sect. 11.3, the social field $V_a(x,t)$ corresponds to the *dynamic force field* which mathematically describes the environmental influences as *perceived* by individuals of subpopulation a.
- If the behaviour x_α of an individual α does not agree with a minimum $\widehat{x}_a$ of the social field $V_a(x_\alpha,t)$, according to (11.9), the social force

$$K_a(x_\alpha,t) = -\nabla_{x_\alpha} V_a(x_\alpha,t) \tag{11.20}$$

will be induced. This is given by the gradient of the social field $V_a(x_\alpha,t)$ and causes a behavioural change according to

$$\frac{dx_\alpha}{dt} \approx K_a(x_\alpha,t) \tag{11.21}$$

which pulls the behaviour in direction of the steepest descent of $V_a(x_\alpha,t)$. Thus the social force (the motivation) $K_a(x_\alpha,t)$ plays the role of LEWIN's *'psychological tension'*.
- The behavioural change dx_α/dt drives the behaviour towards a minimum $\widehat{x}_a$ of the social field $V_a(x_\alpha,t)$. When this minimum is reached ($x_\alpha = \widehat{x}_a$) we have

$$\nabla_{x_\alpha} V_a(x_\alpha,t) = \mathbf{0} \tag{11.22}$$

and, therefore, $K_a(x_\alpha,t) = \mathbf{0}$. As a consequence, the psychological tension $K_a(x_\alpha,t)$ disappears, i.e. it was compensated by the behavioural change. Then, according to (11.9), only accidental behavioural changes take place. The minima $\widehat{x}_a$ of the social field $V_a(x_\alpha,t)$ thus correspond to the behaviours which are in *equilibrium* with the dynamic force field.
- The social fields $V_a(x_\alpha,t)$ of different subpopulations a normally possess different minima $\widehat{x}_a$. Consequently, individuals of different behavioural types a normally feel different motivations $K_a(x_\alpha,t)$ to act which causes different behavioural changes. Therefore, by subpopulations a different types of *personalities* are distinguished.

In the special case where an individual prefers a behaviour $\widehat{x}_a$ we expect behavioural changes in accordance with the formula

$$\frac{dx_\alpha}{dt} \approx \widehat{c}_a(\widehat{x}_a - x_\alpha)\,. \tag{11.23}$$

This corresponds to the social field

$$V_a(x_\alpha, t) \approx \frac{\widehat{c}_a}{2}(\widehat{x}_a - x_\alpha)^2 \,. \tag{11.24}$$

Examples for this case are discussed in [124].

11.4 Computer Simulations

The social force model is able to describe a broad spectrum of social phenomena. It is, for example, suitable to reproduce *pedestrian motion* [123, 124, 134], to model *group dynamics* (cf. pp. 12, 14), or to delineate *opinion formation.*

In the following we will illustrate a few examples for the application of the social force model by computer simulations. For this purpose, let us consider the case of two subpopulations ($A = 2$) and assume a situation where the interesting aspects of behaviour can be characterized by a certain *position* x on a one-dimensional *continuous scale* ($m = 1$, $\boldsymbol{x} \equiv x$). A concrete example for this are *opinion formation processes.* Then, by the two subpopulations, we could distinguish individuals of a conservative and a progressive attitude respectively. Position x would represent the grade of agreement or refusal with regard to a certain political affair (e.g. SDI, nuclear power stations, or the Gulf War). In contrast to Chap. 10 where we distinguished S different opinions i, this time we consider a *one-dimensional opinion continuum.* The various positions of this continuum are represented by 'x' in order to avoid confusions with discrete opinions i.

For one-dimensional behavioural spaces γ_a the integrability conditions (11.15) are automatically fulfilled so that the social fields

$$V_a(x, t) := -\int_{x_0}^{x} dx' \, K_a(x', t) - c_a(t) \tag{11.25}$$

always exist. The parameters $c_a(t)$ can be chosen arbitrarily. We will select the value

$$c_a(t) := \min_x \left(-\int_{x_0}^{x} dx' \, K_a(x', t) \right) \tag{11.26}$$

which sets the absolute minimum of $V_a(x, t)$ to zero.

In the one-dimensional case, according to (6.70), the *formal stationary solution* of the BOLTZMANN-FOKKER-PLANCK equation (11.2) (which is approached in the long-time limit $t \to \infty$) reads

$$P_a^0(x) = P_a^0(x_0) \frac{Q_a(x_0)}{Q_a(x)} \exp\left(2 \int_{x_0}^{x} dx' \, \frac{K_a(x')}{Q_a(x')} \right) . \tag{11.27}$$

Because of the dependence of $K_a(x)$ and $Q_a(x)$ on $P_b^0(y)$ Eqs. (11.27) are only *implicit equations.* However, from (11.27) we can draw the following conclusions:

- If the diffusion coefficients are constant ($Q_a(x) \equiv Q_a$), (11.27) simplifies to

$$P_a^0(x) = P_a^0(x_0) \exp\left[-\frac{2}{Q_a}\Big(V_a(x) + c_a\Big)\right] \tag{11.28}$$

  so that the stationary solution $P_a^0(x)$ is given by the social field $V_a(x)$. In particular, the behavioural distribution $P_a^0(x)$ has its maxima at positions $\widehat{x}_a$ where the social field $V_a(x)$ has its minima. The diffusion constants Q_a regulate the *broadness* of the behavioural distribution $P_a^0(x)$. For $Q_a = 0$ there were no individual behavioural differences and the distribution $P_a^0(x)$ had a sharp maximum at the position of the lowest minimum of $V_a(x)$.
- If the diffusion coefficients $Q_a(x)$ vary with position x, the behavioural distribution $P_a^0(x)$ will be also influenced by the concrete form of $Q_a(x)$. According to (11.27), we expect large probabilities $P_a^0(x)$ at those positions x where the diffusion coefficients $Q_a(x)$ are small. This is illustrated by Figs. 11.4 and 11.7 in which the form of the probability distribution $P_1^0(x)$, in contrast to Fig. 11.3, cannot be understood merely by the social field $V_1(x)$.
- Since the stationary solution $P_a^0(x)$ depends on $V_a(x)$ as well as on $Q_a(x)$, *different* combinations of $V_a(x)$ and $Q_a(x)$ can lead to the *same* probability distribution $P_a^0(x)$ (cf. Fig. 11.6 in the long-time limit).

In the following the simulations will be restricted to the case of imitative and avoidance processes. This has the consequence that the form of the social field $V_a(x,t)$ varies with time t only due to the temporal change of the probability distributions $P_b(y,t)$, i.e. due to behavioural changes of the individuals (cf. (11.5), (11.16a)). For the positions x we will assume a *quasi-continuum* $x \equiv i/20 \in [1/20, 1]$ with the discretization $i \in \{1, 2, \dots, 20\}$. Furthermore, let the transition rates again be of the form (11.3) with

$$R^a(x'|x;t) := \frac{\mathrm{e}^{U^a(x',t)-U^a(x,t)}}{D^a(x',x;t)} \tag{11.29a}$$

(cf. (9.62a)). By means of the choice

$$U^a(x,t) := -\frac{1}{2}\left(\frac{x-x_a}{l_a}\right)^2 - C_a(t) \qquad \text{with} \qquad l_a := \frac{L_a}{20} \tag{11.29b}$$

individuals of subpopulation a prefer position x_a. $1/L_a$ represents a measure for the *resolution* of subpopulation a with respect to its preferred position x_a. The larger L_a the larger is the *indifference* of subpopulation a regarding the position held. Finally, let us assume the relation

$$\frac{\nu_{ab}(t)}{D^a(x',x;t)} := \mathrm{e}^{-|x'-x|/r} \qquad \text{with} \qquad r := \frac{R}{20} \tag{11.29c}$$

for the meeting rates $\nu_{ab}(t)$ and the distance functions $D_a(x', x; t)$. Then position changes occur the less frequently the larger they are, i.e. the larger $|x' - x|$ is. R can be interpreted as a measure for the *'range'* of the interaction.

- If the range R is very small, the diffusion coefficients $Q_a(x, t)$ can be neglected. Consequently, the fluctuations play an unimportant role and the probability distribution $P_a(x, t)$ is mainly given by the social field $V_a(x, t)$ (cf. Fig. 11.3).
- For larger but still small values of R there occur only small position changes to neighbouring positions with a rate worth mentioning. In order to understand the temporal evolution of the probability distribution $P_a(x, t)$ we must take into account the diffusion coefficients $Q_a(x, t)$, then (cf. Figs. 11.4 and 11.7).
- If R exceeds a certain value the second order TAYLOR approximation becomes useless. In order to be able to comprehend the temporal evolution of $P_a(x, t)$, apart from $K_a(x, t)$ and $Q_a(x, t)$, also *higher jump moments* are required (cf. Sect. 6.2). This, however, presupposes a better approximation than that of the BOLTZMANN-FOKKER-PLANCK equations.

In order to avoid mistakes due to the TAYLOR approximation from the very beginning and to be able to use a simple numerical solution method, the calculation of the distributions $P_a(x, t)$, social fields $V_a(x, t)$, and diffusion coefficients $Q_a(x, t)$ was carried out directly via the BOLTZMANN equations (11.1) with the effective transition rates (11.3), (11.29).

Figures 11.3, 11.4, 11.5, 11.6, 11.7 and 11.8 depict the results of computer simulations for various chosen parameters $f^k_{ab}(t)$, L_a, and R. Here the represented functions are scaled in such a way that their maxima reach half of the box height. Again for the *interaction frequencies* $f^k_{ab}(t)$ *of kind* k we use, like in Sect. 10.3.3, the relations

$$f^1_{ab}(t) = f^1_a(t) l_{ab}(t) \tag{11.30a}$$

and

$$f^2_{ab}(t) = f^2_a(t)\Big(1 - l_{ab}(t)\Big) \tag{11.30b}$$

where $l_{ab}(t)$ denotes the *liking* of subpopulation a for subpopulation b ($0 \le l_{ab}(t)$ ≤ 1).

11.4.1 Imitative Processes

In the following let us consider only imitative processes ($f^1_a(t) \equiv 1$, $f^2_a(t) = 0$). Let the preferred positions x_a be at

$$x_1 := \frac{6}{20} \quad \text{and} \quad x_2 := \frac{15}{20} \,. \tag{11.31}$$

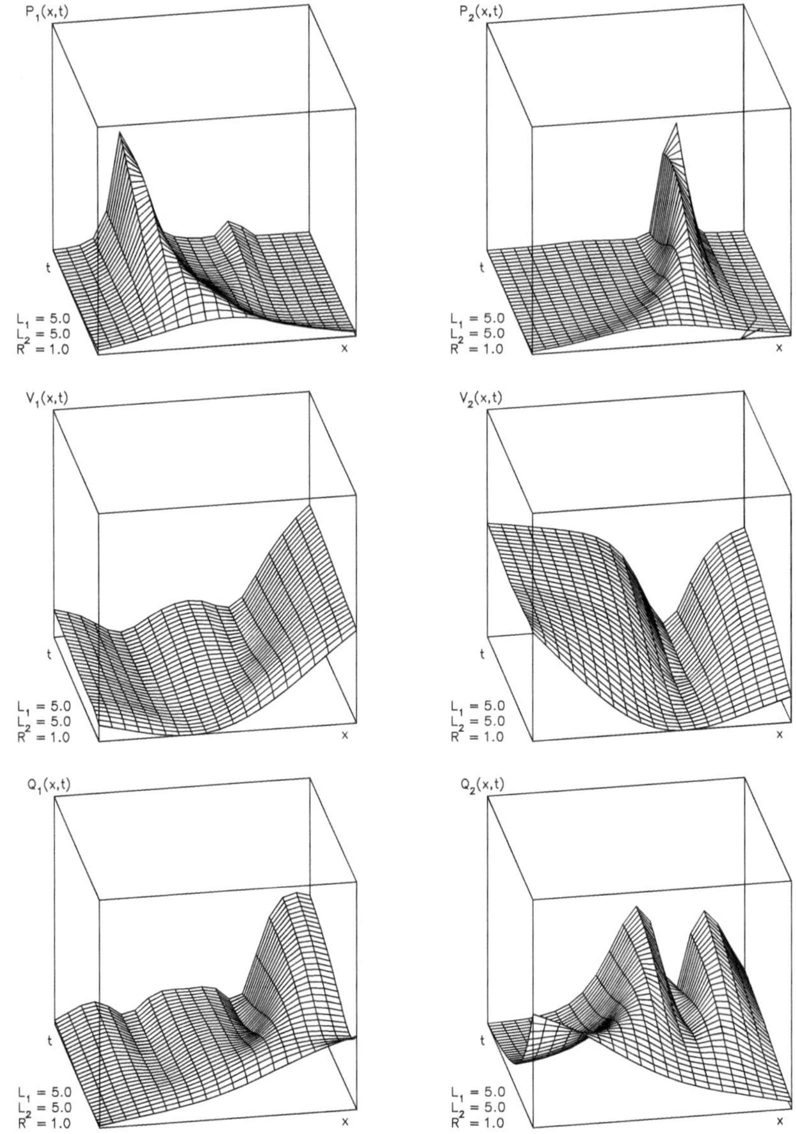

Fig. 11.3 *Opinion distributions* $P_a(x, t)$, *social fields* $V_a(x, t)$, and *diffusion coefficients* $Q_a(x, t)$ for one-sided imitative processes of subpopulation $a = 1$: In the course of time the probability distribution $P_1(x, t)$ develops a second maximum around the position x_2 which is preferred by subpopulation 2. That is, some individuals of subpopulation 1 become persuaded by the preferred position x_2 of subpopulation 2. The maxima of the opinion distributions $P_a(x, t)$ are given by the minima of the corresponding social fields $V_a(x, t)$

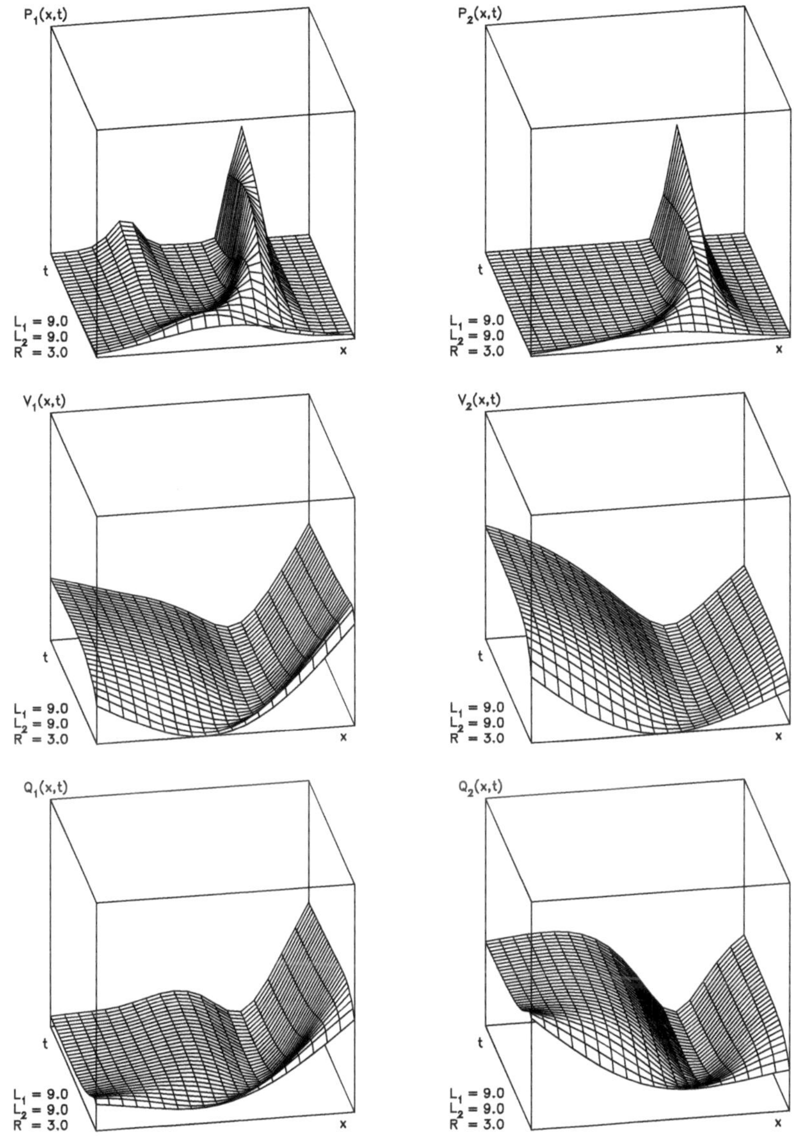

Fig. 11.4 Opinion distributions $P_a(x, t)$, social fields $V_a(x, t)$, and diffusion coefficients $Q_a(x, t)$ for one-sided imitative processes of subpopulation $a = 1$: Since here the indifference L_1 of subpopulation 1 with respect to the position held is *large,* most of its individuals take over the preferred position x_2 of subpopulation 2 (of the *trendsetters*). The opinion distributions $P_a(x, t)$ are not only determined by the corresponding social fields $V_a(x, t)$ but also by the diffusion coefficient $Q_a(x, t)$ (by individual variations of the behavioural changes)

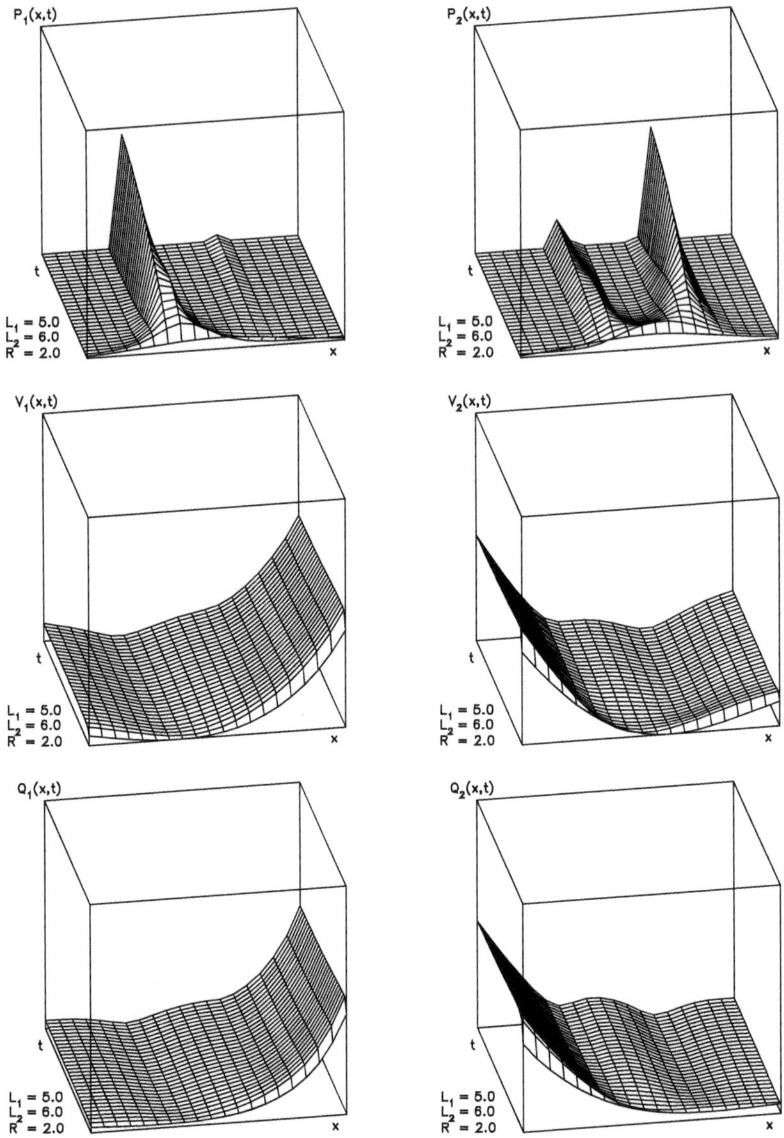

Fig. 11.5 Opinion distributions $P_a(x, t)$, social fields $V_a(x, t)$, and diffusion coefficients $Q_a(x, t)$ for mutual imitative processes of both subpopulations: For a *small* indifference L_a of both subpopulations with respect to their preferred positions x_a, a certain proportion of individuals becomes persuaded by the other subpopulation's preferred position. This is the larger the larger the indifference L_a of subpopulation a is

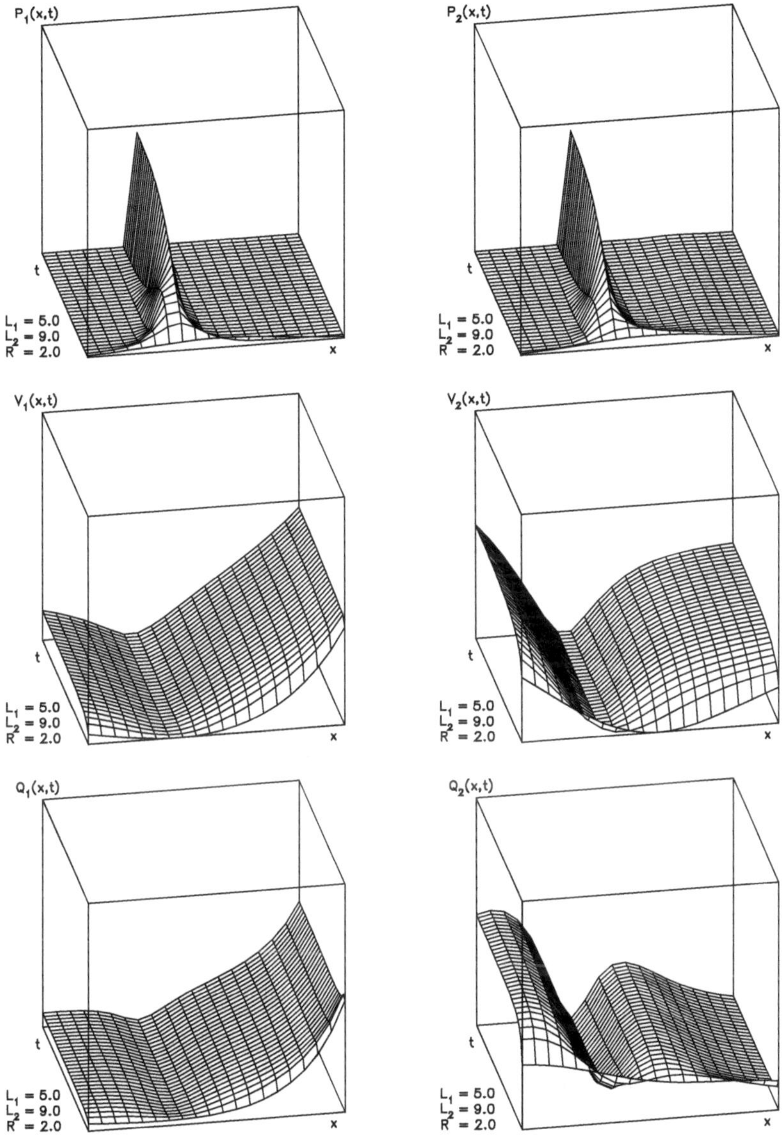

Fig. 11.6 Opinion distributions $P_a(x,t)$, social fields $V_a(x,t)$, and diffusion coefficient $Q_a(x,t)$ for mutual imitative processes of both subpopulations: If the indifference L_a of at least one subpopulation is *large,* the subpopulations will agree by means of persuasion processes on a *common position (compromise)* which is closer to the position x_a that is preferred by the less indifferent subpopulation (here: $a = 1$). Although, after some time, the opinion distributions $P_a(x,t)$ are almost identical the corresponding social fields $V_a(x,t)$ differ considerably due to the difference concerning the preferred positions x_a

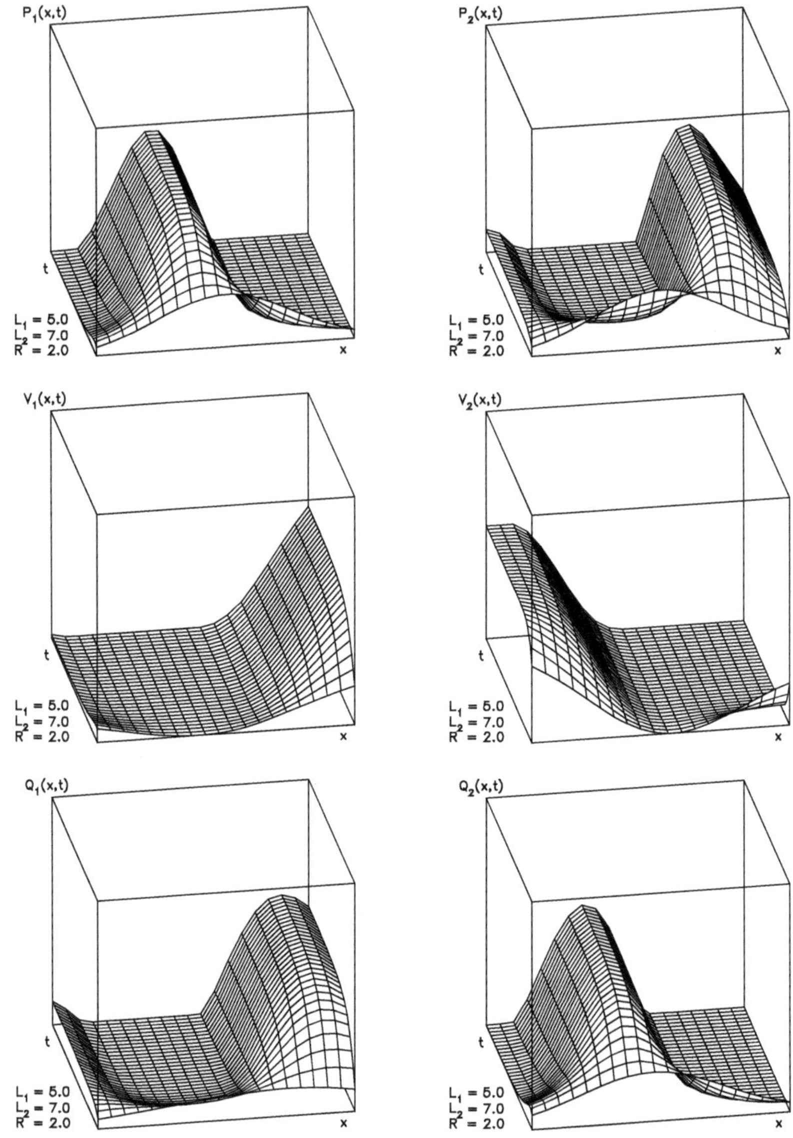

Fig. 11.7 Opinion distributions $P_a(x,t)$, social fields $V_a(x,t)$, and diffusion coefficients $Q_a(x,t)$ for mutual avoidance processes of both subpopulations: The individuals avoid the positions x which are held within the respective other subpopulation. As a consequence, after some time the opinion distributions $P_1(x,t)$ and $P_2(x,t)$ do not overlap any more so that we can observe a *polarization*. In addition, the subpopulation $a = 2$ with the larger indifference L_2 is pushed away from its preferred position x_2

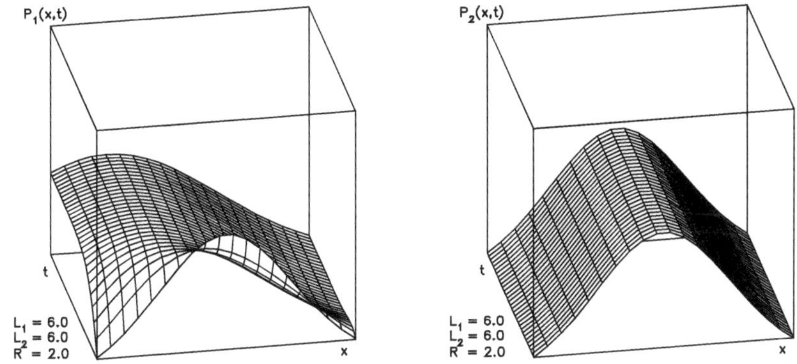

Fig. 11.8 Opinion distributions $P_a(x,t)$ for one-sided avoidance processes of subpopulation 1: Whereas subpopulation 2 keeps its behavioural distribution, the individuals of subpopulation 1 avoid the positions which are held within subpopulation 2. The polarization effect is much smaller than for reciprocal avoidance processes in the case of mutual dislike (cf. Fig. 11.7)

On assumption

$$\Big(l_{ab}(t)\Big) \equiv \Big(f^1_{ab}(t)\Big) := \begin{pmatrix} 1 & 1 \\ 0 & 1 \end{pmatrix} \tag{11.32}$$

individuals of subpopulation $a = 1$ like those of subpopulation $a = 2$ but not vice versa. This implies that subpopulation 2 influences subpopulation 1 in a *one-sided* way.

As expected, for both behavioural distributions $P_a(x,t)$ a maximum occurs at the preferred position x_a. Additionally, a second maximum of $P_1(x,t)$ develops around the preferred position x_2 of subpopulation 2 due to the imitative behaviour of subpopulation 1. This second maximum will be small if the indifference L_1 of subpopulation 1 with respect to changes of the position x is small (cf. Fig. 11.3).

However, for a large indifference L_1 of subpopulation 1 with regard to the position held even the *majority* of individuals imitates the behaviour of subpopulation 2 (cf. Fig. 11.4). One could say that the individuals of subpopulation 2 act as *trend-setters.* This phenomenon is typical for the spreading of a new *fashion.*

Now let us consider the case

$$\Big(l_{ab}(t)\Big) \equiv \Big(f^1_{ab}(t)\Big) := \begin{pmatrix} 1 & 1 \\ 1 & 1 \end{pmatrix} \tag{11.33}$$

where the subpopulations influence each other *mutually* with the same intensity. If the indifference L_a with respect to changes of the position x is small in both subpopulations a, *both* probability distributions $P_a(x,t)$ form *two* maxima. Whereas the higher maximum is located at the position x_a which is preferred by subpopulation a itself, the second maximum is around the preferred position of the *other* subpopulation. It is the higher the larger the indifference L_a is (cf. Fig. 11.5).

However, a *phase transition* occurs if L_a exceeds a certain critical value at least in one subpopulation. Then only *one single* maximum forms so that we observe an entirely different situation. In order to explain this, let us consider Fig. 11.6. Since the indifference L_2 of subpopulation 2 is large, here, its individuals move towards the position of subpopulation 1 due to imitative processes. Nevertheless, also the *majority position* within subpopulation 1 shifts somewhat but, due to the smaller indifference L_1, only a little. After some time, the subpopulations meet each other at one *common position.* Thus we could say that both subpopulations have agreed on a *compromise.* This compromise (the maximum of the distributions $P_a(x,t)$) is closer to the preferred position x_1 of subpopulation 1 which shows the smaller indifference L_1.

The remarkable thing about this compromise is that the behavioural distributions $P_a(x,t)$ agree after some time though the social fields $V_a(x,t)$ and the diffusion coefficients $Q_a(x,t)$ of both subpopulations a differ considerably. The reason for this is that the preferred positions x_a of both subpopulations a are still different.

11.4.2 Avoidance Processes

In the following we will consider only avoidance processes ($f_a^1(t) \equiv 0$, $f_a^2(t) \equiv 1$). With

$$x_1 := \frac{9}{20} \qquad \text{and} \qquad x_2 := \frac{12}{20} \tag{11.34}$$

let us assume that both subpopulations a prefer almost the same position x_a. Fig. 11.7 shows the case

$$\Big(l_{ab}(t)\Big) := \begin{pmatrix} 1 & 0 \\ 0 & 1 \end{pmatrix}, \qquad \text{i.e.} \qquad \Big(f_{ab}^2(t)\Big) \equiv \begin{pmatrix} 0 & 1 \\ 1 & 0 \end{pmatrix}. \tag{11.35}$$

So individuals of different subpopulations do not like each other. This corresponds to a *mutual* influence of both subpopulations on each other. The results of the computer simulations show that

- the individuals avoid the positions x which exist in the other subpopulation,
- the subpopulation $a = 1$ with the smaller indifference $L_1 < L_2$ holds positions which are close to its preferred position x_1 and pushes away the other subpopulation from its preferred position x_2.

Although the initial distribution $P_a(x,0)$ agrees in both subpopulations a, after some time there is almost no overlap of the behavioural distributions $P_1(x,t)$ and $P_2(x,t)$ any more. This is typical for social *polarization phenomena.* Familiar examples for this are the development of *ghettos* or the formation of *hostile groups.*

Figure 11.8 assumes the case that individuals of subpopulation 2 like those of subpopulation 1 and thus do not react on behaviours within subpopulation 1 with avoidance processes:

$$\big(l_{ab}(t)\big) := \begin{pmatrix} 1 & 0 \\ 1 & 1 \end{pmatrix}, \qquad \text{i.e.} \qquad \big(f^2_{ab}(t)\big) \equiv \begin{pmatrix} 0 & 1 \\ 0 & 0 \end{pmatrix}. \tag{11.36}$$

Consequently, the distribution $P_2(x,t)$ does not change with time whereas $P_1(x,t)$ drifts further and further away from the preferred position x_1 due to avoidance processes. Surprisingly enough, the *polarization effect* is much smaller than in Fig. 11.7 according to the considerable overlap of the distributions $P_1(x,t)$ and $P_2(x,t)$. The reason for this is that the gradient of $P_2(x,t)$ is much smaller than in Fig. 11.7 and, in addition, it stays constant. As a consequence, the frequency with which an individual of subpopulation 1 meets disliked individuals of subpopulation 2 can hardly be decreased by a small change of the position x. From this we could draw the conclusion that polarization effects (which often lead to an *escalation*) can be reduced if individuals do not return dislike with dislike.

Chapter 12
Evolutionary Game Theory

12.1 Introduction

For a long time *game theoretical approaches* have been used for a quantitative description of the *competition* or *cooperation* of interacting individuals. These go back to VON NEUMANN and MORGENSTERN [196] and meanwhile they are also empirically validated [46, 230, 233]. In game theory two aspects stand in the centre of interest—the individual attempts to *maximize the personal success* and the *dynamics* resulting from this in connection with the interactions between the individuals.

It turned out that, especially for the understanding of the evolution of cooperation among 'egoists', a repeated interaction and therefore the time dimension is of great importance *('shadow of the future')* [13]. Therefore, a multiple repetition of single game dynamical moves has been suggested *('iterated games')*. Meanwhile, however, the *game dynamical equations* find more and more resonance [137, 139, 256, 271, 303]. These are differential equations and thus continuous in time which will be more realistic at least if there are no separate, clearly defined moves at well-defined points in time. An interesting application of the game dynamical equations is the modelling of the spontaneous *self-organization of behavioural conventions* in the case of equivalent competing behavioural *strategies* (cf. Sect. 12.4.1).

The game dynamical equations originated from the theory of evolution [137, 271, 303] where their application turned out to be very successful. Therefore, in connection with the continuous formulation, we also speak of *evolutionary game theory*. However, in contrast to the theory of evolution where the equations can be derived from the laws of genetics (cf. [138], pp. 43ff.), for their application in the social and behavioural sciences up to now a 'microscopic' foundation is still missing. Such a foundation of the game dynamical equations on the basis of individual behavioural principles will be introduced in Sect. 12.2.4. It connects the game dynamical equations with the BOLTZMANN-like equations for behavioural changes which were developed in Chap. 9. In order to establish this connection we have to interpret $P_a(i, t)$ as the probability with which the *(behavioural) strategy* i is pursued within subpopulation a. Then $p_{ab}(i', j'|i, j; t)$ describes the

D. Helbing, *Quantitative Sociodynamics*, 2nd ed.,
DOI 10.1007/978-3-642-11546-2_12,

probability of two individuals to change their strategies due to pair interactions. Therefore, the remaining problem is to specify the corresponding transition rates $w_{ab}(i', j'|i, j; t) := \nu_{ab}(t) p_{ab}(i', j'|i, j; t)$ in dependence on the expected successes of the strategies.

12.2 Derivation of the Game Dynamical Equations

12.2.1 Payoff Matrix and Expected Success

Let $E_{ab}(i, j;t)$ be the *success* of strategy i for an individual of subpopulation a in an interaction with an individual of subpopulation b pursuing strategy j. Furthermore, let

$$r_{ab}(t) := \frac{\nu_{ab}(t)}{\sum\limits_c \nu_{ac}(t)} \tag{12.1}$$

denote the *relative meeting frequency* of an individual of subpopulation a with individuals of subpopulation b. Then the *expected success* of strategy i is obviously given by

$$E_a(i, t) := \sum_b \sum_j r_{ab}(t) E_{ab}(i, j;t) P_b(j;t) . \tag{12.2a}$$

The matrix

$$\underline{A}_{ab}(t) \equiv \Big(A_{ab}(i, j;t) \Big) := \Big(r_{ab}(t) E_{ab}(i, j;t) \Big) \tag{12.2b}$$

is often called the *payoff matrix.* Normally, the meeting rate $\nu_{ab}(t) = \widehat{\nu}_{ab}(t) N_b$ of an individual of subpopulation a with individuals of subpopulation b is proportional to the relative frequency N_b/N of individuals of subpopulation b:

$$\nu_{ab}(t) = \overline{\nu}(t) \frac{N_b}{N} = \widehat{\nu}(t) N_b . \tag{12.3a}$$

In this case we have

$$r_{ab}(t) = \frac{N_b}{\sum\limits_c N_c} = \frac{N_b}{N} . \tag{12.3b}$$

12.2.2 Customary Derivation

For the temporal evolution of the proportion $P_a(i,t)$ of strategy i within subpopulation a we expect a law of the form

$$\frac{d}{dt} P_a(i,t) = \lambda_a(i,t) P_a(i,t) \tag{12.4}$$

(cf. (10.26a) and Sect. 4.5.2) where the *growth rate* $\lambda_a(i,t)$ depends on the expected success $E_a(i,t)$ of strategy i. Due to the normalization condition, $\lambda_a(i,t)$ must be chosen in such a way that

$$\frac{d}{dt} \sum_i P_a(i,t) = \sum_i \frac{d}{dt} P_a(i,t) \stackrel{!}{=} 0 \tag{12.5}$$

is fulfilled. We immediately obtain a law of the desired form if we assume that the growth rate is proportional to the *difference* between the expected success $E_a(i,t)$ of strategy i and the *average expected success*

$$\langle E_a \rangle := \sum_i E_a(i,t) P_a(i,t) \,, \tag{12.6}$$

i.e.

$$\lambda_a(i,t) := \nu^a(t) \Big[E_a(i,t) - \langle E_a \rangle \Big] \,. \tag{12.7}$$

Then the proportion $P_a(i,t)$ of the strategies i with an expected success $E_a(i,t)$ which exceeds the average $\langle E_a \rangle$ will grow whereas the proportions of the remaining strategies will decrease. The resulting equations

$$\begin{aligned} \frac{d}{dt} P_a(i,t) &= \nu^a(t) P_a(i,t) \Big[E_a(i,t) - \langle E_a \rangle \Big] \\ &= \nu^a(t) P_a(i,t) \Bigg[\sum_{b,j} A_{ab}(i,j;t) P_b(j,t) \\ &\quad - \sum_{b,i',j} P_a(i',t) A_{ab}(i',j;t) P_b(j,t) \Bigg] \end{aligned} \tag{12.8}$$

are known as the *game dynamical equations.*

12.2.3 Fields of Application

Game theory has many interesting applications in economics (e.g. in *oligopoly theory* [269, 280]) as well as in the social and behavioural sciences (e.g. in connection with the *'prisoner's dilemma'*) [12, 13, 139, 175, 196, 234, 252, 253, 283]. However, the game dynamical equations also play an important role in biology *(evolution theory)* where they are often called *replicator equations* [139, 255, 262]. A generalization of (12.8),

$$\frac{d}{dt}P_a(i,t) = \sum_{i'}\Big[w_a(i|i';t)P_a(i',t) - w_a(i'|i;t)P_a(i,t)\Big] \quad (12.9a)$$

$$+ \nu^a(t)P_a(i,t)\Big[E_a(i,t) - \langle E_a\rangle\Big], \quad (12.9b)$$

is known as the *selection-mutation equation* ([139], pp. 249f., [72], pp. 230ff.) or as the EIGEN-FISHER *equation* [60, 78]. Equation (12.9b) can be understood as the effect of *selection* (if we interpret $E_a(i,t)$ as the *fitness* of strategy i). In contrast, (12.9a) describes the effect of spontaneous *mutations.* Whereas the relevance of equation (12.9) for the *theory of evolution* is obvious [61, 72], in game theory the mutation term can be used for modelling *trial and error behaviour* by *spontaneous strategy changes.*

12.2.4 Derivation from the BOLTZMANN-Like Equations

For behavioural changes we have developed BOLTZMANN-like equations (cf. Chap. 9). Consequently, we expect a connection between these equations and the game dynamical equations. A comparison of Eqs. (12.9) with the BOLTZMANN-like equations

$$\frac{d}{dt}P_a(i,t) = \sum_{i'}\Big[w^a(i|i';t)P_a(i',t) - w^a(i'|i;t)P_a(i,t)\Big] \quad (12.10a)$$

with

$$w^a(i|i';t) := w_a(i|i';t) + \nu^a(i,i';t)R^a(i|i';t) \quad (12.10b)$$

and

$$\nu^a(i,i';t) := \sum_b \nu_{ab}(t)\Big[\Big(f^1_{ab}(t)-f^3_{ab}(t)\Big)P_b(i,t)+\Big(f^2_{ab}(t)-f^3_{ab}(t)\Big)P_b(i',t)+f^3_{ab}(t)\Big] \quad (12.10c)$$

shows that the (generalized) game dynamical equations (12.9) can result from these only for

$$\nu^1_{ab}(t) = \nu_{ab}(t)f^1_{ab}(t) := \nu^a(t)\delta_{ab} \qquad \text{and} \qquad f^2_{ab}(t) \equiv 0 \equiv f^3_{ab}(t) . \quad (12.11)$$

Thus for the game dynamical equations, apart from spontaneous transitions, only *imitative processes* are of importance. Avoidance processes or compromising processes have no influence. In order to specify the interaction between the individuals in an appropriate way we argue as follows:

- When an individual meets another individual of *whatever* subpopulation, the expected success $E_a(i,t)$ of the pursued strategy i is evaluated: An individual of subpopulation a meets individuals of subpopulation b at a rate $\nu_{ab}(t)$. With probability $P_b(j,t)$ the interaction partner from subpopulation b pursues strategy j. The success of the individual of subpopulation a in the corresponding interaction will be $E_{ab}(i,j;t)$. Consequently, the evaluation of the expected success

$$E_a(i,t) = \sum_b \sum_j r_{ab}(t) E_{ab}(i,j;t) P_b(j,t) \tag{12.12a}$$

  with

$$r_{ab}(t) = \frac{\nu_{ab}(t)}{\sum_c \nu_{ac}(t)} \tag{12.12b}$$

  is actually possible. Of course, the expected success is normally not *calculated* by the individual but rather it is *estimated* from personal *experience* [228, 229, 251].
- Furthermore, when an individual meets individuals of the *same* subpopulation (who are *cooperative* due to common interests), the expected successes $E_a(i,t)$ of different strategies i are *compared* with each other by *exchange of experiences.* With a certain probability $p^1_{ab}(j|i;t)$ the strategy j of the interaction partner is imitated. The *readiness* $R^a(j|i;t)$ to give up strategy i in favour of strategy j will be the larger the larger the *increase*

$$\Delta_{ji} E_a := E_a(j,t) - E_a(i,t) \tag{12.13}$$

  of the expected success is. However, if an adoption of strategy j were connected with a decrease of the expected success ($\Delta_{ji} E_a < 0$), no imitation would take place at all. In summary, a plausible approach for the readiness to change the strategy from i to j is

$$R^a(j|i;t) := \max\Big(E_a(j,t) - E_a(i,t), 0\Big). \tag{12.14}$$

  It describes an individual *optimization* or *learning process.*
- When an individual meets individuals of *other* subpopulations (who are assumed to behave in a *competitive* way), normally *no* imitative processes will take place: During these interactions the expected success $E_b(j,t)$ of the competitor can at best be *estimated* by *observation* since he/she will not tell his/her experiences. Moreover, there might be

$$E_a(j,t) \not\equiv E_b(j,t) \tag{12.15}$$

due to *different criteria of success.* That is, the expected success $E_b(j,t)$ of strategy j for an individual of subpopulation b does not admit the conclusion about the expected success $E_a(j,t)$ of the same strategy for an individual of subpopulation a. As a consequence, an imitation of the strategy of an individual of subpopulation b by an individual of subpopulation $a(\neq b)$ would be very *risky* since it would probably be connected with a *decrease* of the expected success. Hence the assumption

$$f^1_{ab}(t) \approx 0 \qquad \text{for} \qquad b \neq a \tag{12.16}$$

will normally be justified.

Relation (12.16) also results in cases where the individuals of each subpopulation a *cannot* imitate the strategies of the respective other subpopulations b because these are not available for them (i.e. because of different *strategy sets*).

Now, by inserting (12.11) and (12.14) into (12.10), we obtain the desired *(generalized) game dynamical equations*

$$\begin{aligned}\frac{d}{dt}P_a(i,t) &= \sum_{i'} \Big[w_a(i|i';t)P_a(i',t) - w_a(i'|i;t)P_a(i,t) \Big] \\ &\quad + \nu^a(t)P_a(i,t)\left[E_a(i,t) - \sum_{i'} E_a(i',t)P_a(i',t) \right]\end{aligned} \tag{12.17}$$

because of

$$\max\Big(E_a(i,t)-E_a(i',t),0\Big) - \max\Big(E_a(i',t)-E_a(i,t),0\Big) = E_a(i,t)-E_a(i',t)\,. \tag{12.18}$$

We can conclude that *the game dynamical equations (12.9) result from the* Boltzmann*-like equations for a special case of imitative processes.* In order to define the quantities $\nu^a(t) = \nu_{aa}(t) f^1_{aa}(t)$ and $E_a(i,t)$ in a *unique* way, we finally put

$$f^1_{ab}(t) := \delta_{ab} \qquad \text{which implies} \qquad \nu^a(t) = \nu_{aa}(t) \tag{12.19}$$

so that the rate $\nu^a(t)$ is determined by the meeting rate $\nu_{aa}(t)$ within subpopulation a.

By the way, the approach

$$R^a(j|i;t) := E_a(j,t) \qquad \text{with} \qquad E_a(j,t) \geq 0 \tag{12.20}$$

would also lead to the game dynamical equations (12.9). However, in contrast to (12.14), it would assume a considerable readiness for strategy changes even if these were connected with a *decrease* of expected success. Thus (12.20) is not very plausible.

12.3 Properties of Game Dynamical Equations[1]

In the following we will list some important properties of the game dynamical equations (cf. [139], pp.124ff.).

12.3.1 Non-negativity and Normalization

As it is known from Sect. 4.4.1,

$$P_a(i,t) \geq 0 \qquad \text{and} \qquad \sum_i P_a(i,t) = 1 \qquad \forall t \geq t_0 \tag{12.21}$$

is guaranteed for all BOLTZMANN-like equations (and therefore also for the game dynamical equations) if at a certain time t_0 we have $P_a(i,t_0) \geq 0$ and $\sum_i P_a(i,t_0) = 1$ for all strategies i. The non-negativity and the possibility of normalization are preconditions for an interpretation of the functions $P_a(i,t)$ as *probabilities* or *proportions* of the strategies i.

12.3.2 Formal Solution

For the game dynamical equations (12.9) we can find the *formal solution*

$$P_a(i,t) = \frac{\sum_j M^a_{ij}(t,t_0) P_a(j,t_0)}{\sum_{i',j} M^a_{i'j}(t,t_0) P_a(j,t_0)} \tag{12.22a}$$

with the matrices

$$\underline{M}^a(t,t_0) \equiv \Big(M^a_{ij}(t,t_0)\Big) := \text{T}\exp\Big[(\underline{E}^a + \underline{L}^a), t, t_0\Big] \tag{12.22b}$$

(cf. p. xxiii) and

$$\underline{E}^a(t) \equiv \Big(E^a_{ij}(t)\Big) := \Big(\nu^a(t) E_a(i,t) \delta_{ij}\Big), \tag{12.22c}$$

$$\underline{L}^a(t) \equiv \Big(L^a_{ij}(t)\Big) := \left(w_a(i|j;t) - \delta_{ij} \sum_{i'} w_a(i'|j;t)\right). \tag{12.22d}$$

---

[1] This section can be omitted during a first reading.

This solution satisfies the *initial* and *normalization conditions.* Furthermore, by deriving with respect to time and applying the *quotient rule* (cf. p. xxviii) we obtain

$$
\begin{aligned}
&\frac{d}{dt} P_a(i,t) \\
&= \frac{\left[\sum_{i',j}\Big(E^a_{ii'}(t) + L^a_{ii'}(t)\Big) M^a_{i'j}(t,t_0) P_a(j,t_0)\right] \cdot \left(\sum_{i',j} M^a_{i'j}(t,t_0) P_a(j,t_0)\right)}{\left(\sum_{i',j} M^a_{i'j}(t,t_0) P_a(j,t_0)\right)^2} \\
&\quad - \frac{\left[\sum_{i',j',j}\Big(E^a_{i'j'}(t) + L^a_{i'j'}(t)\Big) M^a_{j'j}(t,t_0) P_a(j,t_0)\right] \cdot \left(\sum_{j} M^a_{ij}(t,t_0) P_a(j,t_0)\right)}{\left(\sum_{i',j} M^a_{i'j}(t,t_0) P_a(j,t_0)\right)^2} \\
&= \sum_{i'} \Big(E^a_{ii'}(t) + L^a_{ii'}(t)\Big) P_a(i',t) - \sum_{i',j'} \Big(E^a_{i'j'}(t) + L^a_{i'j'}(t)\Big) P_a(j',t) P_a(i,t) \\
&= \nu^a(t) E_a(i,t) P_a(i,t) - P_a(i,t) \sum_j \nu^a(t) E_a(j,t) P_a(j,t) \\
&\quad + \sum_{i'} L^a_{ii'}(t) P_a(i',t) - P_a(i,t) \sum_{i',j} L^a_{i'j}(t) P_a(j,t) \\
&= \sum_{i'} L^a_{ii'}(t) P_a(i',t) + \nu^a(t) P_a(i,t)\Big(E_a(i,t) - \langle E_a \rangle\Big) .
\end{aligned}
\tag{12.23}
$$

With (3.100) this means that the formal solution (12.22) fulfills the game dynamical equations (12.9).

12.3.3 Increase of the Average Expected Success in Symmetrical Games

For the *weighted mean*

$$
\langle E \rangle := \sum_a \frac{N_a}{N} \langle E_a \rangle \tag{12.24}
$$

of the average expected successes $\langle E_a(i,t) \rangle$ the *fundamental theorem*

$$
\frac{d\langle E \rangle}{dt} \geq 0 \tag{12.25}
$$

can be proved given $w_a(i'|i;t) = 0$, the time-independence of the successes ($E_{ab}(j,i;t) \equiv E_{ab}(j,i)$), the *symmetry*

$$E_{aa}(i,j) = E_{aa}(j,i)\,, \tag{12.26}$$

and

$$r_{ab}(t) = \frac{N_b}{N} \tag{12.27}$$

(cf. (12.3b)). By interchanging the variables of two interacting individuals α and β we find

$$E_{ab}(i,j) = E_{\alpha\beta}(i,j) = E_{\beta\alpha}(j,i) = E_{ba}(j,i)\,. \tag{12.28}$$

Hence we obtain

$$\begin{aligned}
\frac{d\langle E\rangle}{dt} &= \sum_a \frac{N_a}{N}\left(\sum_i E_a(i,t)\frac{d}{dt}P_a(i,t) + \sum_i P_a(i,t)\frac{d}{dt}E_a(i,t)\right)\\
&= \sum_a \frac{N_a}{N}\sum_i E_a(i,t)\nu^a(t)P_a(i,t)\Big(E_a(i,t) - \langle E_a\rangle\Big)\\
&\quad + \sum_a \frac{N_a}{N}\sum_i P_a(i,t)\sum_{b,j}\frac{N_b}{N}E_{ab}(i,j)\nu^b(t)P_b(j,t)\Big(E_b(j,t) - \langle E_b\rangle\Big) \qquad (12.29)\\
&= \sum_{a,i}\sum_{b,j}\frac{N_a}{N}\frac{N_b}{N}E_{ab}(i,j)P_b(j,t)\nu^a(t)P_a(i,t)\Big(E_a(i,t) - \langle E_a\rangle\Big)\\
&\quad + \sum_{a,i}\sum_{b,j}\frac{N_b}{N}P_b(j,t)\frac{N_a}{N}\underbrace{E_{ba}(j,i)}_{=E_{ab}(i,j)}\nu^a(t)P_a(i,t)\Big(E_a(i,t) - \langle E_a\rangle\Big) \qquad (12.30)\\
&= 2\sum_a \frac{N_a}{N}\nu^a(t)\sum_i P_a(i,t)E_a(i,t)\Big(E_a(i,t) - \langle E_a\rangle\Big)\\
&= 2\sum_a \frac{N_a}{N}\nu^a(t)\Big(\langle (E_a)^2\rangle - \langle E_a\rangle^2\Big)\\
&= 2\sum_a \frac{N_a}{N}\nu^a(t)\sum_i \Big(E_a(i,t) - \langle E_a\rangle\Big)^2 P_a(i,t) \qquad (12.31)\\
&\ge 0\,. \qquad (12.32)
\end{aligned}$$

From (12.29) to (12.30) we interchanged a and b as well as i and j. In the case of one single subpopulation ($A = 1$) relation (12.31) states that *the increase of the average expected success is proportional to the variance of the expected successes.*

12.3.4 Invariant of Motion for Antisymmetrical Games

In the following let us consider the case of one single subpopulation ($A = 1$) and a time-independent *antisymmetrical* success matrix

$$\Big(E(i,j)\Big) = -\Big(E(j,i)\Big) . \tag{12.33}$$

Then we find for the average expected success

$$\langle E(i,t)\rangle = -\langle E(i,t)\rangle = 0 \tag{12.34}$$

and (12.8) simplifies to

$$\frac{d}{dt}P(i,t) = \nu(t)E(i,t)P(i,t) = \nu(t)P(i,t)\sum_j E(i,j)P(j,t) . \tag{12.35}$$

If $P^0 \equiv (P^0(i_1), \ldots, P^0(i_S))^{\text{tr}}$ is a *stationary point* of (12.35) with

$$\sum_j E(i,j)P^0(j) = 0 \qquad \forall i , \tag{12.36}$$

the quantity

$$C := \prod_j P(j,t)^{P^0(j)} \tag{12.37}$$

is time-independent, i.e. an *invariant of motion* due to

$$\begin{aligned} \frac{dC}{dt} &= \sum_i \frac{dP(i,t)}{dt}\frac{P^0(i)}{P(i,t)}\prod_j P(j,t)^{P^0(j)} \\ &= \nu(t)\,C\sum_{i,j} P^0(i)E(i,j)P(j,t) \\ &= -\nu(t)\,C\sum_j P(j,t)\sum_i E(j,i)P^0(i) \\ &= 0 . \end{aligned} \tag{12.38}$$

Because of the invariant of motion (12.37) and normalization condition (12.5), the trajectory $P(t) \equiv (P(i_1,t), \ldots, P(i_S,t))^{\text{tr}}$ moves on a $(S-2)$-dimensional *hypersurface*. Concrete examples for this were discussed in Sect. 10.3.2.

12.3.5 Interrelation with the Lotka-Volterra Equations

The *general* Lotka-Volterra *equations* [172, 173, 282] have the form

$$\frac{dy_i(\tau)}{d\tau} = y_i(\tau)\left(A_{i0} + \sum_{j=1}^{S} A_{ij} y_j(\tau)\right) \qquad (i \in \{1, \ldots, S\}). \tag{12.39}$$

They are particularly often applied in *population ecology* for the description of *food chains (predator-prey systems)* (cf. [91, 109, 139, 219]). $y_i(t) \geq 0$ denotes the *size* of *population* i. The constant A_{i0} describes the rate of change of population i in the absence of other populations whereas the constants A_{ij} reflect the influences of populations j on the temporal change of population i (via predator-prey relations or *symbioses*).

For the Lotka-Volterra equations of S variables $y_1(\tau), \ldots, y_S(\tau)$ it can be shown that they are mathematically *flow-equivalent* to the game dynamical equations of $(S+1)$ variables $P(0,t), P(1,t), \ldots, P(S,t)$ (cf. [139], pp. 134f.). In order to prove this, we first transform the S equations (12.39) with

$$A_{0j} := 0 \qquad \text{and} \qquad y_0(\tau) = y_0(\tau_0) := 1 \tag{12.40a}$$

into the $(S+1)$ equations

$$\frac{d}{d\tau} y_i(\tau) = y_i(\tau) \cdot \left(\sum_{j=0}^{S} A_{ij} y_j(\tau)\right) \qquad (i \in \{0, 1, \ldots, S\}). \tag{12.40b}$$

These equations result from the game dynamical equations

$$\frac{d}{dt} P(i,t) = \nu P(i,t)\Big[E(i,t) - \langle E \rangle\Big] \qquad (i \in \{0, 1, \ldots, S\}) \tag{12.41}$$

by means of invertible variable transformations: Let

$$P(i,t) := \frac{y_i(t)}{\sum\limits_{i=0}^{S} y_i(t)} \tag{12.42}$$

be the *relative population sizes*. Then, due to $y_0(\tau) = 1$, we find

$$P(0,t) = \frac{1}{\sum\limits_{i=0}^{S} y_i(t)} \tag{12.43}$$

and thus

$$\frac{y_i(t)}{y_0(t)} = y_i(t) = \frac{P(i,t)}{P(0,t)} . \tag{12.44}$$

The application of the quotient rule and of (12.41) leads, after a few conversions, to

$$\frac{d}{dt}\left(\frac{P(i,t)}{P(0,t)}\right) = \frac{P(i,t)}{P(0,t)} \cdot \nu\Big[E(i,t) - E(0,t)\Big] . \tag{12.45}$$

(12.45) implies with

$$A_{ij} := \nu A(i,j) \tag{12.46a}$$

and due to

$$E(i,t) = \sum_{j=0}^{S} A(i,j)P(j,t) , \tag{12.46b}$$

$$E(0,t) = \sum_{j=0}^{S} A(0,j)P(j,t) = 0 \tag{12.46c}$$

the relation

$$\frac{d}{dt} y_i(t) = y_i(t) \cdot \left(\sum_{j=0}^{S} A_{ij} P(j,t) \right) = y_i(t) \left(\sum_{j=0}^{S} A_{ij} y_j(t) \right) P(0,t) . \tag{12.47}$$

With the *time scaling (velocity transformation)*

$$\tau(t) := \int_{t_0}^{t} dt' \, P(0,t') \tag{12.48}$$

we finally obtain (12.40).

12.3.6 Limit Cycles and Chaos

The special game dynamical equation (12.8) with $\nu(t) \equiv \nu$ and the payoff matrix

$$\underline{A} := \begin{pmatrix} 0 & 0 & -\kappa & 1 \\ 1 & 0 & 0 & -\kappa \\ -\kappa & 1 & 0 & 0 \\ 0 & -\kappa & 1 & 0 \end{pmatrix} \tag{12.49}$$

has the stationary point $P^0 = (1/4, 1/4, 1/4, 1/4)^{\text{tr}}$. A *linear stability analysis* with $\delta P(i,t) := P(i,t) - P^0(i) \ll 1$ results in

$$\frac{d}{dt}\delta P(i,t) = \sum_j M_{ij}\,\delta P(j,t) \tag{12.50a}$$

where the JACOBIan *matrix*

$$\underline{M} \equiv \left(M_{ij}\right) := \nu \begin{pmatrix} -1+\kappa & -1+\kappa & -1-\kappa & 1+\kappa \\ 1+\kappa & -1+\kappa & -1+\kappa & -1-\kappa \\ -1-\kappa & 1+\kappa & -1+\kappa & -1+\kappa \\ -1+\kappa & -1-\kappa & 1+\kappa & -1+\kappa \end{pmatrix} \tag{12.50b}$$

possesses the eigenvalues

$$\lambda_1 = \frac{\nu}{4}(-1+\kappa)\,, \tag{12.51a}$$

$$\lambda_2 = \frac{\nu}{4}(-1-\kappa)\,, \tag{12.51b}$$

$$\lambda_3 = \frac{\nu}{4}(\kappa+\mathrm{i})\,, \tag{12.51c}$$

$$\lambda_4 = \frac{\nu}{4}(\kappa-\mathrm{i})\,. \tag{12.51d}$$

If $\kappa \in [-0.5, 0.5]$ is varied from negative to positive values, λ_1 and λ_2 will always be negative whereas the complex conjugate pair λ_3, λ_4 will cross the imaginary axis from the left half-plane to the right one. Hence the fixed point P^0 is stable only for $\kappa < 0$. For $\kappa = 0$ one finds a HOPF *bifurcation,* and for small values $\kappa > 0$ one discovers a *limit cycle* (cf. Sect. 7.5 and [138], pp. 178f.).

Game dynamical equations (12.8) can even show a *chaotic* temporal evolution [7, 279]. For example, this is the case for the payoff matrices

$$\underline{A} := \begin{pmatrix} 0.0 & 0.5 & -0.1 & 0.1 \\ 1.1 & 0.0 & -0.6 & 0.0 \\ -0.5 & 1.0 & 0.0 & 0.0 \\ 1.7 & -1.0 & -0.2 & 0.0 \end{pmatrix} \tag{12.52}$$

and

$$\underline{A} := \begin{pmatrix} 0.000 & 0.063 & 0.000 & 0.437 \\ 0.537 & 0.000 & -0.036 & -0.001 \\ -0.535 & 0.380 & 0.000 & 0.655 \\ 0.536 & -0.032 & -0.004 & 0.000 \end{pmatrix} \tag{12.53}$$

(cf. [248]).

12.4 Stochastic Version of the Game Dynamical Equations

Due to the fact that the game dynamical equations are, according to Sect. 12.2.4, a special case of the BOLTZMANN-like equations we can, in connection with Chap. 5, immediately formulate a stochastic version of the game dynamical equations (cf. [72], pp. 207ff., [58], pp. 239ff.). This is given by the *configurational master equation*

$$\frac{d}{dt}P(n,t) = \sum_{n'}\Big[w(n|n';t)P(n',t) - w(n'|n;t)P(n,t)\Big] \tag{12.54a}$$

with the transition rates

$$w(n'|n;t) = \begin{cases} \widehat{w}_a(i'|i;n;t)n_i^a & \text{if } n' = n_{i'i}^{aa} \\ \widehat{w}_{ab}(i',j'|i,j;n;t)n_i^a n_j^b & \text{if } n' = n_{i'j'ij}^{abab} \\ 0 & \text{otherwise} \end{cases} \tag{12.54b}$$

(cf. (5.50)) ($\widehat{w}_a = w_a, \widehat{w}_{ab} \stackrel{N_a \gg 1}{=} w_{ab}/N_b$). Here for $\widehat{w}_{ab}(i',j'|i,j;n;t)$ we use the approach (9.52a) for imitative processes:

$$\begin{aligned}\widehat{w}_{ab}(i',j'|i,j;n;t) := \widehat{\nu}_{ab}(t)f^1_{ba}(t)R^a(i|j;n;t)\delta_{ii'}\delta_{ij'}(1-\delta_{ij}) \\ + \widehat{\nu}_{ab}(t)f^1_{ab}(t)R^a(j|i;n;t)\delta_{jj'}\delta_{ji'}(1-\delta_{ij})\,.\end{aligned} \tag{12.54c}$$

Particularly, for the special imitative processes of game dynamics we have

$$\widehat{\nu}_{ab}(t)N_b f^1_{ab}(t) = \nu_{ab}(t)f^1_{ab}(t) := \nu^a(t)\delta_{ab} \tag{12.54d}$$

and

$$R^a(j|i;n;t) := \max\Big(E_a(j,n;t) - E_a(i,n;t), 0\Big) \tag{12.54e}$$

with

$$E_a(i,n;t) := \sum_b\sum_j A_{ab}(i,j;t)\frac{n_j^b}{N_b} \tag{12.54f}$$

(cf. (12.11), (12.14), and (12.2)).

The stochastic formulation of the game dynamical equations allows, on the one hand, the calculation of the *probability* $P(n,t)$ with which possible socioconfigurations n (i.e. strategy distributions) occur. On the other hand, it allows the derivation of *covariance equations* which are necessary

- for the calculation of the *reliability of game dynamical results* which is given by the *standard deviations* $\sqrt{\sigma_{aa}}$,
- for the determination of the *period of time* during which game dynamical equations (12.9) can be interpreted as *approximate mean value equations* (cf. Sect. 5.3).

Moreover, *corrected mean value* and *covariance equations* (cf. Sect. 5.4) become derivable. For this purpose, let us introduce the following abbreviations:

$$L^{a-}_{ii'} := w_a(i|i';n;t) - \delta_{ii'} \sum_j w_a(j|i';n;t)\,, \tag{12.55a}$$

$$L^{a+}_{ii'} := w_a(i|i';n;t) + \delta_{ii'} \sum_j w_a(j|i';n;t)\,, \tag{12.55b}$$

$$M^{a-}_{ii'} := \nu^a(t)\Big[R^a(i|i';n;t) - R^a(i'|i;n;t)\Big]\,, \tag{12.55c}$$

$$M^{a+}_{ii'} := \nu^a(t)\Big[R^a(i|i';n;t) + R^a(i'|i;n;t)\Big] \tag{12.55d}$$

Then, with (5.15) and (5.24), we obtain

$$\frac{\partial\langle n^a_i\rangle}{\partial t} = \sum_{i'}\left(L^{a-}_{ii'}\langle n^a_{i'}\rangle + M^{a-}_{ii'}\frac{\langle n^a_i\rangle\langle n^a_{i'}\rangle}{N_a}\right) \tag{12.56a}$$

$$\begin{aligned}
&+ \frac{s}{2}\sum_{b,c}\sum_{j,j'}\sigma^{bc}_{jj'}\sum_{i'}\frac{\partial^2 L^{a-}_{ii'}}{\partial\langle n^b_j\rangle\partial\langle n^c_{j'}\rangle}\langle n^a_{i'}\rangle \\
&+ \frac{s}{2}\sum_{b,c}\sum_{j,j'}\sigma^{bc}_{jj'}\sum_{i'}\frac{\partial^2 M^{a-}_{ii'}}{\partial\langle n^b_j\rangle\partial\langle n^c_{j'}\rangle}\frac{\langle n^a_i\rangle\langle n^a_{i'}\rangle}{N_a} \\
&+ s\sum_b\sum_{i',j}\sigma^{ab}_{i'j}\left(\frac{\partial L^{a-}_{ii'}}{\partial\langle n^b_j\rangle} + \frac{\partial M^{a-}_{ii'}}{\partial\langle n^b_j\rangle}\frac{\langle n^a_i\rangle}{N_a}\right) \\
&+ s\sum_b\sum_j\sigma^{ab}_{ij}\left(M^{a-}_{ij}\frac{\delta_{ab}}{N_a} + \sum_{i'}\frac{\partial M^{a-}_{ii'}}{\partial\langle n^b_j\rangle}\frac{\langle n^a_{i'}\rangle}{N_a}\right)
\end{aligned} \tag{12.56b}$$

as *approximate mean value equations* ($s = 0$) and *corrected mean value equations* ($s = 1$) respectively (cf. (5.16) and (5.47)). The game dynamical equations correspond to the *approximate* mean value equations (12.56a). However, according to Sect. 7.1, they can also be interpreted as equations for the *most probable individual behaviour.* The interpretation as approximate mean value equations is only possible as long as the variances σ_{aa} are small (cf. (5.20)). In doubtful cases one should rather use the corrected mean value equations (12.56) which themselves depend on the covariances σ_{ab} and thus take into account the *fluctuations* which result from the stochasticity of the described processes. However, the corrected mean value equations require the solution of the covariance equations. As *approximate covariance equations* ($s = 0$) and *corrected covariance equations* ($s = 1$) respectively, we find (cf. (5.22) and (5.48)):

$$
\begin{aligned}
&\frac{\partial \sigma_{i\,i'}^{ab}}{\partial t} \\
&= \delta_{ab}\left[\delta_{ii'}\sum_j\left(L_{ij}^{a+}\langle n_j^a\rangle + M_{ij}^{a+}\frac{\langle n_i^a\rangle\langle n_j^a\rangle}{N_a}\right)\right. \\
&\qquad \left. -\left(w_a(i|i';n;t)\langle n_{i'}^a\rangle + w_a(i'|i;n;t)\langle n_i^a\rangle + M_{ii'}^{a+}\frac{\langle n_i^a\rangle\langle n_{i'}^a\rangle}{N_a}\right)\right] \\
&\quad + \sum_{c,k}\sigma_{i\,k}^{ac}\left[\sum_j\left(\frac{\partial L_{i'j}^{b-}}{\partial\langle n_k^c\rangle}\langle n_j^b\rangle + \frac{\partial M_{i'j}^{b-}}{\partial\langle n_k^c\rangle}\frac{\langle n_{i'}^b\rangle\langle n_j^b\rangle}{N_b}\right) + \delta_{bc}\left(L_{i'k}^{b-} + M_{i'k}^{b-}\frac{\langle n_{i'}^b\rangle}{N_b}\right)\right] \\
&\quad + \sum_{c,k}\sigma_{i'k}^{bc}\left[\sum_j\left(\frac{\partial L_{ij}^{a-}}{\partial\langle n_k^c\rangle}\langle n_j^a\rangle + \frac{\partial M_{ij}^{a-}}{\partial\langle n_k^c\rangle}\frac{\langle n_i^a\rangle\langle n_j^a\rangle}{N_a}\right) + \delta_{ac}\left(L_{ik}^{a-} + M_{ik}^{a-}\frac{\langle n_i^a\rangle}{N_a}\right)\right] \\
&\quad + \sigma_{i\,i'}^{ab}\sum_j\left(M_{ij}^{a-}\frac{\langle n_j^a\rangle}{N_a} + M_{i'j}^{b-}\frac{\langle n_j^b\rangle}{N_b}\right)
\end{aligned}
\tag{12.57a}
$$

$$
\begin{aligned}
&\quad + \delta_{ab}\left\{\delta_{ii'}\left[\frac{s}{2}\sum_{c,d}\sum_{k,k'}\sigma_{k\,k'}^{cd}\sum_j\left(\frac{\partial^2 L_{ij}^{a+}}{\partial\langle n_k^c\rangle\partial\langle n_{k'}^d\rangle}\langle n_j^a\rangle + \frac{\partial^2 M_{ij}^{a+}}{\partial\langle n_k^c\rangle\partial\langle n_{k'}^d\rangle}\frac{\langle n_i^a\rangle\langle n_j^a\rangle}{N_a}\right)\right.\right. \\
&\quad + s\sum_{c,k}\sum_j\sigma_{j\,k}^{ac}\left(\frac{\partial L_{ij}^{a+}}{\partial\langle n_k^c\rangle} + \frac{\partial M_{ij}^{a+}}{\partial\langle n_k^c\rangle}\frac{\langle n_i^a\rangle}{N_a}\right) \\
&\quad \left. + s\sum_{c,k}\sigma_{i\,k}^{ac}\left(\sum_j\frac{\partial M_{ij}^{a+}}{\partial\langle n_k^c\rangle}\frac{\langle n_j^a\rangle}{N_a} + M_{ik}^{a+}\frac{\delta_{ac}}{N_a}\right)\right] \\
&\quad - \left[\frac{s}{2}\sum_{c,d}\sum_{k,k'}\sigma_{k\,k'}^{cd}\left(\frac{\partial^2 w_a(i,i';t)}{\partial\langle n_k^c\rangle\partial\langle n_{k'}^d\rangle}\langle n_{i'}^a\rangle + \frac{\partial^2 w_a(i'|i;t)}{\partial\langle n_k^c\rangle\partial\langle n_{k'}^d\rangle}\langle n_i^a\rangle\right)\right. \\
&\quad + \frac{s}{2}\sum_{c,d}\sum_{k,k'}\sigma_{k\,k'}^{cd}\frac{\partial^2 M_{ii'}^{a+}}{\partial\langle n_k^c\rangle\partial\langle n_{k'}^d\rangle}\frac{\langle n_i^a\rangle\langle n_{i'}^a\rangle}{N_a} \\
&\quad + s\sum_{c,k}\left(\sigma_{i\,k}^{ac}\frac{\partial w_a(i'|i;t)}{\partial\langle n_k^c\rangle} + \sigma_{i'k}^{ac}\frac{\partial w_a(i|i';t)}{\partial\langle n_k^c\rangle}\right) \\
&\quad \left.\left. + s\sum_{c,k}\left(\sigma_{i\,k}^{ac}\frac{\partial M_{ii'}^{a+}}{\partial\langle n_k^c\rangle}\frac{\langle n_{i'}^a\rangle}{N_a} + \sigma_{i'k}^{ac}\frac{\partial M_{ii'}^{a+}}{\partial\langle n_k^c\rangle}\frac{\langle n_i^a\rangle}{N_a}\right) + \sigma_{i\,i'}^{aa}\frac{M_{ii'}^{a+}}{N_a}\right]\right\}.
\end{aligned}
\tag{12.57b}
$$

An example for the application of these equations and of the game dynamical methods is given in the next section.

12.4.1 Self-Organization of Behavioural Conventions for the Case of Two Equivalent Competing Strategies

In the following let us consider the case of one single subpopulation ($A = 1$). Then the subscripts and superscripts a, b, c can be omitted. (A socio-biological example for game dynamical equations for two subpopulations can be found in [139], pp. 137ff.)

If the considered individuals have the choice between only two strategies $i \in \{1, 2\}$, this implies

$$n_2 = N - n_1 \tag{12.58}$$

where N denotes the number of individuals. Furthermore, for *equivalent strategies* the payoff matrix $\underline{A}$ is *symmetrical* and of the form

$$\underline{A} \equiv \Big(A(i, j)\Big) := \begin{pmatrix} \widehat{A} + \widehat{B} & \widehat{B} \\ \widehat{B} & \widehat{A} + \widehat{B} \end{pmatrix} . \tag{12.59}$$

By scaling the time we can achieve

$$\nu(t) \equiv \nu . \tag{12.60}$$

With

$$w(i'|i; n; t) := \widehat{W} \tag{12.61}$$

we will assume the *simplest* case for the transition rates describing spontaneous strategy changes.

An example for the above mentioned situation is the *avoidance behaviour of pedestrians* (cf. Fig. 12.1): In crowds of pedestrians who walk into opposite directions, avoidance manoeuvres fail the less frequently (i.e. have the larger expected success) the larger the number n_i of pedestrians is who swerve to the same side i. Thus in the payoff matrix (12.59) there is $\widehat{A} > 0$ because it is advantageous to apply the *same* strategy as the interaction partner. Consequently, we expect a *symmetry breaking* via a *phase transition.* In fact, according to experience, in Germany and some other countries pedestrians prefer the right-hand side ($i := 1$) [204, 205]. With the same probability, however, a preference of the left-hand side ($i := 2$) could have

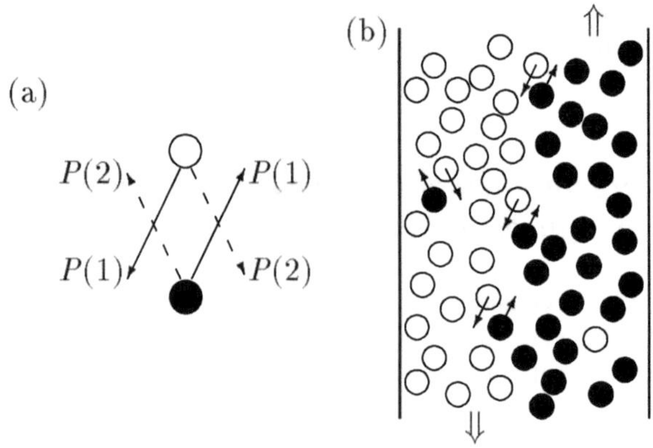

Fig. 12.1 (**a**) Avoidance problem of pedestrians: For pedestrians who walk in opposite directions it will be advantageous if they swerve both to the right (solid arrows) or both to the left (dashed arrows). Otherwise they are forced to stop in order to avoid a collision. The probability $P(1)$ to swerve to the right is (in Germany and some other countries) normally greater than the probability $P(2) = 1 - P(1)$ to swerve to the left. (**b**) In dense pedestrian crowds *lanes* develop in which one of the two opposite walking directions (represented by black and white circles respectively) dominates. So the frequency of avoidance manoeuvres is reduced. These are indicated by arrows

won out which is a priori completely equivalent. Since in some countries this case seems also to be given, an explanation of the symmetry breaking by the asymmetry of the human body must be discarded.

Some people argue that the reason for the asymmetrical avoidance behaviour of pedestrians could be the habits regarding road traffic. In this case, however, a similar model as above would be required to explain why in *some* countries left-hand traffic is prescribed (e.g. in Great Britain) and in *others* right-hand traffic (e.g. on European mainland). Here we can assume that the legislator adapts to once established norms or conventions. Very probably these conventions first developed at pedestrian traffic since the avoidance problem is much older there than at vehicular traffic.

Apart from this, in many countries of the world (e.g. India) vehicular traffic plays a completely subordinate role compared to pedestrian traffic, not to mention that the fewest pedestrians have a driving licence there. At least in these countries it is improbable that pedestrians behave according to the perspective of a car driver.

The model for the self-organization of behavioural conventions is not limited to the avoidance problem. It can rather be applied to a number of other cases where it is advantageous for the majority to choose the same behavioural alternative. Some examples are discussed on pp. 4, 5.

For the above described situation the stochastic version of the game dynamical equations has the explicit form

$$\begin{aligned}\frac{d}{dt}P(n_1, n_2; t) = &\left[w(1|2) + \nu R(1|2; n_1 - 1, n_2 + 1; t)\frac{n_1 - 1}{N}\right](n_2 + 1) \\ &\times P(n_1 - 1, n_2 + 1; t) \\ &+ \left[w(2|1) + \nu R(2|1; n_1 + 1, n_2 - 1; t)\frac{n_2 - 1}{N}\right](n_1 + 1) \\ &\times P(n_1 + 1, n_2 - 1; t) \\ &- \left[w(1|2) + \nu R(1|2; n_1, n_2; t)\frac{n_1}{N}\right] n_2 P(n_1, n_2; t) \\ &- \left[w(2|1) + \nu R(2|1; n_1, n_2; t)\frac{n_2}{N}\right] n_1 P(n_1, n_2; t) \end{aligned} \tag{12.62}$$

with

$$R(j|i; n; t) := \max\Big(E(j, n; t) - E(i, n; t), 0\Big) \tag{12.63}$$

and

$$E(i, n; t) := \sum_j A(i, j)\frac{n_j}{N}. \tag{12.64}$$

For the calculation of the related mean value equations (12.56) and covariance equations (12.57) we require

$$L^-_{ii'} = \widehat{W} - \delta_{ii'} S\widehat{W}, \tag{12.65a}$$

$$L^+_{ii'} = \widehat{W} + \delta_{ii'} S\widehat{W}, \tag{12.65b}$$

$$M^-_{ii'} = \nu\Big[E(i, t) - E(i', t)\Big], \tag{12.65c}$$

$$M^+_{ii'} = \nu\Big|E(i, t) - E(i', t)\Big| \tag{12.65d}$$

with

$$E(i, t) = \sum_j A(i, j)P(j, t). \tag{12.66}$$

In particular, the approximate mean value equations (and thus the game dynamical equations) have, due to $P(2, t) = 1 - P(1, t)$, the explicit form

$$\frac{d}{dt}P(i, t) = -2\left(P(i, t) - \frac{1}{2}\right)\left[\widehat{W} + \nu\widehat{A}P(i, t)\Big(P(i, t) - 1\Big)\right]. \tag{12.67}$$

Obviously, $P(i, t) \equiv P^0(i) = 1/2$ is a stationary solution of (12.67) which describes a situation in which both strategies are chosen with the same frequency. This solution will be only stable if the parameter

$$\kappa := 1 - \frac{4\widehat{W}}{\nu \widehat{A}} \tag{12.68}$$

fulfills the condition $\kappa < 0$, i.e. if spontaneous strategy changes are so frequent that no strategy can win out.

For $\kappa > 0$ the stationary solution $P^0(i) = 1/2$ is unstable and the game dynamical equations (12.67) can be represented in the form

$$\frac{d}{dt}P(i,t) = -2\left(P(i,t) - \frac{1}{2}\right)\left(P(i,t) - \frac{1+\sqrt{\kappa}}{2}\right)\left(P(i,t) - \frac{1-\sqrt{\kappa}}{2}\right). \tag{12.69}$$

Consequently, for $\kappa > 0$ there exist two additional stationary solutions $P^0(i) = (1+\sqrt{\kappa})/2$ and $P^0(i) = (1-\sqrt{\kappa})/2$ which are stable. That is, either strategy 1 or strategy 2 will gain a majority of $100 \cdot \sqrt{\kappa}$ percent which essentially depends on the (random) initial condition $P(1, t_0) = 1 - P(2, t_0)$. This majority will be the larger the smaller the rate $\widehat{W}$ of spontaneous strategy changes is.

For $\kappa = \kappa_0 := 0$ there obviously occurs a *phase transition.* This becomes particularly clear in Fig. 12.2 which shows, for $\widehat{A} := 1$ and various values of κ, the temporal course of the probability $P(n_1, n_2; t) \equiv P(n_1, N - n_1; t)$ with which n_1 individuals pursue strategy 1 and $n_2 = N - n_1$ individuals apply strategy 2. (For illustrative reasons we started from a *broad* initial distribution which corresponds to a situation in which the initial state is not exactly known. With the unrealistically small number of $N = 40$ individuals we achieve that the distribution $P(n_1, N - n_1; t)$ *remains* relatively broad in the course of time. With increasing population size N the maxima of the corresponding probability distribution become more and more distinct. In addition, note that Figs. 12.2 and 12.6 are scaled in such a way that the maximum of $P(n_1, N - n_1; t)$ reaches half of the box height.)

From Fig. 12.2 we can conclude the following:

- For $\kappa < 0$ the probability distribution $P(n_1, n_2; t)$ will have only *one* maximum at $n_1 = n_2 = N/2$: That is, the individuals are very likely to pursue both strategies with the same frequency.
- For $\kappa = \kappa_0 = 0$ there occurs the *phase transition.* Due to *critical fluctuations* (cf. Sect. 7.4) the probability distribution $P(n_1, n_2; t)$ becomes very *broad.*
- For $\kappa > 0$ the probability distribution $P(n_1, n_2; t)$ becomes *multimodal:* It is very likely that one of the strategies will gain the majority of individuals. In the short term the initial condition decides which strategy has the advantage. However, since for $t \to \infty$ the maxima are located symmetrically with respect to $N/2$, in the long term both strategies have the same chance to win over the majority (cf. Sect. 7.3).

  For growing values of κ the maxima of the probability distribution $P(n_1, n_2; t)$ which describe the chance that one of the strategies will win out become more and more distinct.

Figures 12.3, 12.4 and 12.5 compare the *exact* course of the mean value $\langle n_1 \rangle \equiv \langle n_1 \rangle_t$ and the variance $\sigma_{11} \equiv \sigma_{11}(t)$ of n_1 (—) with the solutions of the corresponding

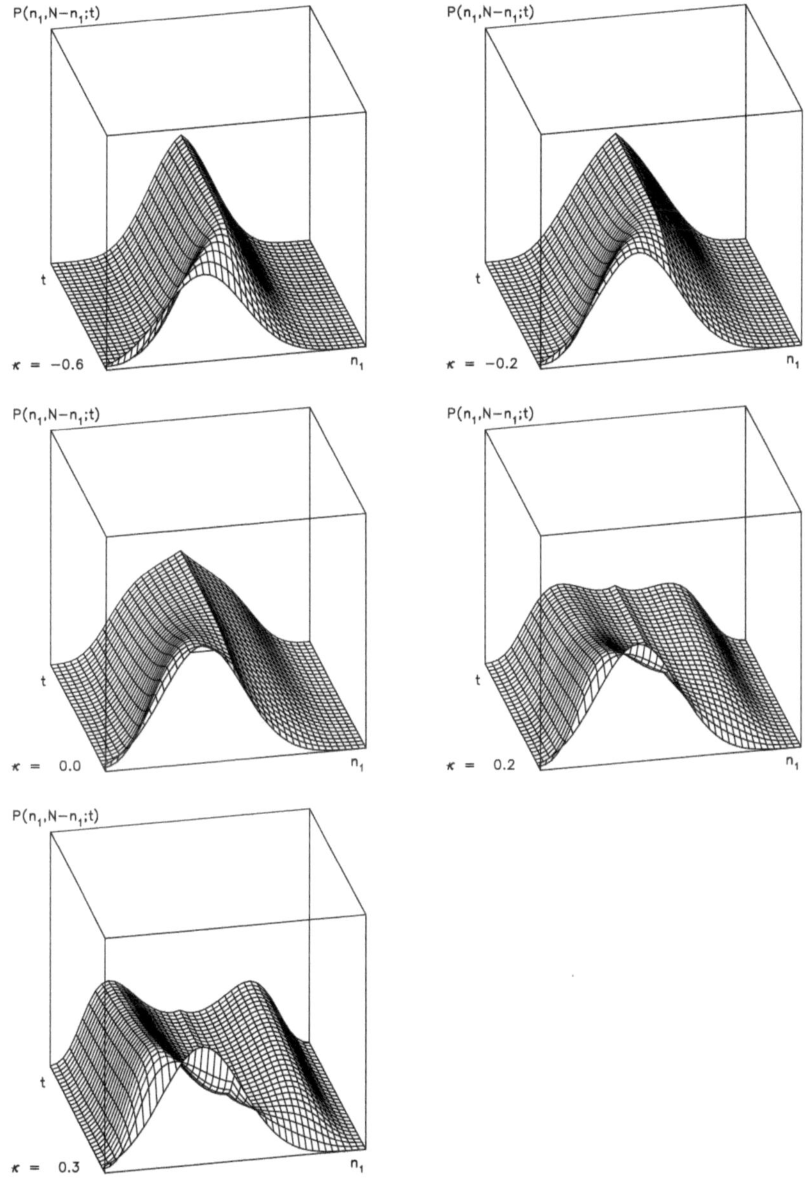

Fig. 12.2 Illustration of the temporal evolution of the configurational distribution $P(n_1, N - n_1; t)$ according to the stochastic version of the game dynamical equations for various values of the control parameter κ. $P(n_1, N - n_1; t)$ denotes the probability with which n_1 individuals pursue strategy 1 and $n_2 = N - n_1$ individuals strategy 2. For $\kappa = 0$ a *phase transition* takes place: Whereas for $\kappa < 0$ very probably both strategies are applied with roughly the same frequency, for $\kappa > 0$ one of the strategies will win out with a high probability, i.e. a *behavioural convention* will arise by social *self-organization*

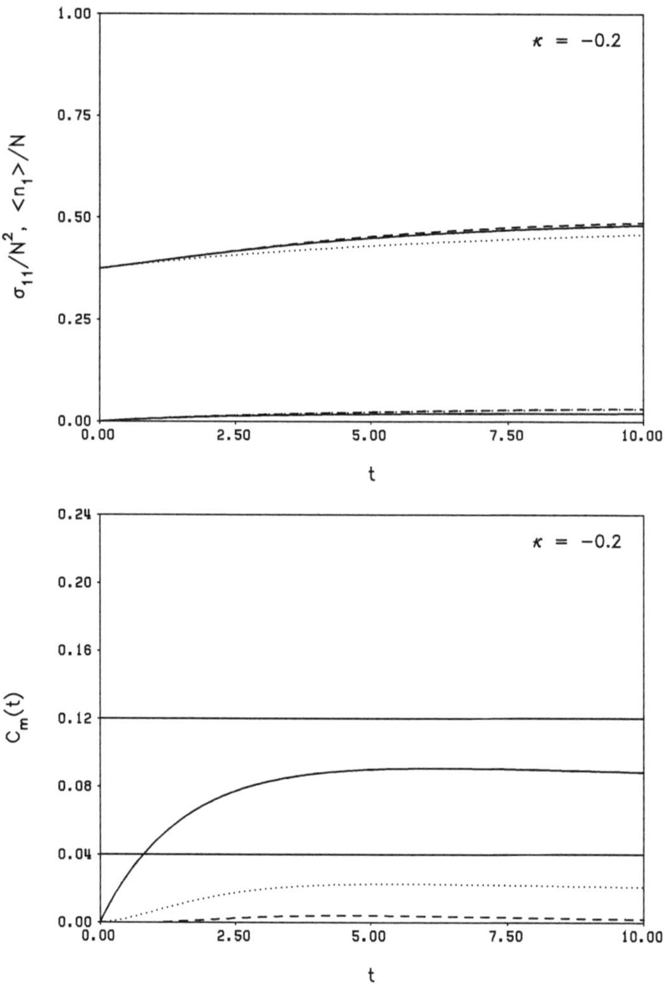

Fig. 12.3 *Above*: Exact (—), approximate (· · ·), and corrected (– –) mean values (*upper curves*) and variances (*lower curves*) for a small unimodal distribution ($\kappa < 0$; cf. Fig. 12.2): The simulation results for the approximate equations are acceptable, those for the corrected equations very well. *Below*: *Relative central moments* $C_m(t)$ for the stochastic version of the game dynamical equations (— : $m = 2$; – – : $m = 3$; · · · : $m = 4$)

approximate mean value and variance equations (· · · ; cf. (12.56a), (12.57a)) as well as with the solutions of the *corrected* mean value and variance equations (– – ; cf. (12.56), (12.57)).

- For $\kappa \ll \kappa_0 = 0$ the approximate as well as the corrected mean value and variance equations provide good approximations (cf. Fig. 12.3).
- Since the probability distribution $P(n_1, N - n_1; t)$ becomes broader for growing values of κ, the solutions of the corresponding approximate and corrected

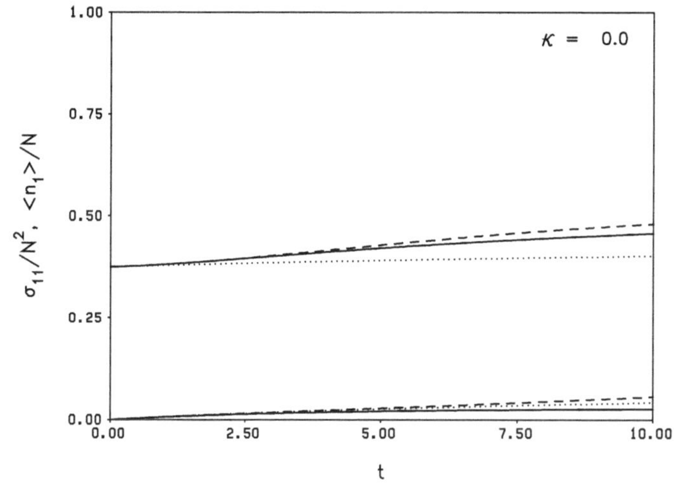

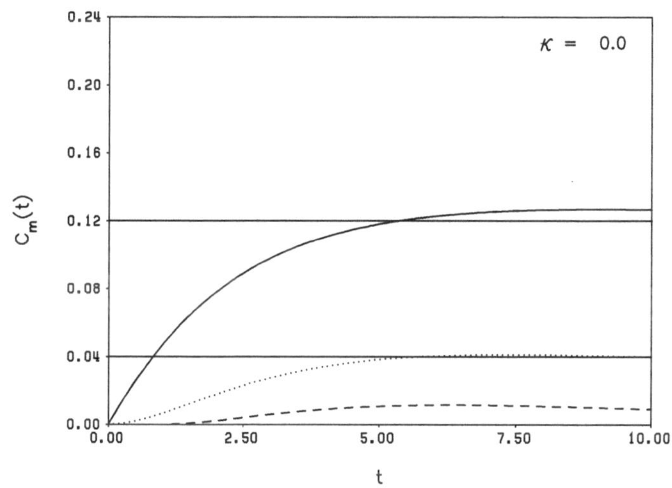

Fig. 12.4 *Above*: The approximate ($\cdots$) and corrected (– –) mean values (*upper curves*) and variances (*lower curves*) show, for a broad unimodal distribution ($\kappa = 0$; cf. Fig. 12.2), deviations from the exact mean value and variance respectively (—). As expected, the results of the approximate equations are worse than those of the corrected ones. *Below*: Relative central moments $C_m(t)$ for the stochastic version of the game dynamical equations (—: $m = 2$; – –: $m = 3$; $\cdots$: $m = 4$). A comparison with the figure above shows that the *approximate* mean value and variance start moving off from the exact course when $C_2(t)$ exceeds 0.04 whereas for the *corrected* mean value and variance this is true as soon as we have $C_4(t) > 0.04$ and $C_2(t) > 0.12$ respectively. The approximate equations are valid for a considerably shorter time interval than the corrected equations

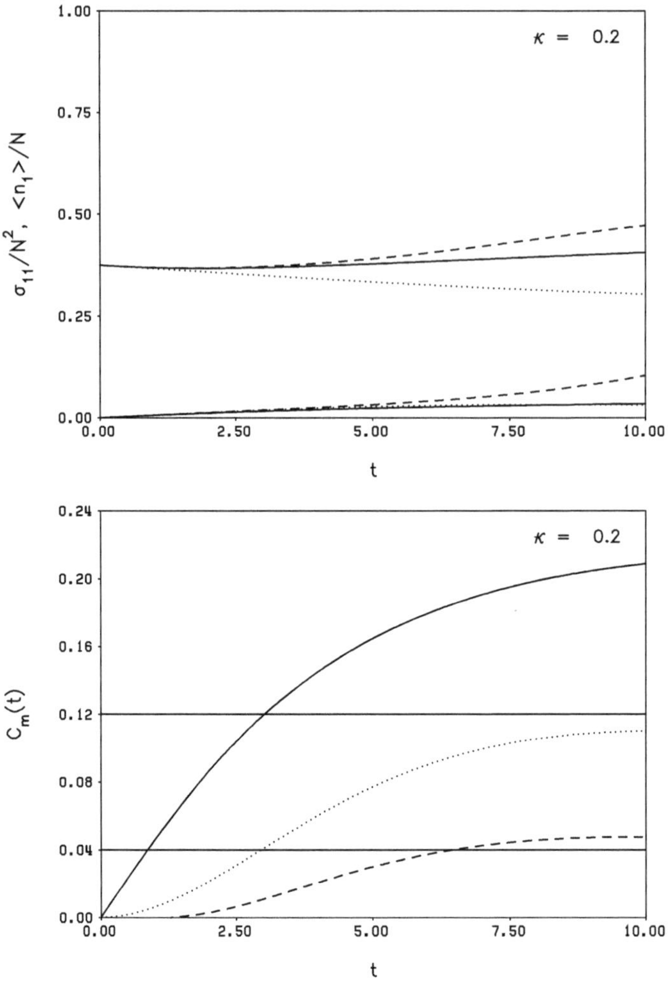

Fig. 12.5 Above: For a multi-modal distribution ($\kappa > 0$; cf. Fig. 12.2) it is only for a short time period that the approximate ($\cdots$) and corrected (– –) mean values (*upper curves*) and variances (*lower curves*) are a valid approximation of the exact mean value and variance respectively (—). However, the approximate equations still describe the *most probable individual behaviour. Below*: Relative central moments $C_m(t)$ for the stochastic version of the game dynamical equations (—: $m = 2$; – –: $m = 3$; $\cdots$: $m = 4$). A comparison with the figure above shows again that the *approximate* mean value and variance deviate from the exact course for $C_2(t) > 0.04$, the *corrected* mean value and variance, in comparison, for $C_4(t) > 0.04$ and $C_2(t) > 0.12$ respectively. As expected, the approximate equations are valid for a considerably shorter time interval than the corrected equations

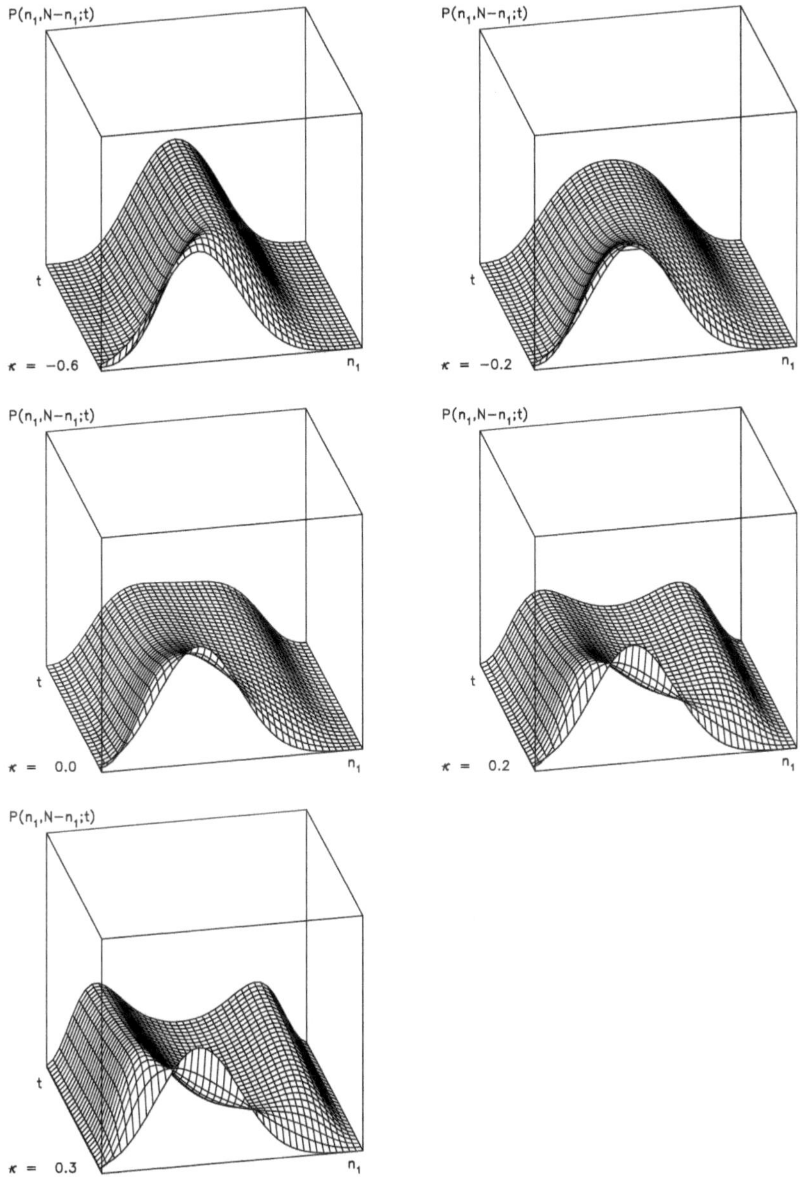

Fig. 12.6 Illustration of the temporal evolution of the configurational distribution $P(n_1, N - n_1; t)$ for various values of the control parameter κ according to the stochastic version of the *modified game dynamic equations*. Like in Fig. 12.2, for $\kappa = 0$ there occurs a *phase transition* which describes the self-organization of a behavioural convention. This time, the modified transition rates lead to smooth distributions and take into account the incompleteness of information during the decision on a strategy change

equations deviate more and more from the exact solution, as expected. Because of (5.42), (5.43), they are valid only for the period of time during which the *relative central moments*

$$C_m(t) \equiv C^{a_1 \dots a_m}_{i_1 \dots i_m}(t) := \frac{\left\langle (n^{a_1}_{i_1} - \langle n^{a_1}_{i_1} \rangle) \cdot \ldots \cdot (n^{a_m}_{i_m} - \langle n^{a_m}_{i_m} \rangle) \right\rangle}{\langle n^{a_1}_{i_1} \rangle \cdot \ldots \cdot \langle n^{a_m}_{i_m} \rangle} \tag{12.70}$$

fulfill the condition

$$|C_m(t)| \stackrel{!}{\leq} \varepsilon' \ll 1 \tag{12.71a}$$

for certain well-defined values of m. The decisive values for m are

- $m = 2$ for the approximate equations (cf. (5.42))
- $m = 3$ and $m = 4$ for the corrected equations (cf. (5.43)).

As we can see, the curves of the *approximate* and *corrected* equations begin to deviate from the exact curves roughly at the time t when, for the first time, we have $|C_m(t)| > 0.04$ for the decisive values of m. Hence we put

$$\varepsilon' := 0.04\,. \tag{12.71b}$$

Unfortunately, the examination of the validity of condition (12.71) will be impossible for $C_m(t)$ with $m > 2$ if we do not evaluate master equation (12.62) but only the corrected mean value and covariance equations. From Figs. 12.3, 12.4, 12.5 and 12.7, however, we can see that condition (12.71) will be fulfilled for the decisive values of m if we have

$$|C_2(t)| = \left| \frac{\sigma^{ab}_{ij}(t)}{\langle n^a_i \rangle \langle n^b_j \rangle} \right| \leq \varepsilon \qquad \text{with} \qquad \varepsilon := 0.12\,. \tag{12.72}$$

- Comparing (12.71) with (12.72) we expect that the *corrected* equations are valid for a significantly longer time period than the approximate equations (cf. Figs. 12.4 and 12.5). The corrected equations have explanatory power even for a rather broad probability distribution $P(n_1, N - n_1; t)$ but they fail when the probability distribution becomes *multimodal.* This phase transition is preceded by a considerable increase of the covariances $\sigma^{ab}_{ij}(t)$ connected with a violation of condition (12.72) in the course of time t.

Nevertheless, the *approximate* mean value equation is relevant also for those values of κ and those times t for which it lost its applicability as *mean value equation*—since, according to Sect. 7.1, it approximately agrees with the equation for the *most probable behaviour* of individuals.

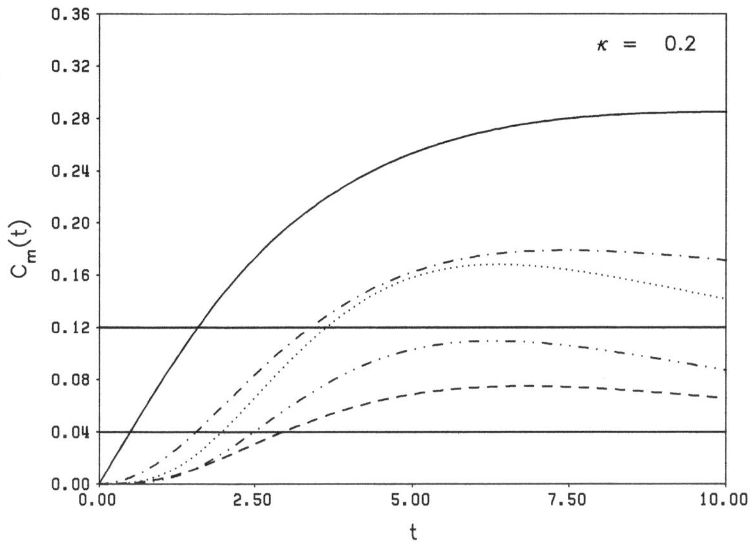

Fig. 12.7 Relative central moments $C_m(t)$ for the stochastic version of the *modified* game dynamical equations (—: $m = 2$; – –: $m = 3$; – · –: $m = 4$; – · · : $m = 5$; · · · : $m = 6$). The corrected mean value and variance equations become invalid as soon as there is $C_m(t) > 0.04$ for any $m > 2$ or $C_2(t) > 0.12$. Due to the dependence of the exact mean value equation on moments of arbitrarily large order m, approximations for this are valid only for a short time interval

12.4.1.1 Modified Game Dynamical Equations

Note that the probability distributions $P(n_1, N - n_1; t)$ in Fig. 12.2 show a curious crease at $n_1 = N/2$. This has its origin in the discontinuity of the derivative of $R^a(j|i; n; t)$ with respect to $(E_a(j, n; t) - E_a(i, n; t))$ which results from the dependence on the maximum function (12.54e). With the approach

$$R^a(j|i; n; t) := \frac{e^{U^a(j,n;t) - U^a(i,n;t)}}{D^a(j, i; t)} \tag{12.73a}$$

that was motivated in Chap. 9 and the *expected utility*

$$U^a(i, n; t) := E_a(i, n; t) = \sum_b \sum_j A_{ab}(i, j; t) \frac{n_j^b}{N_b} \tag{12.73b}$$

we can obtain results that will turn out more satisfactory than those of (12.54e). This approach *also* guarantees that transitions from strategy i to strategy j are particularly frequent if the expected increase of success $\Delta_{ji} E_a$ (or increase of utility $\Delta_{ji} U^a$) is large. If a *decrease* of success (or utility) is to be expected ($\Delta_{ji} E_a = \Delta_{ji} U^a < 0$), transitions will be rare. According to the discussion of Sect. 9.2.1, approach (12.73), in contrast to (12.54e), takes into account that during a decision on a strategy change the *information* about the exact value of the expected utility U^a is usually *incomplete*. The decisions rather base on an *estimate*

$$\begin{aligned} U_*^a(i, n; t) &:= \sum_b \sum_j A_{ab}^*(i, j; t) \frac{n_j^{b*}(t)}{N_b} \\ &= \sum_b \sum_j A_{ab}(i, j; t) \frac{n_j^b(t)}{N_b} + \varepsilon_a'(i, n; t) \\ &= U^a(i, n; t) + \varepsilon_a'(i, n; t) \end{aligned} \tag{12.74}$$

of the expected utility $U^a(i, n; t)$ which is evaluated from the successes of the preceding interactions. The random variations $\varepsilon_a'(i, n; t)$ of the estimated utility $U_*^a(i, n; t)$ result from the accidental variations of the relative frequencies

$$r_{ab}^*(t) \frac{n_j^{b*}(t)}{N_b} \tag{12.75}$$

with which an individual of subpopulation a meets individuals of subpopulation b with strategy j (and possibly from other uncertainties).

In addition to this, by variation of the distance functions $D^a(j, i; t)$, approach (12.73) allows to take into account the effort $S^a(j, i; t) = \ln D^a(j, i; t)$ which is connected with a strategy change from i to j. In the case of small utility differences $\Delta_{ji} U^a$ and $D^a(j, i; t) := 2$, by means of a TAYLOR approximation, we again arrive at the game dynamical equations (12.9). From Fig. 12.6 we find for this case that the modified approach (12.73) yields similar results than (12.54e), especially a phase transition at $\kappa = 0$ (cf. Fig. 12.2). However, a disadvantage of approach (12.73) is that, in order to guarantee the validity of the corrected mean value equation, condition (12.71) must be fulfilled for all $m > 2$ according to (5.43). This condition is difficult to check and, due to the fact that the exact mean value equation depends on moments of arbitrary order m, violated early (cf. Fig. 12.7).

Chapter 13
Determination of the Model Parameters from Empirical Data

13.1 Introduction

The decisive measure of the quality and applicability of a model is the comparison with empirical data. Hence we require strategies for the determination of the model parameters as $U_a(x,t)$ or $D_a(x',x;t)$ from suitable data material. These mostly base on the *method of least squares.* Since available data material is not always *complete,* we will develop useful methods even for such cases.

After the model parameters have been determined, a *sensitivity analysis* should be carried out. A test for the *significance* of the model parameters normally allows a *model reduction.* Furthermore, from the viewpoint of interpretation it is of interest on which *factors* the utility and distance functions of the considered behavioural changes depend. A corresponding decomposition with respect to *explanatory variables* is possible by means of a *regression analysis* in connection with a *factor analysis* or, alternatively, a *ranking regression analysis.* This is especially important for the *prognosis* of the future behaviour of the investigated system. Examples for the decomposition of utility functions into interpretable contributions will be discussed for *purchase pattern* and *voting behaviour.*

13.2 The Case of Complete Data

In the following let us mark empirical data, in contrast to model variables, by a subscript or superscript 'e'. We speak of *'complete data'* of a considered social process if we know all the following quantities

$$P_a^{\mathrm{e}}(x,t_i) \qquad \forall x\,, \tag{13.1a}$$

$$w_a^{\mathrm{e}}(x'|x;t_i) \qquad \forall x' \neq x\,, \tag{13.1b}$$

$$\nu_{ab}^{\mathrm{e}}(t_i)\,, \qquad \text{and} \tag{13.1c}$$

$$p_{ab}^{k,\mathrm{e}}(x',y'|x,y;t_i) \qquad \forall (x',y') \neq (x,y) \tag{13.1d}$$

D. Helbing, *Quantitative Sociodynamics*, 2nd ed.,
DOI 10.1007/978-3-642-11546-2_13, © Springer-Verlag Berlin Heidelberg 2010

for a sequence of times

$$t_i \in \{t_0, \dots, t_n\} . \tag{13.2}$$

Here the time intervals $\Delta t_i := t_i - t_{i-1}$ must be considerably smaller than the reciprocals of the individual transition rates of behavioural changes, i.e.

$$\sum_{x'(\neq x)} w_a^{\mathrm{e}}(x'|x; t_i) \ll \frac{1}{\Delta t_i} \tag{13.3a}$$

and

$$\sum_k \sum_{x'(\neq x)} \sum_y \sum_{y'} \frac{\nu_{ab}^{\mathrm{e}}(t_i)}{N_b} p_{ab}^{k,\mathrm{e}}(x', y'|x, y; t_i) \ll \frac{1}{\Delta t_i} \tag{13.3b}$$

(cf. (9.5)).

Now a fitting of the model rates

$$w_a(x'|x; t_i) , \qquad w_{ab}^k(x', y'|x, y; t_i) \tag{13.4}$$

can be achieved by minimizing the GAUSS*ian error function*

$$\begin{aligned} F := &\sum_i \sum_a \sum_{x \neq x'} \Big[w_a^{\mathrm{e}}(x'|x; t_i) P_a^{\mathrm{e}}(x, t_i) - w_a(x'|x; t_i) P_a^{\mathrm{e}}(x, t_i) \Big]^2 \\ &+ \sum_i \sum_k \sum_{a,b} \sum_{(x,y)\neq(x',y')} \Big[\nu_{ab}^{\mathrm{e}}(t_i) p_{ab}^{k,\mathrm{e}}(x', y'|x, y; t_i) P_a^{\mathrm{e}}(x, t_i) P_b^{\mathrm{e}}(y, t_i) \\ &- w_{ab}^k(x', y'|x, y; t_i) P_a^{\mathrm{e}}(x, t_i) P_b^{\mathrm{e}}(y, t_i) \Big]^2 \end{aligned} \tag{13.5}$$

under variation of the model parameters (*method of least squares;* [19, 81, 88]). However, the *maximum likelihood method* could also be used [82, 149].

For the model rates let us again choose approaches of the form

$$w_a(x'|x; t_i) := \nu_a(t_i) \frac{\mathrm{e}^{U_a(x',t_i) - U_a(x,t_i)}}{D_a(x', x; t_i)} \tag{13.6}$$

and

$$\begin{aligned} w_{ab}^1(x', y'|x, y; t_i) = &\ \nu_{ab}(t_i) p_{ba}^1(x|y; t_i) \delta_{xx'} \delta_{xy'} (1 - \delta_{xy}) \\ &+ \nu_{ab}(t_i) p_{ab}^1(y|x; t_i) \delta_{yy'} \delta_{yx'} (1 - \delta_{xy}) , \end{aligned} \tag{13.7a}$$

$$
\begin{aligned}
w^2_{ab}(x', y'|x, y; t_i) = {} & \nu_{ab}(t_i) p^2_{ab}(x|x; t_i) p^2_{ba}(y'|x; t_i)\delta_{xy}\delta_{xx'}(1-\delta_{xy'}) \\
& +\nu_{ab}(t_i) p^2_{ab}(x'|y; t_i) p^2_{ba}(y|y; t_i)\delta_{xy}\delta_{yy'}(1-\delta_{yx'}) \\
& +\nu_{ab}(t_i) p^2_{ab}(x'|x; t_i) p^2_{ba}(y'|x, t_i)\delta_{xy}(1-\delta_{xx'}) \\
& \times(1-\delta_{xy'}) ,
\end{aligned}
\tag{13.7b}
$$

$$
\begin{aligned}
& w^3_{ab}(x', y'|x, y; t_i) \\
& \quad = \nu_{ab}(t_i) p^3_{ab}(x|x; t_i) p^3_{ba}(y'|y; t_i)\delta_{xx'}(1-\delta_{xy})(1-\delta_{yy'})(1-\delta_{xy'}) \\
& \quad +\nu_{ab}(t_i) p^3_{ab}(x'|x, t_i) p^3_{ba}(y|y; t_i)\delta_{yy'}(1-\delta_{xy})(1-\delta_{xx'})(1-\delta_{yx'}) \\
& \quad +\nu_{ab}(t_i) p^3_{ab}(x'|x; t_i) p^3_{ba}(y'|y, t_i)(1-\delta_{xy})(1-\delta_{xx'})(1-\delta_{yy'}) \\
& \quad \times(1-\delta_{yx'})
\end{aligned}
\tag{13.7c}
$$

with

$$
p^k_{ab}(x'|x; t_i) := f^k_{ab}(t_i) \frac{\mathrm{e}^{U^a(x',t_i)-U^a(x,t_i)}}{D^a(x', x; t_i)}
\tag{13.8}
$$

(cf. Chap. 9). Additionally, for reasons of uniqueness we demand

$$
\frac{1}{S}\sum_x U_a(x, t_i) \stackrel{!}{=} 0 \stackrel{!}{=} \frac{1}{S}\sum_x U^a(x, t_i)
\tag{13.8a}
$$

and

$$
\frac{1}{S(S-1)}\sum_{x\neq x'} D_a(x', x; t_i) \stackrel{!}{=} 1 \stackrel{!}{=} \frac{1}{S(S-1)}\sum_{x\neq x'} D^a(x', x; t_i)
\tag{13.8b}
$$

(cf. (9.37), (9.61)). Then the method of least squares will allow a determination of the utility functions $U_a(x, t_i)$, $U^a(x, t_i)$, the distance functions $D_a(x', x; t_i)$, $D^a(x', x; t_i)$, and the interaction frequencies $f^k_{ab}(t)$. A minimization of the error function F with respect to the model parameters $\nu_a(t_i)$, $\nu^k_{ab}(t_i) := \nu^{\mathrm{e}}_{ab}(t_i) f^k_{ab}(t_i)$, $U_a(x, t_i)$, $U^a(x, t_i)$, $D_a(x', x; t_i)$, and $D^a(x', x; t_i)$ is in general carried out numerically, e.g. by means of the *method of steepest descent* [19, 62, 80].

If $P_a^{\mathrm{e}}(x,t)$ is unknown, instead of using (13.28), we can determine the model parameters by applying the error function

$$F' := \sum_i \sum_a \sum_{x \neq x'} \Big[w_a^{\mathrm{e}}(x'|x;t_i) - w_a(x'|x;t_i) \Big]^2 + \sum_i \sum_k \sum_{a,b} \sum_{(x,y)\neq(x',y')} \Big[\nu_{ab}^{\mathrm{e}}(t_i) p_{ab}^{k,\mathrm{e}}(x',y'|x,y;t_i) - w_{ab}^k(x',y'|x,y;t_i) \Big]^2 . \tag{13.9}$$

However, the transition rates are not weighted with the magnitude of their contribution to $dP_a(x,t)/dt$, then. Hence this method is not as suitable as a basis for *prognoses*.

13.3 The Case of Incomplete Data

For financial reasons and due to special difficulties with the data collection of pair interaction rates, one is often confronted with a poor data situation. Therefore, let us assume in the following that all we know are the quantities

$$P_a^{\mathrm{e}}(x,t_i) \qquad \forall x , \tag{13.10a}$$

$$w_{\mathrm{e}}^a(x'|x;t_i) \qquad \forall x' \neq x \tag{13.10b}$$

for certain times $t_i \in \{t_0, \dots, t_n\}$. This time we must presuppose for $\Delta t_i = t_i - t_{i-1}$ that

$$\sum_{x'(\neq x)} w_{\mathrm{e}}^a(x'|x;t_i) \ll \frac{1}{\Delta t_i} \tag{13.11}$$

is fulfilled (cf. (9.5)). We will try to describe $w_{\mathrm{e}}^a(x'|x;t_i)$ for $x' \neq x$ by the model approach

$$w^a(x'|x;t_i) := \nu_a(t_i) \frac{\mathrm{e}^{U_a(x',t_i)-U_a(x,t_i)}}{D_a(x',x;t_i)} + \nu^a(x',x;t_i) \frac{\mathrm{e}^{U^a(x',t_i)-U^a(x,t_i)}}{D^a(x',x;t_i)} \tag{13.12a}$$

from Chap. 9 where

$$\nu^a(x',x;t_i) := \sum_b \Big[\Big(\nu_{ab}^1(t_i) - \nu_{ab}^3(t_i) \Big) P_b^{\mathrm{e}}(x',t_i) + \Big(\nu_{ab}^2(t_i) - \nu_{ab}^3(t_i) \Big) P_b^{\mathrm{e}}(x,t_i) + \nu_{ab}^3(t_i) \Big] . \tag{13.12b}$$

In order to determine the functions $\nu_a(t_i)$, $\nu^k_{ab}(t_i)$, $U_a(x,t_i)$, $U^a(x,t_i)$, $D_a(x',x;t_i)$, and $D^a(x',x;t_i)$ in a unique way, we again demand that the constraints (13.8) are fulfilled. Due to linear dependences of some model parameters, instead of $\nu^1_{ab}(t_i)$, $\nu^2_{ab}(t_i)$, and $\nu^3_{ab}(t_i)$, one can only determine

$$\overline{\nu^1_{ab}}(t_i) := \nu^1_{ab}(t_i) - \nu^3_{ab}(t_i)\,, \tag{13.13a}$$

$$\overline{\nu^2_{ab}}(t_i) := \nu^2_{ab}(t_i) - \nu^3_{ab}(t_i)\,, \tag{13.13b}$$

and

$$\overline{\nu^3_a}(t_i) := \sum_b \nu^3_{ab}(t_i)\,. \tag{13.13c}$$

However, since for $A = 1$ we have

$$\nu^3(t_i) = \overline{\nu^3}(t_i) \tag{13.14a}$$

and

$$\nu^1(t_i) = \overline{\nu^1}(t_i) + \overline{\nu^3}(t_i)\,, \tag{13.14b}$$

$$\nu^2(t_i) = \overline{\nu^2}(t_i) + \overline{\nu^3}(t_i)\,, \tag{13.14c}$$

this problem will not occur if only one single subpopulation a is considered ($A = 1$) which is normally the case.

Now our task is to determine the functions $\nu_a(t_i)$, $\overline{\nu^1_{ab}}(t_i)$, $\overline{\nu^2_{ab}}(t_i)$, $\overline{\nu^3_a}(t_i)$, $U_a(x,t_i)$, $U^a(x,t_i)$, $D_a(x',x;t_i)$, and $D^a(x',x;t_i)$ from the $(n+1)AS(S-1)$ data values $w^a_e(x'|x;t_i)$ $(x' \neq x)$ with $t_i \in \{t_0,\dots,t_n\}$, $x \in \{x_1,\dots,x_S\}$, $a \subset \{1,\dots,A\}$. Table 13.1 shows how many independent quantities belong to the different functions.

Obviously, model (13.12) has

$$(n+1)A\Big[2 + 2A + 2(S-1) + S(S-1) - 2\Big] \tag{13.15}$$

independent model parameters, i.e. it is totally underdeterminate by the $(n+1)$ $AS(S-1)$ empirical data $w^a_e(x'|x;t_i)$. Thus we introduce the additional assumption

$$D^a(x',x;t) \equiv D_a(x',x;t)\,. \tag{13.16}$$

Then there remain

$$(n+1)A\left[2 + 2A + 2(S-1) + \frac{S(S-1)}{2} - 1\right] \tag{13.17}$$

Table 13.1 Number of independent variables of the empirical data and the functions occuring in the behavioural models

Function	Constraints	Number of indep. variables
$w^{\text{e}}_a(x'\|x;t_i)$	$x' \neq x$	$+(n+1)AS^2$
		$-(n+1)AS$
	Altogether	$+(n+1)AS(S-1)$
$\nu_a(t_i)$	$\overset{!}{\geq} 0$	$+(n+1)A$
$\overline{\nu^3_a}(t_i)$	$\overset{!}{\geq} 0$	$+(n+1)A$
$\overline{\nu^1_{ab}}(t_i)$	$\overset{!}{\geq} -\nu^3_{ab}(t_i)$	$+(n+1)A^2$
$\overline{\nu^2_{ab}}(t_i)$	$\overset{!}{\geq} -\nu^3_{ab}(t_i)$	$+(n+1)A^2$
$U_a(x,t_i)$	$\frac{1}{S}\sum_x U_a(x,t) \overset{!}{=} 0$	$+(n+1)AS$
		$-(n+1)A$
	Altogether	$+(n+1)A(S-1)$
$U^a(x,t_i)$	$\frac{1}{S}\sum_x U^a(x,t) \overset{!}{=} 0$	$+(n+1)A(S-1)$
$D_a(x',x;t_i)$	$\geq 0,\ x' \neq x$	$+(n+1)AS(S-1)$
	$D_a(x,x';t_i) = D_a(x',x;t_i)$	$-(n+1)A\frac{S(S-1)}{2}$
	$\frac{1}{S(S-1)}\sum_{x\neq x'} D_a(x',x;t_i) \overset{!}{=} 1$	$-(n+1)A$
	Altogether	$+(n+1)A\left(\frac{S(S-1)}{2}-1\right)$
$D^a(x',x;t_i)$	Analogously	$+(n+1)A\left(\frac{S(S-1)}{2}-1\right)$

independent model parameters. Assumption (13.16) is very plausible since it implies that the *distance* between (dissimilarity of) two behaviours x and x' agrees for spontaneous behavioural changes and behavioural changes due to pair interactions. *On the basis of the distances $D^a(x',x;t) \equiv D_a(x',x;t)$ one can construct an abstract behavioural space with the method of multidimensional scaling* [161, 301]. This describes a part of the *psychological topology (psychological structure)* of the individuals of subpopulation a since the distances reflect which behaviours are more or less related (compatible) (cf. [212], Chap. 7). With the dependence on a, the distance functions $D_a(x',x;t)$ distinguish different psychological structures (personality profiles) which could be denoted as *'characters'* and manifest themselves as different behavioural types.

The model parameters can only be uniquely determined from the empirical data on the condition

$$2A + 2S + \frac{S(S-1)}{2} - 1 \stackrel{!}{\leq} S(S-1), \quad \text{i.e.} \quad S \stackrel{!}{\geq} \frac{5 + \sqrt{17 + 16A}}{2}. \tag{13.18}$$

So, for the case of one single subpopulation ($A = 1$), at least $S = 6$ behaviours must be distinguished.

If the data differentiate less behaviours, we can, for instance, make the further simplification

$$U^a(x, t) \equiv U_a(x, t). \tag{13.19}$$

This assumes that the readiness $R^a(x'|x; t)$ for behavioural changes from x to x' due to pair interactions is the same as the corresponding readiness $R_a(x'|x; t)$ for spontaneous transitions. One consequence of (13.19) is that, instead of $\nu_a(t)$ and $\overline{\nu_a^3}(t)$, we can only determine

$$\overline{\nu_a}(t) := \nu_a(t) + \overline{\nu_a^3}(t) \tag{13.20}$$

so that even in the case of one single subpopulation ($A = 1$) the contributions by spontaneous transitions and compromising processes are no longer separable. However, since the resulting model

$$\begin{aligned} w^a(x'|x; t_i) := \frac{\mathrm{e}^{U_a(x',t_i) - U_a(x,t_i)}}{D_a(x', x; t_i)} \Bigg[\overline{\nu_a}(t_i) + \sum_b \Big(\overline{\nu_{ab}^1}(t_i) P_b^{\mathrm{e}}(x', t_i) \\ + \overline{\nu_{ab}^2}(t_i) P_b^{\mathrm{e}}(x, t_i) \Big) \Bigg] \end{aligned} \tag{13.21}$$

has only

$$(n+1)A \left[1 + 2A + (S-1) + \frac{S(S-1)}{2} - 1 \right] \tag{13.22}$$

independent model parameters, these can be determined already for

$$S \stackrel{!}{\geq} \frac{3 + \sqrt{1 + 16A}}{2}. \tag{13.23}$$

That is, for one single subpopulation ($A = 1$) we must distinguish at least $S = 4$ behaviours and for two subpopulations ($A = 2$) at least $S = 5$ behaviours.

A remarkable aspect of Eq. (13.21) is that it *implicitly* takes into account compromising processes by *effective rates* $\overline{\nu_a}(t_i)$, $\overline{\nu_{ab}^1}(t_i)$, and $\overline{\nu_{ab}^2}(t_i)$ for spontaneous, imitative, and avoidance processes respectively. As a consequence, *a separate discussion of compromising processes is often not necessary.*

A further drastic simplification is, for example, given by the assumption of *time-independence* of the distance functions $D_a(x', x; t)$:

$$D_a(x', x; t) = e^{S_a(x',x,t)} \equiv e^{S_a(x',x)} = D_a(x', x) . \tag{13.24}$$

Then there remain

$$(n+1)A\Big[1 + 2A + (S-1)\Big] + A\left[\frac{S(S-1)}{2} - 1\right] \tag{13.25}$$

independent model parameters which can be determined in the case of a sufficient number $(n+1) \gg 1$ of time steps if

$$S \overset{!}{>} 1 + \sqrt{1+2A} . \tag{13.26}$$

In general it is not reasonable to choose a model which contains as many parameters as one just can determine from the data. On the one hand, the *explanatory power* of the single parameters is small, then. On the other hand, surplus parameters often describe only *accidental variations* of the data instead of the significant *systematic interrelations* one is interested in (cf. Sect. 13.4.2).

13.3.1 Parameter Estimation[1]

Now the particular way of proceeding for the determination of the model parameters will be discussed for the model

$$w^a(x'|x; t_i) := \nu_a(t_i) \frac{e^{U_a(x',t_i) - U_a(x,t_i)}}{D_a(x', x; t_i)} + \nu^a(x', x; t_i) \frac{e^{U^a(x',t_i) - U^a(x,t_i)}}{D_a(x', x; t_i)} , \tag{13.27a}$$

$$\nu^a(x', x; t_i) := \sum_b \left[\overline{\nu^1_{ab}}(t_i) P^e_b(x', t_i) + \overline{\nu^2_{ab}}(t_i) P^e_b(x, t_i) \right] + \overline{\nu^3_a}(t_i) . \tag{13.27b}$$

An optimal estimation of the parameters can be achieved (similar to the procedure in [294], pp. 33ff.) by minimizing the error function

$$\begin{aligned} F := \sum_{a'} \sum_{\hat{x} \neq \hat{x}'} \sum_{i'} \frac{1}{2} \Bigg\{ & w^{a'}_e(\widehat{x}'|\widehat{x}; t_{i'}) P^e_{a'}(\widehat{x}, t_{i'}) \\ & - \frac{1}{D_{a'}(\widehat{x}', \widehat{x}; t_{i'})} \Big[\nu_{a'}(t_{i'}) e^{U_{a'}(\hat{x}',t_{i'}) - U_{a'}(\hat{x},t_{i'})} \\ & + \nu^{a'}(\widehat{x}', \widehat{x}; t_{i'}) e^{U^{a'}(\hat{x}',t_{i'}) - U^{a'}(\hat{x},t_{i'})} \Big] P^e_{a'}(\widehat{x}, t_{i'}) \Bigg\}^2 \end{aligned} \tag{13.28}$$

---

[1] This section can be omitted during a first reading.

under variation of $\nu_a(t_i)$, $\overline{\nu_a^3}(t_i)$, $\overline{\nu_{ab}^1}(t_i)$, $\overline{\nu_{ab}^2}(t_i)$, $U_a(x, t_i)$, $U^a(x, t_i)$, and $D_a(x', x; t_i)$. During this minimization procedure we must take into account constraints (13.8), the condition of positivity (9.32), and symmetry conditions (9.30b). However, it is possible to include these constraints directly in the formulation of the optimization problem: In order to guarantee the positivity of $D_a(x', x; t)$ we will use, like in Chapter 9, the representation

$$D_a(x', x; t) := \mathrm{e}^{S_a(x', x; t)} \tag{13.29a}$$

(cf. (9.32)). The symmetry conditions (9.30b) can be obtained by the representation

$$S_a(x', x; t) := G_a(x', x; t) + G_a(x, x'; t) \,. \tag{13.29b}$$

Constraints (13.8b) can be replaced by introducing LAGRANGE *multipliers* $\lambda_a(t_i)$ and minimizing the function

$$F' := F - \sum_{a'} \sum_{i'} \lambda_{a'}(t_{i'}) \left[\sum_{\hat{x} \neq \hat{x}'} \left(\mathrm{e}^{G_{a'}(\hat{x}', \hat{x}; t_{i'}) + G_{a'}(\hat{x}, \hat{x}'; t_{i'})} - 1 \right) \right] \tag{13.30}$$

instead of F. Here representation (13.29) must be used in F. As solutions of the minimization problem we find a *solution spectrum* of optimal model parameters in dependence of the still undetermined LAGRANGE multipliers $\lambda_a(t_i)$. Now we choose the LAGRANGE multipliers $\lambda_a(t_i)$ just in such a way that the constraints (13.8b) are satisfied. The minimum of F' agrees with the minimum of F, then (cf. (13.30)).

The minimum of F' is characterized by a vanishing of the *total differential*

$$\begin{aligned}
\delta F' = &\sum_a \sum_x \sum_i \left[\frac{\delta F}{\delta U_a(x, t_i)} \right] \delta U_a(x, t_i) \\
&+ \sum_a \sum_x \sum_i \left[\frac{\delta F}{\delta U^a(x, t_i)} \right] \delta U^a(x, t_i) \\
&+ \sum_a \sum_{x \neq x'} \sum_i \left[\frac{\delta F}{\delta G_a(x, x', t_i)} \right. \\
&\left. - \lambda_a(t_i) \mathrm{e}^{G_a(x', x; t_i) + G_a(x, x'; t_i)} \right] \delta G_a(x, x'; t_i) \\
&+ \sum_a \sum_i \left[\frac{\delta F}{\delta \nu_a(t_i)} \right] \delta \nu_a(t_i)
\end{aligned}$$

$$+\sum_{a}\sum_{i}\left[\frac{\delta F}{\delta\overline{\nu_a^3}(t_i)}\right]\delta\overline{\nu_a^3}(t_i)$$

$$+\sum_{a,b}\sum_{i}\left[\frac{\delta F}{\delta\overline{\nu_{ab}^1}(t_i)}\right]\delta\overline{\nu_{ab}^1}(t_i)$$

$$+\sum_{a,b}\sum_{i}\left[\frac{\delta F}{\delta\overline{\nu_{ab}^2}(t_i)}\right]\delta\overline{\nu_{ab}^2}(t_i)\,. \tag{13.31}$$

Except for the restrictions

$$\delta U_a(x_S,t_i) = -\sum_{x'(\neq x_S)} \delta U_a(x',t_i)\,, \tag{13.32a}$$

$$\delta U^a(x_S,t_i) = -\sum_{x'(\neq x_S)} \delta U^a(x',t_i) \tag{13.32b}$$

given by the constraints (13.8a), all model parameters can be varied independently of each other. Inserting (13.32) into (13.31) and demanding $\delta F' \stackrel{!}{=} 0$ leads to the *minimum conditions*

$$\frac{\delta F}{\delta U_a(x,t_i)} - \mu_a(t_i) \stackrel{!}{=} 0 \qquad (x \neq x_S)\,, \tag{13.33a}$$

$$\frac{\delta F}{\delta U^a(x,t_i)} - \mu^a(t_i) \stackrel{!}{=} 0 \qquad (x \neq x_S)\,, \tag{13.33b}$$

$$\frac{\delta F}{\delta G_a(x,x';t_i)} - \lambda_a(t_i)\mathrm{e}^{G_a(x',x;t_i)+G_a(x,x';t_i)} \stackrel{!}{=} 0\,, \tag{13.33c}$$

$$\frac{\delta F}{\delta \nu_a(t_i)} \stackrel{!}{=} 0\,, \tag{13.33d}$$

$$\frac{\delta F}{\delta\overline{\nu_a^3}(t_i)} \stackrel{!}{=} 0\,, \tag{13.33e}$$

$$\frac{\delta F}{\delta\overline{\nu_{ab}^1}(t_i)} \stackrel{!}{=} 0\,, \tag{13.33f}$$

$$\frac{\delta F}{\delta\overline{\nu_{ab}^2}(t_i)} \stackrel{!}{=} 0 \tag{13.33g}$$

with

$$\mu_a(t_i) := \frac{\delta F}{\delta U_a(x_S, t_i)}, \tag{13.34a}$$

$$\mu^a(t_i) := \frac{\delta F}{\delta U^a(x_S, t_i)}. \tag{13.34b}$$

Using the abbreviations

$$f^a(x', x; t_i) := \Big(w_{\mathrm{e}}^a(x'|x; t_i) - \mathrm{e}^{-S_a(x',x;t_i)} g^a(x', x; t_i)\Big)\Big(P_a^{\mathrm{e}}(x, t_i)\Big)^2, \tag{13.35a}$$

$$g^a(x', x; t_i) := \nu_a(t_i)\mathrm{e}^{U_a(x',t_i)-U_a(x,t_i)} + \nu^a(x', x; t_i)\mathrm{e}^{U^a(x',t_i)-U^a(x,t_i)}, \tag{13.35b}$$

the *functional derivatives* are given by

$$\frac{\delta F}{\delta U_a(x, t_i)} = -\sum_{x'(\neq x)} \mathrm{e}^{-S_a(x',x;t_i)} \nu_a(t_i)\Big[f^a(x, x'; t_i)\mathrm{e}^{U_a(x,t_i)-U_a(x',t_i)} - f^a(x', x; t_i)\mathrm{e}^{U_a(x',t_i)-U_a(x,t_i)}\Big], \tag{13.36a}$$

$$\frac{\delta F}{\delta U^a(x, t_i)} = -\sum_{x'(\neq x)} \mathrm{e}^{-S_a(x',x;t_i)}\Big[f^a(x, x'; t_i)\nu^a(x, x'; t_i) \times \mathrm{e}^{U^a(x,t_i)-U^a(x',t_i)} - f^a(x', x; t_i) \times \nu^a(x', x; t_i)\mathrm{e}^{U^a(x',t_i)-U^a(x,t_i)}\Big], \tag{13.36b}$$

$$\frac{\delta F}{\delta G_a(x', x; t_i)} = \mathrm{e}^{-S_a(x',x;t_i)}\Big[f^a(x', x; t_i)g^a(x', x; t_i) + f^a(x, x'; t_i)g^a(x, x'; t_i)\Big], \tag{13.36c}$$

$$\frac{\delta F}{\delta \nu_a(t_i)} = -\sum_{x\neq x'} f^a(x', x; t_i)\mathrm{e}^{-S_a(x',x;t_i)}\mathrm{e}^{U_a(x',t_i)-U_a(x,t_i)}, \tag{13.36d}$$

$$\frac{\delta F}{\delta \overline{\nu_a^3}(t_i)} = -\sum_{x\neq x'} f^a(x', x; t_i)\mathrm{e}^{-S_a(x',x;t_i)}\mathrm{e}^{U^a(x',t_i)-U^a(x,t_i)}, \tag{13.36e}$$

$$\frac{\delta F}{\delta \overline{\nu_{ab}^1}(t_i)} = -\sum_{x\neq x'} f^a(x', x; t_i)\mathrm{e}^{-S_a(x',x;t_i)}\mathrm{e}^{U^a(x',t_i)-U^a(x,t_i)} \times P_b^{\mathrm{e}}(x', t_i), \tag{13.36f}$$

$$\frac{\delta F}{\delta \overline{\nu_{ab}^2}(t_i)} = -\sum_{x\neq x'} f^a(x', x; t_i)\mathrm{e}^{-S_a(x',x;t_i)}\mathrm{e}^{U^a(x',t_i)-U^a(x,t_i)} \times P_b^{\mathrm{e}}(x, t_i). \tag{13.36g}$$

Obviously we have

$$\sum_x \frac{\delta F}{\delta U_a(x,t_i)} = -\sum_{x\neq x'} \nu_a(t_i)\mathrm{e}^{-S_a(x',x;t_i)} f^a(x,x';t_i)\mathrm{e}^{U_a(x,t_i)-U_a(x',t_i)}$$
$$+\sum_{x\neq x'} \nu_a(t_i)\mathrm{e}^{-S_a(x',x;t_i)} f^a(x',x;t_i)\mathrm{e}^{U_a(x',t_i)-U_a(x,t_i)}$$
$$= 0 \tag{13.37a}$$

and analogously

$$\sum_x \frac{\delta F}{\delta U^a(x,t_i)} = 0\,, \tag{13.37b}$$

i.e.

$$\mu_a(t_i) = \frac{1}{S}\sum_x \frac{\delta F}{\delta U_a(x,t_i)} = 0\,, \tag{13.38a}$$

$$\mu^a(t_i) = \frac{1}{S}\sum_x \frac{\delta F}{\delta U^a(x,t_i)} = 0\,. \tag{13.38b}$$

Hence conditions (13.33a) and (13.33b) can be rewritten in the form

$$\frac{\delta F}{\delta U_a(x,t_i)} \stackrel{!}{=} 0\,, \tag{13.39a}$$

$$\frac{\delta F}{\delta U^a(x,t_i)} \stackrel{!}{=} 0\,. \tag{13.39b}$$

Next we must determine the LAGRANGE multipliers $\lambda_a(t_i)$. By taking into account constraints (13.8b), from (13.33c) and (13.29b) we get the relations

$$\lambda_a(t_i) \stackrel{!}{=} \lambda_a(t_i)\frac{1}{S(S-1)}\sum_{x\neq x'} \mathrm{e}^{S_a(x',x;t_i)} \stackrel{!}{=} \frac{1}{S(S-1)}\sum_{x\neq x'} \frac{\delta F}{\delta G_a(x,x';t_i)}\,. \tag{13.40}$$

Inserting (13.40) into (13.33c), we find

$$\sum_{x\neq x'} \frac{\delta F}{\delta G_a(x,x';t_i)} - \left(\frac{1}{S(S-1)}\sum_{x\neq x'} \frac{\delta F}{\delta G_a(x,x';t_i)}\right)\sum_{x\neq x'} \mathrm{e}^{S_a(x',x;t_i)} = 0 \tag{13.41}$$

and, as desired,

$$\frac{1}{S(S-1)}\sum_{x\neq x'} \mathrm{e}^{S_a(x',x;t_i)} = 1 \tag{13.42}$$

if only conditions (13.33c) are fulfilled.

Another possibility to take into account constraints (13.8b) is to minimize the error function F with respect to $\overline{\nu_a^3}(t_i)$, $\overline{\nu_{ab}^1}(t_i)$, $\overline{\nu_{ab}^2}(t_i)$, $U_a(x,t_i)$, $U^a(x,t_i)$, and $D_a(x',x;t_i)$ using representations (13.29). In contrast, $\nu_a(t_i)$ must be kept constant. Without loss of generality we can put $\nu_a(t_i) \equiv 1$. Then, instead of (13.33c), we have the *minimum condition*

$$\frac{\delta F}{\delta G_a(x,x';t_i)} \stackrel{!}{=} 0 \,. \tag{13.43}$$

For the *unscaled* distances $D_a(x',x;t_i)$ we obtain, from (13.43) and (13.36c), the useful relations

$$D_a(x',x;t_i) = \mathrm{e}^{S_a(x',x;t_i)} = \frac{n_a(x',x;t_i)}{d_a(x',x;t_i)} \tag{13.44a}$$

(cf. (13.29a)) with

$$n_a(x',x;t_i) := \Big(g^a(x',x;t_i)P_a^{\mathrm{e}}(x,t_i)\Big)^2 + \Big(g^a(x,x';t_i)P_a^{\mathrm{e}}(x',t_i)\Big)^2 \tag{13.44b}$$

and

$$\begin{aligned} d_a(x',x;t_i) := {} & w_{\mathrm{e}}^a(x'|x;t_i)g^a(x',x;t_i)\Big(P_a^{\mathrm{e}}(x,t_i)\Big)^2 \\ & + w_{\mathrm{e}}^a(x|x';t_i)g^a(x,x';t_i)\Big(P_a^{\mathrm{e}}(x',t_i)\Big)^2 . \end{aligned} \tag{13.44c}$$

If the error function F is minimal for the parameters $U_a(x,t_i)$, $U^a(x,t_i)$, $D_a(x',x;t_i)$, $\nu_a(t_i) \equiv 1$, $\overline{\nu_a^3}(t_i)$, $\overline{\nu_{ab}^1}(t_i)$, and $\overline{\nu_{ab}^2}(t_i)$, the same is, according to model approach (13.27), the case for the *scaled parameters*

$$U_a^{\mathrm{e}}(x,t_i) := U_a(x,t_i) - C_a(t_i) \,, \tag{13.45a}$$

$$U_{\mathrm{e}}^a(x,t_i) := U^a(x,t_i) - C^a(t_i) \,, \tag{13.45b}$$

$$D_a^{\mathrm{e}}(x',x;t_i) := \frac{D_a(x',x;t_i)}{D_a(t_i)} \,, \tag{13.45c}$$

$$\nu_a^{\mathrm{e}}(t_i) := \frac{\nu_a(t_i)}{D_a(t_i)} \,, \tag{13.45d}$$

$$\overline{\nu_a^{3\mathrm{e}}}(t_i) := \frac{\overline{\nu_a^3}(t_i)}{D_a(t_i)} \,, \tag{13.45e}$$

$$\overline{\nu_{ab}^{1e}}(t_i) := \frac{\overline{\nu_{ab}^{1}}(t_i)}{D_a(t_i)}, \tag{13.45f}$$

$$\overline{\nu_{ab}^{2e}}(t_i) := \frac{\overline{\nu_{ab}^{2}}(t_i)}{D_a(t_i)}. \tag{13.45g}$$

In order to fix the scaled parameters in a unique way, we again demand that $U_a^{\rm e}(x, t_i)$, $U_{\rm e}^a(x, t_i)$, and $D_a^{\rm e}(x', x; t_i)$ satisfy the constraints (13.8). This leads to the relations

$$C_a(t_i) := \frac{1}{S} \sum_x U_a(x, t_i), \tag{13.46a}$$

$$C^a(t_i) := \frac{1}{S} \sum_x U^a(x, t_i), \tag{13.46b}$$

and

$$D_a(t_i) := \frac{1}{S(S-1)} \sum_{x \neq x'} D_a(x', x; t_i). \tag{13.47}$$

$C_a(t_i)$ and $C^a(t_i)$ have the meaning of *mean utilities* whereas $D_a(t_i)$ represents some kind of a *unit of distance.*

Since an analytical solution of the minimum conditions (13.33), (13.36) is not possible, for the explicit determination of the model parameters from empirical data we must use numerical, computer-aided methods. Especially suited for this problem is the *method of steepest descent* [19, 62, 80]. The estimated parameters should be carefully checked for their plausibility because the minimization problem possibly has several (relative) minima. That is, the solution found can depend on the *initial approximations* for the parameters. Among other things, a reasonable solution must yield the *non-negativity* of the parameters $\nu_a(t_i)$ and $\overline{\nu_a^3}(t_i)$.

13.3.2 Model Reduction

The model for behavioural changes contains a lot of variables from which certainly not all are *significant* for the description of the system. Thus a systematic simplification of the model (i.e. a *model reduction*) is desirable (cf. [190], pp. 461ff.). For example, we can test the *hypothesis* $H_0: \nu_{ab}^3 \equiv 0$ that compromising processes yield no significant contribution to the considered behavioural changes. This is done by means of a statistical *significance test* (in *linear* cases e.g. with the F*-test*: cf. [190], pp. 663ff.) which checks whether certain parameters can explain an essential part of the variance of the quantities that are to be described.

13.4 Migration in West Germany

In the following we will illustrate the method for the determination of the model parameters from Sect. 13.3 by an example. It is not easy to find suitable data for that since data usually only exist for the temporal evolution of a certain behavioural distribution $P_a^{\mathrm{e}}(x, t_i)$ but not for the corresponding transition rates $w_{\mathrm{e}}^a(x'|x; t_i)$. However, for some countries all necessary data are available for *migration,* i.e. for removals from one region to another (cf. [294]). In this case the behaviour $x \equiv x$ describes in which region $x \in \{1, \ldots, S\}$ out of S distinguished regions an individual lives.

Here the results for migration in West Germany will be presented. West Germany was subdivided into ten federal states and West Berlin (cf. Fig. 13.1 and Table 13.2).

With respect to migration there can be assumed a more or less homogeneous population so that there is $A = 1$ and the subscripts/superscripts a, b can be omitted.

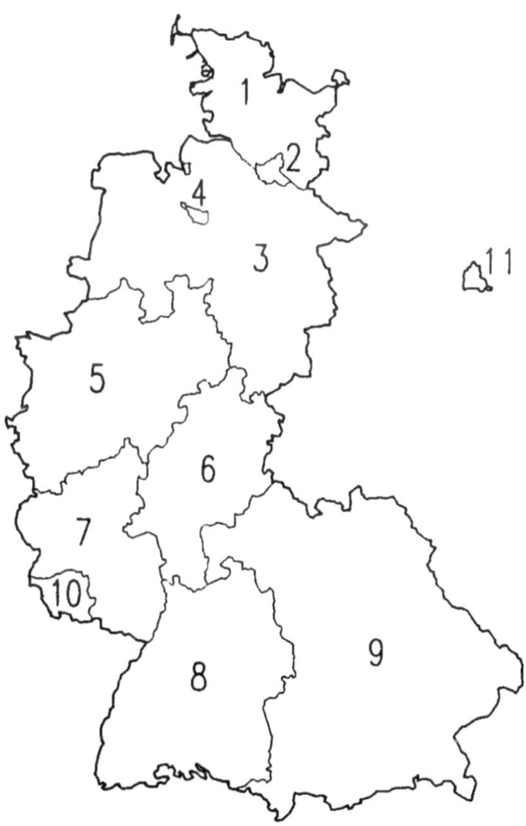

Fig. 13.1 Subdivision of West Germany into eleven federal states

Table 13.2 The eleven federal states of West Germany, their names, and the symbols that will be used for their graphical representation

Symbol	Region x	Name
$+$	1	Schleswig-Holstein
$\odot$	2	Hamburg
$\triangle$	3	Lower Saxony
$\square$	4	Bremen
$\diamond$	5	North Rhine-Westphalia
$\oplus$	6	Hesse
$\bowtie$	7	Rhineland-Palatinate
$\times$	8	Baden-Württemberg
$\triangledown$	9	Bavaria
$\backslash$	10	Saarland
$/$	11	Berlin

Hence the corresponding migration model reads

$$w(x'|x;t_i) := \frac{1}{D^{\mathrm{e}}(x',x;t_i)}\Big[\nu^{\mathrm{e}}(t_i)\mathrm{e}^{U^{\mathrm{e}}(x',t_i)-U^{\mathrm{e}}(x,t_i)} + \Big(\nu^{1\mathrm{e}}(t_i)P^{\mathrm{e}}(x',t_i) + \nu^{2\mathrm{e}}(t_i)P^{\mathrm{e}}(x,t_i)\Big)\mathrm{e}^{U_{\mathrm{e}}(x',t_i)-U_{\mathrm{e}}(x,t_i)}\Big], \tag{13.48}$$

if the rather improbable compromising processes are neglected.

The data for the $S = 11$ regions can be found in the *Statistische Jahrbücher* which are edited in Wiesbaden (Germany) on an annual basis. In the following the $n+1 = 26$ years $t_0 = 1960$, $t_1 = 1961$ to $t_{26} = 1985$ will be examined. Figure 13.2 shows the time-dependent *population distribution* $P^{\mathrm{e}}(x,t_i)$ whereas Fig. 13.3 depicts some examples for the temporal variation of the *effective transition rates* $w_{\mathrm{e}}(x'|x;t_i)$.

13.4.1 First Model Reduction

Applying the method of steepest descent, the minimization of error function (13.28) provides the result

$$\nu^{2\mathrm{e}}(t_i) \ll \nu^{\mathrm{e}}(t_i) < \nu^{1\mathrm{e}}(t_i) \qquad \text{or} \qquad \nu^{2\mathrm{e}}(t_i) \approx 0\,. \tag{13.49}$$

Thus avoidance processes are negligible from an empirical point of view though theoretically they could play a role (e.g. for individuals who move away from a

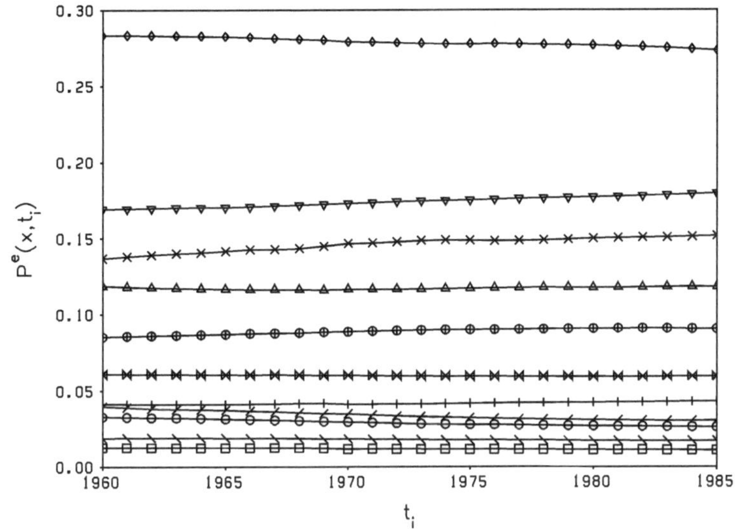

Fig. 13.2 Time-dependent population distribution over the various federal states of West Germany

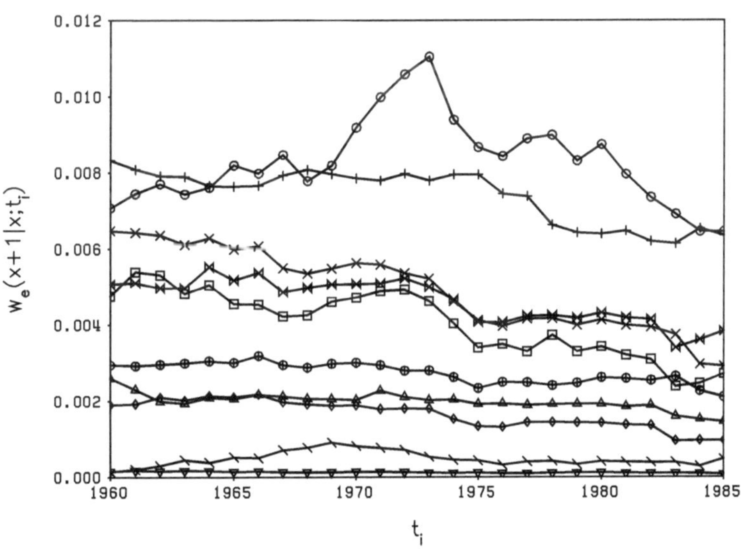

Fig. 13.3 Temporal dependence of some effective transition rates of migration in West Germany

town because they do not bear its population density any longer). Consequently, the migration model reduces to

$$w(x'|x;t_i) := \frac{1}{D^e(x',x;t_i)}\Big[\nu^e(t_i)e^{U^e(x',t_i)-U^e(x,t_i)} + \nu^{1e}(t_i)P^e(x',t_i)e^{U_e(x',t_i)-U_e(x,t_i)}\Big]. \qquad (13.50)$$

The optimal parameters for this model according to the method of steepest descent are illustrated in Figs. 13.4, 13.5 and 13.6. Figure 13.4 shows the rate $\nu^e(t_i)$ of *'spontaneous removals'* as well as the rate $\nu^{1e}(t_i)$ of *'imitative removals'* which are induced by personal contacts to the destination region. The utility functions $U^e(x,t_i)$ for spontaneous behavioural changes and the utility functions $U_e(x,t_i)$ for imitative processes are depicted in Figs. 13.5 and 13.6.

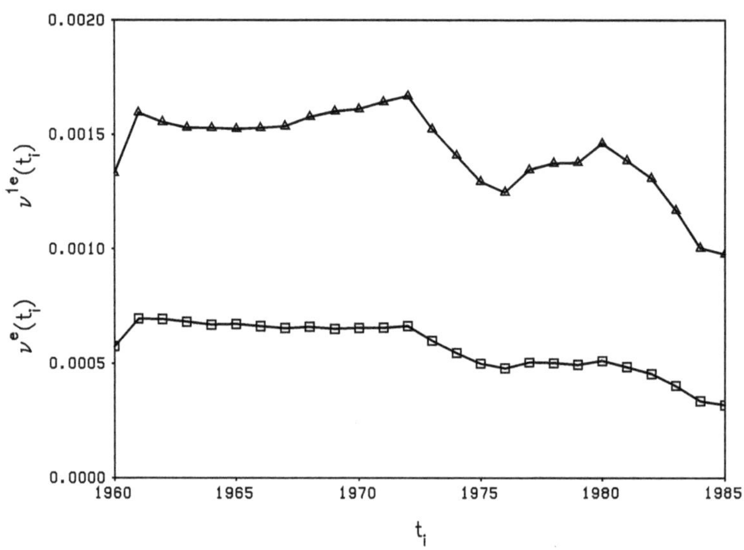

Fig. 13.4 Rate $\nu^e(t_i)$ of spontaneous removals (□) and rate $\nu^{1e}(t_i)$ of imitative processes (△) according to the migration data for West Germany

13.4.2 Second Model Reduction

The irregularity of the utility functions $U_e(x,t_i)$ indicates that these might reflect accidental fluctuations of the migration data. Thus we will examine the *correlation coefficient*

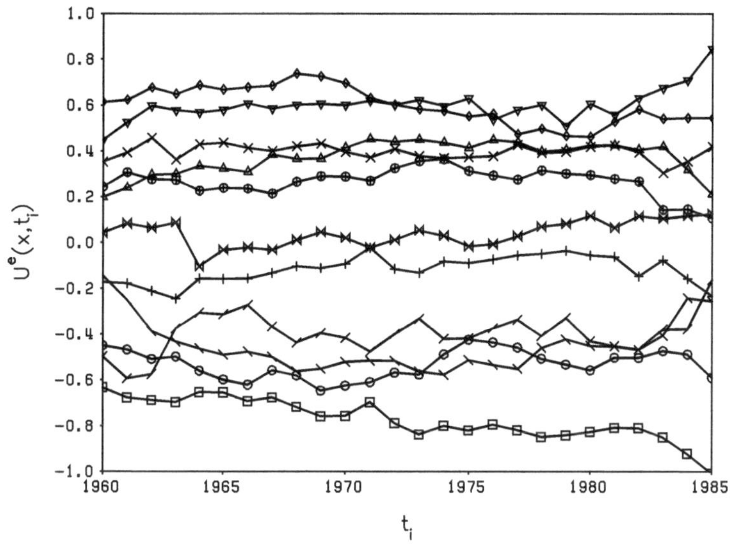

Fig. 13.5 Utility functions for *spontaneous* removals from one federal state of West Germany to another

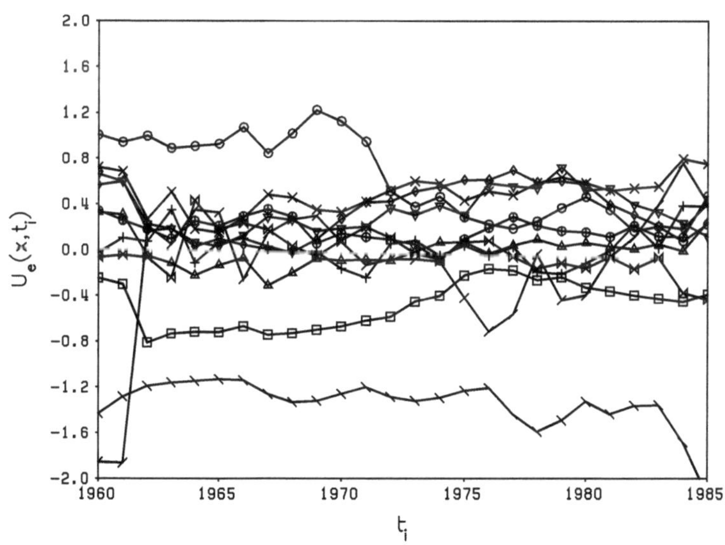

Fig. 13.6 Utility functions for removals from one federal state of West Germany to another due to *imitative* processes

$$r := \frac{\sum_i \sum_{x \neq x'} \Big(w(x'|x;t_i) - W\Big)\Big(w_e(x'|x;t_i) - W_e\Big)}{\sqrt{\sum_i \sum_{x \neq x'} \Big(w(x'|x;t_i) - W\Big)^2} \sqrt{\sum_i \sum_{x \neq x'} \Big(w_e(x'|x;t_i) - W_e\Big)^2}} \tag{13.51}$$

between the empirical transition rates $w_e(x'|x;t_i)$ and the model rates $w(x'|x;t_i)$ for various models $w(x'|x;t_i)$. Here we introduced the *average rates*

$$W := \frac{1}{(n+1)S(S-1)} \sum_i \sum_{x \neq x'} w(x'|x;t_i) \tag{13.52a}$$

and

$$W_e := \frac{1}{(n+1)S(S-1)} \sum_i \sum_{x \neq x'} w_e(x'|x;t_i) \,. \tag{13.52b}$$

Apart from the correlation coefficient r, the quantity

$$\frac{\sum_i \sum_{x \neq x'} \Big(w(x'|x;t_i) - W\Big)^2}{\sum_i \sum_{x \neq x'} \Big(w_e(x'|x;t_i) - W_e\Big)^2} \tag{13.53}$$

is also of particular interest. It describes how many percent of the *empirical variance*

$$\frac{1}{(n+1)S(S-1)} \sum_i \sum_{x \neq x'} \Big(w_e(x'|x;t_i) - W_e\Big)^2 \tag{13.54}$$

of the data are explained by a model $w(x'|x;t_i)$.

Whereas model (13.50) has a correlation of $r = 0.9997$ and reconstructs 100.0% of the empirical variance, the contribution

$$w(x'|x;t_i) := \frac{\nu^e(t_i)}{D^e(x',x;t_i)} e^{U^e(x',t_i) - U^e(x,t_i)} \tag{13.55}$$

describing *spontaneous* transitions reproduces 73.6% of the empirical variance with a correlation of $r = 0.981$. Here it is essential that, during the calculation of (13.53) for contribution (13.55), we used the same model parameters as were determined for the total model (13.50). The contribution

$$w(x'|x;t_i) := \frac{\nu^{le}(t_i)}{D^e(x',x;t_i)} P^e(x',t_i) e^{U_e(x',t_i) - U_e(x,t_i)} \tag{13.56}$$

representing *imitative* processes has a correlation of $r = 0.694$ and explains only 5.2% of the empirical variance. Hence contribution (13.56) is practically insignificant, and the migration model can be reduced to the form (13.55) which, interestingly enough, agrees with the migration model of WEIDLICH and HAAG (cf. [294]).

For the *reduced* model (13.55) once more a parameter optimization is carried out by means of the method of steepest descent. This yields the *mobility rate* $\nu^e(t_i)$ depicted in Fig. 13.7 and the utility functions $U^e(x, t_i)$ illustrated in Fig. 13.8. The *effective distances* $D^e(x', x; t_i)$ can be calculated via formula (13.44). They are not only a measure for the average *geographical distances* but also for the *transaction costs* (e.g. removal costs) and for the differences of mentality, language, etc. which influence the removal behaviour.

A comparison with Figs. 13.4, 13.5 and 13.6 shows that the optimal parameter values of the reduced model (13.55) are different from those of the total model (13.50). For example, the utility functions $U^e(x, t_i)$ for spontaneous transitions now reflect also tendencies which in the total model were delineated by the utility functions $U_e(x, t_i)$ for imitative processes (compare the curves for Berlin (/)).

13.4.3 Comparison of the WEIDLICH-HAAG Model and the Generalized Gravity Model

The correlation of the WEIDLICH-HAAG model (13.55) with the empirical migration rates $w_e(x'|x; t_i)$ for the parameters depicted in Figs. 13.7 and 13.8 is $r = 0.996$. Hence this model reproduces the empirical data almost perfectly.

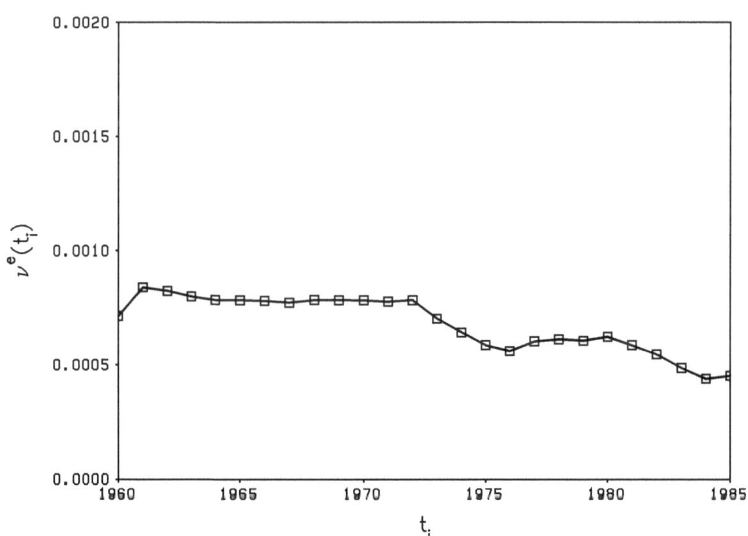

Fig. 13.7 Mobility rate for spontaneous removals in West Germany according to the migration model of WEIDLICH and HAAG

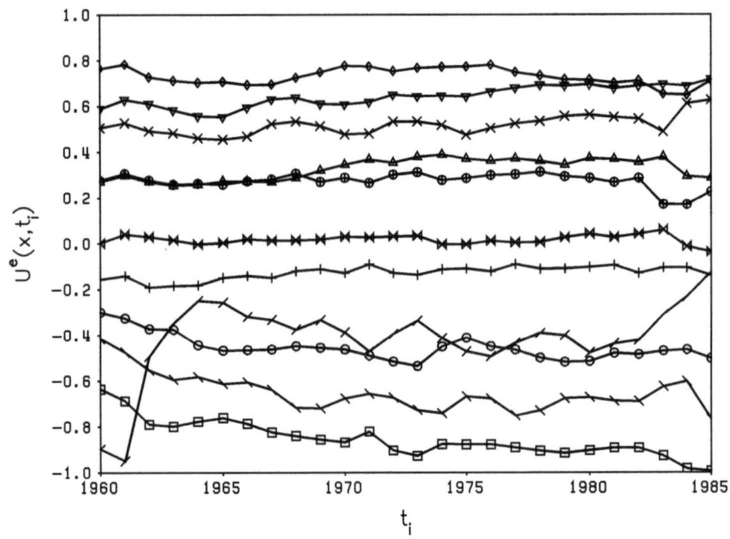

Fig. 13.8 Utility functions of the single federal states of West Germany for the migration model of WEIDLICH and HAAG. Apparent is the extreme dependence of the utility of West Berlin (/) on changes of the *political situation*. For example, a remarkable utility increase is observed after the erection of the Berlin wall in 1961

Astonishingly enough, apart from the WEIDLICH-HAAG model there exist *other* models which have the same number of parameters and show exactly the same correlation with the empirical data. This applies, for example, to the *generalized gravity model*

$$w'(x'|x;t_i) := \frac{\nu^{\mathrm{le}}(t_i)}{D_{\mathrm{e}}(x',x;t_i)} P^{\mathrm{e}}(x',t_i) \mathrm{e}^{U_{\mathrm{e}}(x',t_i) - U_{\mathrm{e}}(x,t_i)} \tag{13.57}$$

(cf. (9.69)) because of the following reason: If the model parameters of the generalized gravity model are calculated via the formulas

$$U_{\mathrm{e}}(x,t_i) := U^{\mathrm{e}}(x,t_i) - \ln\sqrt{P^{\mathrm{e}}(x,t_i)} + \frac{1}{S}\sum_{x'} \ln\sqrt{P^{\mathrm{e}}(x',t_i)} \tag{13.58a}$$

and

$$\frac{\nu^{\mathrm{le}}(t_i)}{D_{\mathrm{e}}(x',x;t_i)} := \frac{1}{\sqrt{P^{\mathrm{e}}(x,t_i) P^{\mathrm{e}}(x',t_i)}} \frac{\nu^{\mathrm{e}}(t_i)}{D^{\mathrm{e}}(x',x;t_i)} \tag{13.58b}$$

from the optimal model parameters for the WEIDLICH-HAAG model, we have

$$w'(x'|x;t_i) \equiv w(x'|x;t_i) \,. \tag{13.59}$$

Whereas the WEIDLICH-HAAG model bases on *spontaneous* removals, the generalized gravity model supposes *imitative* processes. Obviously, apart from the correlation coefficient r we require another criterion in order to decide which of the two models describes migration in the most adequate way. For this purpose, we utilize the circumstance that the effective distances $D^{\text{e}}(x', x; t_i)$ and $D_{\text{e}}(x', x; t_i)$ respectively between two regions x and x' should be nearly independent of time t_i. In particular, they should not depend on the population distribution $P^{\text{e}}(x, t_i)$ which, however, is inevitably the case for one of both models due to relation (13.58b). Thus we will prefer that model for which the corresponding distances $D^{\cdot\cdot}_{\cdot\cdot}(x', x; t_i)$ possess the smallest *average variance*

$$\frac{1}{(n+1)S(S-1)} \sum_i \sum_{x \neq x'} \left[D^{\cdot\cdot}_{\cdot\cdot}(x', x; t_i) - \left(\frac{1}{n+1} \sum_j D^{\cdot\cdot}_{\cdot\cdot}(x', x; t_j) \right) \right]^2 \tag{13.60}$$

in the course of time. This criterion decides clearly in favour of the generalized gravity model.

Figure 13.9 depicts the mobility rate $\nu^{1\text{e}}(t_i)$ for the generalized gravity model (13.57). The corresponding utility functions $U_{\text{e}}(x, t_i)$ are illustrated in Fig. 13.10. A comparison with Fig. 13.8 shows that the utility functions $U_{\text{e}}(x, t_i)$ of the generalized gravity model possess a considerably smaller variance than the utility functions $U^{\text{e}}(x, t_i)$ of the WEIDLICH-HAAG model. This can be interpreted in such a way that the utility functions $U^{\text{e}}(x, t_i)$ of the WEIDLICH-HAAG model still depend on

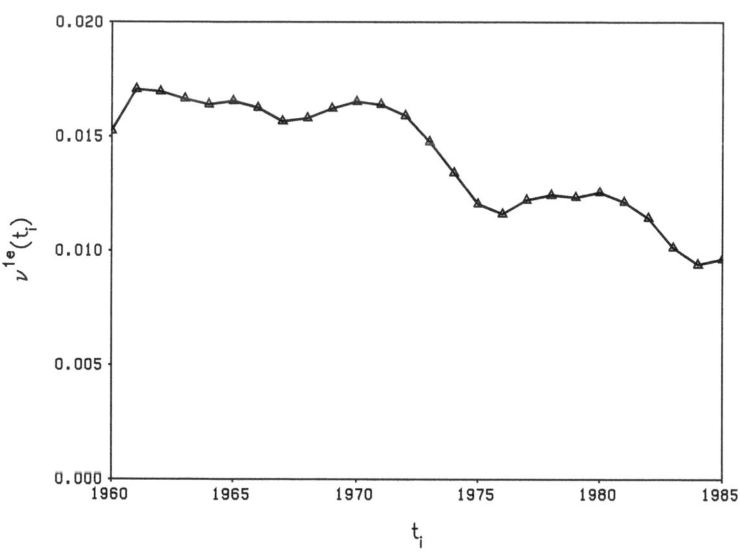

Fig. 13.9 Mobility rate for removals in West Germany according to the generalized gravity model

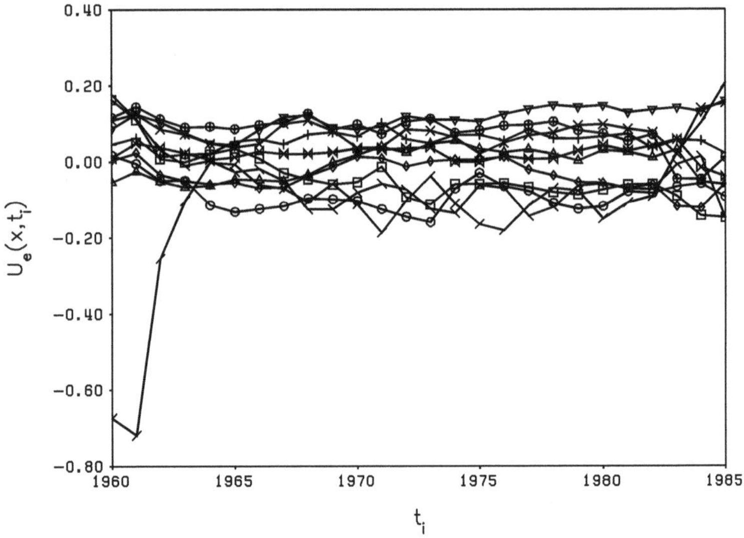

Fig. 13.10 Utility functions of the single federal states of West Germany for the generalized gravity model

the population distribution $P^{\rm e}(x, t_i)$ (cf. [294], pp. 74ff.) which seems not to be the case any more for the utility functions $U_{\rm e}(x, t_i)$ of the gravity model.

13.4.4 Third Model Reduction

The replacement of the time-dependent distances $D_{\rm e}(x', x; t_i)$ by *time-independent effective distances*

$$D_*(x', x) := \frac{1}{n+1} \sum_i D_{\rm e}(x', x; t_i) \tag{13.61a}$$

allows a further model reduction. The remaining model reads

$$w(x'|x; t_i) := \frac{\nu^{1*}(t_i)}{D_*(x', x)} P^{\rm e}(x', t_i) {\rm e}^{U_{\rm e}(x', t_i) - U_{\rm e}(x, t_i)} . \tag{13.61b}$$

Then the optimal value for the mobility rate $\nu^{1*}(t_i)$ is given by

$$\nu^{1*}(t_i) \stackrel{!}{=} \nu^{1{\rm e}}(t_i) \frac{\displaystyle\sum_{x \neq x'} \frac{1}{D_{\rm e}(x', x; t_i)}}{\displaystyle\sum_{x \neq x'} \frac{1}{D_*(x', x)}} \tag{13.61c}$$

since this relation best fulfils the demand

$$\frac{\nu^{1*}(t_i)}{D_*(x',x)} \stackrel{!}{=} \frac{\nu^{1e}(t_i)}{D_e(x',x;t_i)} . \tag{13.62}$$

The *improved mobility rate* $\nu^{1*}(t_i)$ is depicted in Fig. 13.11. Moreover, Table 13.3 lists the time-independent effective distances $D_*(x',x)$.

For the description of the $S(S-1)(n+1) = 2860$ effective transition rates $w^e(x'|x;t_i)$ the final model (13.61b) requires only $(S+1)(n+1)+S(S-1)/2 = 367$ parameters. That is, we have achieved a data reduction to $367/2860 = 12.8\%$. For

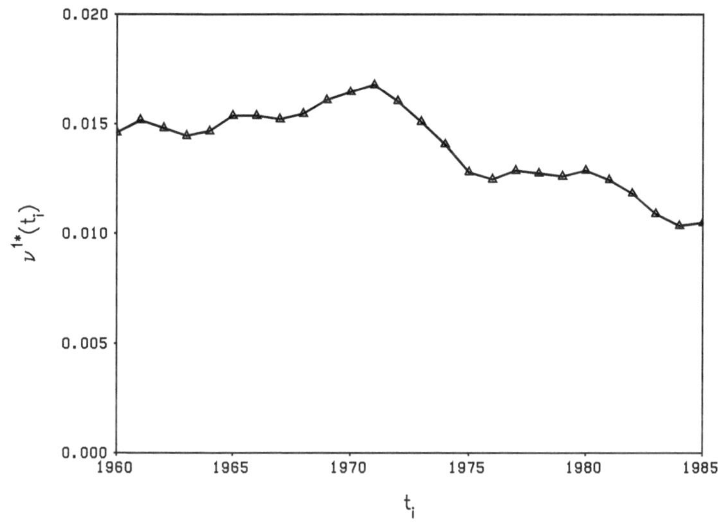

Fig. 13.11 Improved mobility rate of removals in West Germany for the generalized gravity model with *time-independent* effective distances

Table 13.3 Time-independent effective distances $D_*(x',x)$ between the eleven federal states of West Germany according to the generalized gravity model

x	1	2	3	4	5	6	7	8	9	10	11
1	–	0.05	0.33	0.23	0.91	0.99	1.27	1.19	1.53	2.01	0.42
2	0.05	–	0.22	0.26	1.20	0.94	1.78	1.32	1.58	2.52	0.43
3	0.33	0.22	–	0.08	0.68	0.77	1.51	1.40	1.86	2.49	0.44
4	0.23	0.26	0.08	–	0.93	0.85	1.40	1.18	1.65	2.18	0.46
5	0.91	1.20	0.68	0.93	–	0.86	0.58	1.36	1.73	1.53	0.77
6	0.99	0.94	0.77	0.85	0.86	–	0.29	0.59	0.79	0.76	0.54
7	1.27	1.78	1.51	1.40	0.58	0.29		0.46	1.19	0.15	0.90
8	1.19	1.32	1.40	1.18	1.36	0.59	0.46	–	0.52	0.68	0.69
9	1.53	1.58	1.86	1.65	1.73	0.79	1.19	0.52	–	1.67	0.75
10	2.01	2.52	2.49	2.18	1.53	0.76	0.15	0.68	1.67	–	1.15
11	0.42	0.43	0.44	0.46	0.77	0.54	0.90	0.69	0.75	1.15	–

the description of one region the model needs, on average, only $367/[S(n+1)] = 1.28$ parameters per year. Despite this, model (13.61) has a very good correlation of $r = 0.98$.

13.5 Evaluation of Empirically Obtained Results

13.5.1 Sensitivity Analysis

After determining the model parameters from empirical data, a *sensitivity analysis* of the resulting model with respect to small variations of the parameters should be carried out [34, 268, 307]. This is necessary in order to realize whether a possible deviation of the estimated value of a parameter

$$\Phi_k(t_i) \in \left\{ \nu_a(t_i),\, \overline{\nu_a^3}(t_i),\, \overline{\nu_{ab}^1}(t_i),\, \overline{\nu_{ab}^2}(t_i),\, U_a(x,t_i),\, U^a(x,t_i),\, D_a(x',x;t_i) \right\} \tag{13.63}$$

from its actual value would have an essential influence on the model behaviour.

The way of proceeding in a sensitivity analysis is as follows: First, the temporal evolution of $P_a(x,t)$ is calculated according to the BOLTZMANN-like equations (9.62) for the estimated model parameters $\Phi_k \equiv \Phi_k(t_i)$. Then one separately varies each model parameter Φ_k by a small amount $\delta\Phi_k$ and computes the *corresponding* temporal evolution $P'_a(x,t)$. A suitable measure for the *dissimilarity* of $P_a(x,t)$ and $P'_a(x,t)$ is provided by THEIL's *inequality coefficient* [272]

$$T_a(x,t) := \frac{\sqrt{\displaystyle\int_{t_0}^{t} dt' \left(P'_a(x,t') - P_a(x,t') \right)^2}}{\sqrt{\displaystyle\int_{t_0}^{t} dt' \left(P'_a(x,t') \right)^2} + \sqrt{\displaystyle\int_{t_0}^{t} dt' \left(P_a(x,t') \right)^2}}\,, \tag{13.64}$$

which lies between 0 and 1 and deviates the more from zero the more different $P_a(x,t)$ and $P'_a(x,t)$ are. As the *absolute sensitivity* of the model with respect to a change of parameter Φ_k we define the quantity

$$S_k^a(x,t) := \frac{\delta P_a(x,t)}{\delta \Phi_k} \tag{13.65}$$

whereas with the term *'relative sensitivity'* we mean

$$R_k^a(x,t) := \frac{\delta P_a(x,t)}{P_a(x,t)} \frac{\Phi_k}{\delta \Phi_k} \tag{13.66}$$

where $\delta P_a(x,t) := P'_a(x,t) - P_a(x,t)$ [307].

13.5.2 Decomposition of the Utility Functions with Respect to Explanatory Variables

The utility functions usually split up into different contributions:

$$U_a(x, t_i) = \sum_{l=1}^{L} \widehat{c}_a^l(x) \widehat{U}_a^l(x, t_i), \tag{13.67a}$$

$$U^a(x, t_i) = \sum_{l=1}^{L} \widehat{c}_l^a(x) \widehat{U}_a^l(x, t_i). \tag{13.67b}$$

Here $\widehat{U}_a^l(x, t_i)$ represent different *influencing factors* (which are often independent of x and a). By a decomposition with regard to influencing factors we try to express the time-dependence of the utility functions $U_a(x, t_i)$, $U^a(x, t_i)$ by a smaller number of easily measurable quantities $\widehat{U}_a^l(x, t_i)$. The estimation of the constants $\widehat{c}_a^l(x)$ and $\widehat{c}_l^a(x)$ is carried out by means of a *regression method* (cf. [190], pp. 637ff., [81], pp. 415ff., and [82], pp. 494ff.). It presupposes the knowledge of the utility functions and influencing factors for $(n+1) \geq ASL$ times t_i.

Often the temporal evolutions of different influencing factors are *correlated* with each other. Thus one tries to express the utility functions $U_a(x, t)$, $U^a(x, t)$ by *uncorrelated* and, in addition, *interpretable explanatory variables*

$$U_a^l(x, t_i) := \sum_j M_{lj}^a(x) \widehat{U}_a^j(x, t_i). \tag{13.68}$$

Then we have

$$U_a(x, t_i) = \sum_{l=1}^{L} c_a^l(x) U_a^l(x, t_i), \tag{13.69a}$$

$$U^a(x, t_i) = \sum_{l=1}^{L} c_l^a(x) U_a^l(x, t_i) \tag{13.69b}$$

with

$$c_a^l(x) := \sum_j \widehat{c}_a^j(x) (\underline{M}^a)_{jl}^{-1}, \tag{13.70a}$$

$$c_l^a(x) := \sum_j \widehat{c}_j^a(x) (\underline{M}^a)_{jl}^{-1}. \tag{13.70b}$$

Suitable transformation matrices $\underline{M}^a \equiv \left(M_{ij}^a\right)$ can be determined by a *factor analysis* (cf. [149], pp. 401ff.).

A decomposition of the utility functions with respect to explanatory variables is also possible in one single step by means of the *ranking regression analysis* (cf. [238, 294]).

13.5.3 Prognoses

A particularly interesting topic is the *prognosis* of behavioural changes [97, 155]. The simplest prognosis technique would certainly be an *extrapolation* of the temporal evolution of $P_a(x,t)$. By doing this, however, the knowledge of the processes that take place in the system is not utilized. This can, especially due to the *nonlinearity* of the dynamics, easily lead to incorrect prognoses. A more reliable prognosis method bases on the extrapolation of the *explanatory* variables. $P_a(x,t)$ is then calculated in accordance with the BOLTZMANN-like equations (9.62).

13.6 Examples for Decompositions of Utility Functions

13.6.1 Purchase Pattern

The decision on the purchase of a product $x \equiv i \in \{1,\dots,S\}$ of a certain kind of good will decisively depend on the utility $U_a(i,t)$ of the product for the purchaser. From the point of view of *sales strategy* it is of special interest which factors determine the utility of a product. The empirical data that are necessary for this are collected in *market analyses* and *surveys*. In the following let us assume that the utility functions $U_a(i,t)$ and $U^a(i,t)$ have already been estimated from the data (cf. Sects. 13.2 and 13.3).

First, it is reasonable to split off the influence of the number $n_i^a(t)$ of purchasers from the transition rates:

$$U_a(i,t) \equiv U_a(i,n;t) := \delta_a(i,t) + \sum_b \kappa_{ab} n_i^b(t) - \sum_{b,c} \sigma_{abc} n_i^b(t) n_i^c(t) \,. \qquad (13.71)$$

The terms

$$\sum_b \kappa_{ab} n_i^b(t) - \sum_{b,c} \sigma_{abc} n_i^b(t) n_i^c(t) \qquad (13.72)$$

describe the *feedback effect* of the purchasers' decisions on the purchase decisions. The single contributions to (13.71) can be interpreted as follows:

- κ_{ab} are *agglomeration parameters*. For $\kappa_{ab} > 0$ individuals of subpopulation a will tend towards purchasing a product i the more frequently the more individuals of subpopulation b decide in favour of it. So *trend* and *fashion effects* can be taken into account. (With regard to clothing, for example, for the customer one important aspect is to receive attention which represents an essential contribution to the utility.)

- σ_{abc} are *saturation parameters.* For $\sigma_{abc} > 0$ they reflect the avoidance of products which are widely spread and thus no longer individual or original.
- The *inclination parameters* $\delta_a(i, t)$ delineate the part of the utility of a product which is independent of the behaviour of other purchasers, i.e. they describe the degree of preference for or refusal of a product i by individuals of subpopulation a. $\delta_a(i, t)$ is determined by the *purchaser interests* and *product characteristics.*

The objective of the following discussion is to decompose the inclination parameters $\delta_a(i, t)$ into the various factors which are relevant for the purchase decision. For this purpose, let us consider L different *characteristics* l of the products i, e.g. price, standard/service, quantity, perishability/durability, colour, and shape. Moreover, let $d_a^l(i, t)$ with

$$0 \leq d_a^l(i, t) \leq 1 \tag{13.73}$$

be a measure for the difference between the *'ideal'* and *real* characteristic $l \in \{1, \ldots, L\}$ of product $i \in \{1, \ldots, S\}$ for individuals of subpopulation a. The quantity $d_a^l(i, t)$ can be collected by means of a survey using a *rating scale* similar to the *semantic differential* (cf. Sect. 8.2.6) and will further be denoted as *dissatisfaction* with respect to characteristic l. We expect that the inclination towards a product i decreases with growing dissatisfaction $d_a^l(i, t)$. This effect will be the stronger the larger the *relevance* r_a^l of characteristic l for individuals of subpopulation a. Hence a plausible approach is

$$\delta_a(i, t) := \pi_a(i) - \sum_{l=1}^{L} r_a^l d_a^l(i, t) - C_a(t) . \tag{13.74}$$

Here $C_a(t)$ is a *scaling parameter* (cf. (13.8a)) and $\pi_a(i)$ with

$$\sum_i \pi_a(i) \stackrel{!}{=} 0 \tag{13.75}$$

denotes the *prestige* which a product i has among subpopulation a independently of the L rated characteristics l.

In summary, the model for the utility functions $U_a(i, t)$ has the form

$$U_a(i, t) := \pi_a(i) + \sum_b \kappa_{ab} n_i^b(t) - \sum_{b,c} \sigma_{abc} n_i^b(t) n_i^c(t) - \sum_{l=1}^{L} r_a^l d_a^l(i, t) - C_a(t) . \tag{13.76a}$$

For the utility functions $U^a(i,t)$ that are related to *pair interactions* we choose an analogous approach:

$$U^a(i,t) := \pi^a(i) + \sum_b \kappa^{ab} n_i^b(t) - \sum_{b,c} \sigma^{abc} n_i^b(t) n_i^c(t) - \sum_{l=1}^{L} r_l^a d_a^l(i,t) - C^a(t) \tag{13.76b}$$

with

$$\sum_i \pi^a(i) \stackrel{!}{=} 0 \,. \tag{13.77}$$

By distinguishing (13.76a) and (13.76b), we take into account that a characteristic l possibly has another relevance at *pair interactions* (i.e. conversations) than at *spontaneous* opinion changes ($r_a^l(t) \neq r_l^a(t)$).

For the determination of the coefficients $\pi_a(i)$, κ_{ab}, σ_{abc}, r_a^l, $\pi^a(i)$, κ^{ab}, σ^{abc}, and r_l^a we must carry out a *regression analysis.* A certain characteristic l will only yield an essential contribution to the utilities $U_a(i,t)$ or $U^a(i,t)$ of a product i if

1. the dissatisfaction with respect to l is not negligible (i.e. there is not $d_a^l(i,t) \ll 1$),
2. the relevances r_a^l and r_l^a of the characteristic l are large,
3. the relevances r_a^l and r_l^a are *significantly* different from zero (i.e. the variations of the terms $r_a^l d_a^l(i,t)$ and $r_l^a d_a^l(i,t)$ can explain an essential part of the variations of $U_a(i,t)$ and $U^a(i,t)$).

As a consequence of the first criterion, from the economist's point of view it is not very reasonable to improve product characteristics which are already close to the purchasers' wishes (unless the expectations of the purchasers can be raised by clever advertising).

In order to take into account also those individuals who *do not* decide on the purchase of one of the products $i \in \{1, \dots, S\}$, an additional state $i = 0$ can be introduced. For the corresponding utility functions we choose

$$U_a(0,t) := \pi_a(0) + \sum_b \kappa_{ab} n_0^b(t) - \sum_{b,c} \sigma_{abc} n_0^b(t) n_0^c(t) - C_a(t) \,, \tag{13.78a}$$

$$U^a(0,t) := \pi^a(0) + \sum_b \kappa^{ab} n_0^b(t) - \sum_{b,c} \sigma^{abc} n_0^b(t) n_0^c(t) - C^a(t) \tag{13.78b}$$

since, this time, it is not the product characteristics which play a role but the feedback effects.

13.6.2 Voting Behaviour

In the case of a voter's decision (electoral decision) among several parties the states $i \in \{1, \dots, S\}$ denote just the S parties which can be elected. By $i = 0$ we take into account the possibility to vote for *none* of the parties. Analogously to Sect. 13.6.1, from the utility functions we first split off the dependence on the numbers $n_i^a(t)$ of voters (the *feedback effects* or influences of *indirect interactions*):

$$U_a(i,t) := \delta_a(i,t) + \sum_b \kappa_{ab} n_i^b(t) - \sum_{b,c} \sigma_{abc} n_i^b(t) n_i^c(t)\,. \tag{13.79a}$$

- This time the *agglomeration parameters* κ_{ab} reflect that a party i is the more preferred the more individuals vote for it. On the one hand, the reason for this might be that people believe that a big party is capable of exerting more power to make its political goals come true. On the other hand, individuals orientate themselves by the behaviour of their *social environment* especially if it is a widely disseminated behaviour *(bandwaggon effect)*. It is particularly the so-called *people's parties* with a high proportion of *loyal voters* which profit from this fact.
- The *saturation parameters* σ_{abc} describe a reduction of the bandwaggon effect in cases where a party already has a huge number n_i^a of followers. This reduction results from the attempt to avoid the predominance of one single party i. (A well working democracy requires a *balance of power,* in particular, an opposition that is not too weak.)
- The *inclination parameters* $\delta_a(i,t)$ delineate the degree of preference for or refusal of party i by subpopulation a which is independent of the number of its followers. $\delta_a(i,t)$ is determined by the *voters' interests* and the *goals of the parties (party programmes).*

This time the inclination parameters will be decomposed in the sense of a *competence model:*

$$\delta_a(i,t) := \pi_a(i) + \sum_l c_a^l D_a^l(t) k_a^l(i,t) - \widehat{C}_a(t)\,. \tag{13.79b}$$

According to (13.79b), a party $i \in \{1, \dots, S\}$ will be the more preferred with respect to an affair $l \in \{1, \dots, L\}$ the higher its *competence* $k_a^l(i,t)$ and the stronger the *need for action*

$$H_a^l(t) := c_a^l D_a^l(t) \tag{13.80}$$

in this affair from the standpoint of subpopulation a.

- With

$$0 \le k_a^l(i,t) \le 1 \tag{13.81}$$

the competence $k_a^l(i,t)$ is demanded to lie in the interval [0, 1]. It can be determined by surveys using suitable *rating scales.*

- c_a^l represents a measure for the *importance* of affair l for individuals of subpopulation a.
- $D_a^l(t)$ with

$$0 \le D_a^l(t) \le 1 \tag{13.82}$$

is a measure for the *discrepancy* between the *'ideal'* and *real* state of society from the viewpoint of subpopulation a with regard to affair l. Relevant affairs l are, for example, the topics unemployment, purchasing power, standard of living, housing market, leisure time, ecology, abortion, or armament.

- $\pi_a(i)$ with

$$\sum_i \pi_a(i) \stackrel{!}{=} 0 \tag{13.83}$$

reflects the *prestige* of a party i within subpopulation a.

- $\widehat{C}_a(t)$ denotes a *scaling parameter* (cf. (13.8a)).

Equation (13.79) implies that a party i can, with respect to affair l, gain political attractiveness in various ways:

- The party can increase its competence $k_a^l(i,t)$ with respect to affair l.
- It can profit from an increase of the discrepancy $D_a^l(t)$ between the 'ideal' and real state of society which is possibly caused by a deterioration of the *social situation.* However, a growth of discrepancy $D_a^l(t)$ is also achievable by raising the expectations about the 'ideal' state of the society. (If the 'ideal' state with respect to a certain affair l is reached ($D_a^l(t) = 0$) there will be no longer a need for action $H_a^l(t)$. Thus, with respect to this affair there will exist no reason to prefer party i to another party i'.)
- The party can also try to gain public interest in affairs l in which it is especially competent, i.e. to increase the importance c_a^l of these affairs. In addition, it can try to play down or to hush up affairs where it is incompetent.

The approach and interpretation concerning the utility functions $U^a(i,t)$ for *pair interactions* is completely analogous:

$$U^a(i,t) := \delta^a(i,t) + \sum_b \kappa^{ab} n_i^b(t) - \sum_{b,c} \sigma^{abc} n_i^b(t) n_i^c(t)\,, \tag{13.84a}$$

$$\delta^a(i,t) := \pi^a(i) + \sum_l c_l^a D_a^l(t) k_a^l(i,t) - \widehat{C}^a(t)\,, \tag{13.84b}$$

$$\sum_i \pi^a(i) \stackrel{!}{=} 0\,. \tag{13.84c}$$

By distinguishing (13.79) and (13.84) we take into account that it is possible for an affair l to have another importance at *pair interactions* (i.e. discussions) than at *spontaneous* opinion changes ($c^l_a(t) \neq c^a_l(t)$).

Finally, like in (13.78), we assume for the utility to vote for *none* of the parties

$$U_a(0,t) := \pi_a(0) + \sum_b \kappa_{ab} n^b_0(t) - \sum_{b,c} \sigma_{abc} n^b_0(t) n^c_0(t) - \widehat{C}_a(t)\,, \qquad (13.85a)$$

$$U^a(0,t) := \pi^a(0) + \sum_b \kappa^{ab} n^b_0(t) - \sum_{b,c} \sigma^{abc} n^b_0(t) n^c_0(t) - \widehat{C}^a(t)\,. \qquad (13.85b)$$

An alternative model for voting behaviour would, analogously to Sect. 13.6.1, be provided by approach (13.76) which expresses the preference for a certain party i by the *dissatisfaction* $d^l_a(i,t)$ of the voters of subpopulation a with the programmes of the parties i with respect to different affairs l. A comparison shows that model (13.79), (13.84) can be transformed into model (13.76) by means of the identifications

$$r^l_a(t) := c^l_a D^l_a(t)\,, \qquad (13.86a)$$

$$r^a_l(t) := c^a_l D^l_a(t)\,, \qquad (13.86b)$$

$$d^l_a(i,t) := 1 - k^l_a(i,t)\,, \qquad (13.86c)$$

$$C_a(t) := \widehat{C}_a(t) - \sum_l c^l_a D^l_a(t)\,, \qquad (13.86d)$$

$$C^a(t) := \widehat{C}^a(t) - \sum_l c^a_l D^l_a(t)\,. \qquad (13.86e)$$

Accordingly, the relevances $r^l_a(t)$ and $r^a_l(t)$ of an affair l are given by the needs for action $c^l_a(t)D^l_a(t)$ and $c^a_l(t)D^l_a(t)$ respectively. Moreover, the competence $k^l_a(i,t) = 1 - d^l_a(i,t)$ of party i with respect to affair l from the viewpoint of subpopulation a decreases with growing dissatisfaction $d^l_a(i,t)$, i.e. with growing distance between the party programme and the political expectations of subpopulation a.

Obviously, the identifications (13.86) are quite plausible so that the models (13.76) and (13.79), (13.84) are compatible. However, in the case of sufficient data, the competence model (13.79), (13.84) should be preferred because, with (13.86a), (13.86b), it additionally describes the time-dependence of the relevances $r^l_a(t)$, $r^a_l(t)$.

The decomposition of the utility functions $U_a(i,t)$ and $U^a(i,t)$ is especially interesting with regard to *prognoses*. However, the prognosis of *election results* is connected with some complications [52]:

- The voters' decisions depend on the *expected* election results, especially on predictable *coalitions*.
- Since the voters make a decision not only for a short time but for a whole *parliamentary term*, they include in their decision procedure the *expectations* about the future change of the discrepancies $D_a^l(t)$ and of the dissatisfaction $d_a^l(i,t)$ with respect to the different affairs l.
- Of great importance are also political *events of the day* which are hardly predictable.

In the case of purchase decisions comparable complications are normally unimportant. Consequently, for these better results can be expected.

13.6.3 Gaps in the Market and Foundations of New Parties

On the one hand, after the introduction of a new product or the foundation of a new party $i := S+1$ at the time t_0, new prognoses for the most probable purchase pattern or voting behaviour might become necessary. On the other hand, however, our models from Sects. 13.6.1 and 13.6.2 allow the estimation of the *achievable market share* of a new product or the *attainable number of followers* of a new party. With this aim in view, we calculate the solution $P_a(S+1,t)$ of BOLTZMANN-like equations (9.62) on the basis of the predicted temporal evolution of the utility functions $U_a(i,t)$ and $U^a(i,t)$. In order to take into account the *non-eligibility* of the product or the party $S+1$ before the time t_0 of appearance, we formally set

$$U_a(S+1,t) := -\infty \quad \text{and} \quad U^a(S+1,t) := -\infty \quad \text{for } t < t_0 . \tag{13.87}$$

For reasons of uniqueness and continuity of the utility functions we choose the constraints

$$\sum_{i=0}^{S} U_a(i,t) \stackrel{!}{=} 0 , \qquad \sum_{i=0}^{S} U^a(i,t) \stackrel{!}{=} 0 \qquad \text{for } t \le t_0 \tag{13.88a}$$

and

$$\sum_{i=0}^{S+1} U_a(i,t) \stackrel{!}{=} U_a(S+1,t_0) , \quad \sum_{i=0}^{S+1} U^a(i,t) \stackrel{!}{=} U^a(S+1,t_0) \quad \text{for } t > t_0 . \tag{13.88b}$$

The result

$$P(S+1,t) = \sum_a \frac{N_a}{N} P_a(S+1,t) \tag{13.89}$$

for the achievable market share or voter proportion $P(S+1,t)$ at the time t allows a decision whether a new product or party has a chance to establish.

One can even carry out analyses about what characteristics a new product should have or which political programme a new party should pursue in order to gain an optimal market or voter proportion $P(S+1,t)$ at a certain time t. The determination of *gaps in the market* or in the *political landscape* takes into account

1. the *expectations* (needs and desires) of purchasers or voters and their *satisfaction (coverage)* by existing products or parties i via $d_a^l(i,t)$,
2. the *sizes* N_a of the subpopulations having these expectations by (13.89),
3. the *relevances* of different characteristics or affairs l via r_a^l and r_l^a.

References

1. Aguda BD, Larter R, Clarke BL (1989) Dynamic elements of mixed-mode oscillations and chaos in a peroxidase-oxidase model network. J Chem Phys 90(8):4168–4175
2. Allen P (1976) Population dynamics and evolution. In: Jantsch E, Waddington CH (eds) Evolution and consciousness. Human systems in transition. Addison-Wesley, Reading, MA
3. Allen PM, Engelen G, Sanglier M (1986) Towards a general dynamic model of the spatial evolution of urban systems. In: Hutchinson B, Batty M (eds) Advances in urban systems modelling. Elsevier, Amsterdam
4. Allen PM, Sanglier M (1979) A dynamic model of a central place system. Geogr Anal 11:256ff
5. Allen PM, Sanglier M (1981) Urban evolution, self-organization and decision making. Environ Plan A 13:167
6. Allen PM, Sanglier M, Engelen G, Boon F (1984) Evolutionary spatial models of urban and regional systems. Sistemi Urbani 1
7. Arneodo A, Coullet P, Tresser C (1980) Occurrence of strange attractors in threedimensional Volterra equations. Phys Lett A 79(4):259–263
8. Arnold L, Lefever R (eds) (1981) Stochastic nonlinear systems in physics, chemistry and biology. Springer, Berlin
9. Arthur B (1989) Competing technologies, increasing returns, and lock-in by historical events. Econ J 99:116–131
10. Ashcroft NW, Mermin ND (1976) Solid state physics. Holt-Saunders, Tokyo, pp 715–718
11. Atkins PW (1986) Physical chemistry, Chaps. 28, 29, 3rd edn. Oxford University Press, Oxford
12. Axelrod R (1984) The evolution of cooperation. Basic Books, New York, NY
13. Axelrod R, Dion D (1988) The further evolution of cooperation. Science 242:1385–1390
14. Babloyantz A, Destexhe A (1987) Strange attractors in the human cortex. In: Rensing L, van der Heiden U, Mackey MC (eds) Temporal disorder in human oscillatory systems. Springer, Heidelberg
15. Babloyantz A, Nicolis C, Salazar M (1985) Evidence of chaotic dynamics of brain activity during the sleep cycle. Phys Lett A 111:152
16. Bai-Lin H (ed) (1984) Chaos. World Scientific, Singapore
17. Barnsley M (1988) Fractals everywhere. Academic Press, Boston
18. Bartholomew DJ (1967) Stochastic models for social processes. Wiley, London
19. Beightler CS, Phillips DT, Wilde DJ (1979) Foundations of optimization, 2nd edn. Prentice-Hall, Englewood Cliffs, NJ, pp 220–223, 228–233
20. Berberian SK (1992) Linear algebra. Oxford University Press, Oxford
21. Bernoulli D (1954) Specimen theorae norae de mensura sortis. Econometrica 22:23–26 (trans: Sommer L)
22. von Bertalanffy L (1968) General system theory. Braziller, New York, NY

D. Helbing, *Quantitative Sociodynamics*, 2nd ed.,
DOI 10.1007/978-3-642-11546-2,

23. Binder K (ed) (1979) Monte Carlo methods in statistical physics, Chap. 1.3.1: Various Monte Carlo realizations of the same master equation. Springer, Berlin.
24. Binney JJ, Dowrick NJ, Fisher AJ, Newman MEJ (1992) The theory of critical phenomena, Chap. 6: Mean-field theory. Clarendon Press, Oxford
25. Bogoliubov NN (1962) In: Uhlenbeck GE, deBoer J (eds) Studies in statistical mechanics, vol 1. North-Holland, Amsterdam
26. Boltzmann L (1964) Lectures on gas theory. University of California, Berkeley, CA
27. Brand H, Schenzle A (1981) Exactly soluble multidimensional Fokker-Planck equations with nonlinear drift. Phys Lett A 81(6):321–325
28. Braun M (1983) Differential equations and their applications, Chap. 3.10: Equal roots, 3rd edn. Springer, Berlin
29. Braun N (1993) Socially embedded exchange. Lang, Frankfurt/Main
30. Brehm JW (1966) A theory of psychological reactance. Academic Press, New York, NY
31. Brehm JW (1972) Responses to loss of freedom: a theory of psychological reactance. General Learning Press, Morristown, NJ
32. Brehm SS (1976) The application of social psychology to clinical practice. Wiley, New York, NY
33. Buckley W (1967) Sociology and modern systems theory. Prentice-Hall, Englewood Cliffs, NJ
34. Burns J (1973) Application of control theory to system dynamics models of social systems. PhD thesis, Purdue University
35. Burnstein E (1967) Sources of cognitive bias in the representation of simple social structures: balance, minimal change, positivity, reciprocity, and the respondent's own attitude. J Pers Soc Psychol 7:36–48
36. Burt RS (1982) Towards a structural theory of action: network models of social structure, perception and action. Academic Press, New York, NY
37. Cartwright D, Harary F (1956) Structural balance: a generalization of Heider's theory. Psychol Rev 63:277–293
38. Chapman S, Cowling TG (1970) The mathematical theory of non-uniform gases, 3rd edn. Cambridge University Press, New York, NY
39. Christaller W (1980) Die zentralen Orte in Süddeutschland. Wissenschaftliche Buchgesellschaft, Darmstadt, p 71
40. Cohen EGD (1968) The kinetic theory of dense gases. In: Cohen EGD (ed) Fundamental problems in statistical mechanics II. Wiley, New York, NY
41. Cohen EGD (1969) The kinetic theory of moderately dense gases. In: Hanley HJM (ed) Transport phenomena in fluids. Dekker, New York, NY
42. Coleman J (1973) The mathematics of collective action. Heinemann Educational Books, London
43. Coleman JS (1964) Introduction to mathematical sociology. The Free Press of Glencoe, New York, NY
44. Coleman JS (1990) Foundations of social theory. Belknap (Harvard University), Cambridge, MA
45. Collet P, Eckmann J-P (1980) Iterated maps on the interval as dynamical systems. Birkhäuser, Basel
46. Colman AM (1984) Game theory and experimental games. Pergamon Press, Oxford
47. Czerwon H-J, Der R, Haberlandt R (1977) Quantum statistical theory of chemical reactions. I. General formulation and lowest-order results. Mol Phys 33(6):1743–1756
48. Daly AJ, Zachary S (1978) Improved multiple choice models. In: Hensher DA, Dalvi MQ (eds) Determinants of travel choice. Saxon House, Westmead
49. Davydov AS (1976) Quantum mechanics, § 75: Self-consistent Hartree-Fock field, 2nd edn. Pergamon Press, Oxford
50. Diekmann A (1992) The log-logistic distribution as a model for social diffusion processes. J Sci Ind Res 51:285–290

51. Diekmann A, Mitter P (eds) (1984) Stochastic modelling of social processes. Academic Press, Orlando, FL
52. Dinkel RH (1989) Landtagswahlen unter dem Einfluß der Bundespolitik: Die Erfahrungen der letzten Legislaturperioden. In: Falter JW, Rattinger H, Troitzsch KG (eds) Wahlen und politische Einstellungen. Lang, Frankfurt/Main
53. Domencich TA, McFadden D (1975) Urban travel demand. A behavioral analysis. North-Holland, Amsterdam, pp 61–69
54. Dorfman JR (1981) In: Raveché HJ (ed). Perspectives in statistical physics. North-Holland, Amsterdam
55. Douglas PH (1934) The theory of wages. Macmillan, New York, NY, pp 132–135
56. Ebeling W (1991) Dynamics of competition and valuation in non-physical systems. In: Sydow A (ed) Handbook on computational systems analysis. Elsevier, Amsterdam
57. Ebeling W (1991) Stochastic models of competition processes in non-physical systems. Syst Anal Model Simul 8:3
58. Ebeling W, Engel A, Feistel R (1990) Physik der Evolutionsprozesse. Akademie, Berlin
59. Ebeling W, Jiménez-Montaño MA, Bruckner E, Scharnhorst A (1992) A stochastic model of technological evolution. In: Haag G, Mueller U, Troitzsch KG (eds) Economic evolution and demographic change. Formal models in social sciences. Springer, Berlin
60. Eigen M (1971) The selforganization of matter and the evolution of biological macromolecules. Naturwissenschaften 58:465
61. Eigen M, Schuster P (1979) The hypercycle. Springer, Berlin
62. Eiselt HA, Pederzoli G, Sandblom C-L (1987) Continuous optimization models, Chap. 2.2.1: The gradient (steepest descent) method. De Gruyter, Berlin
63. Empacher N (1992) Die Wegintegrallösung der Mastergleichung. PhD thesis, University of Stuttgart
64. Enskog D (1917) The kinetic theory of phenomena in fairly rare gases. PhD thesis, Uppsala
65. Erdmann G, Fritsch B (Jul 1985) Synergismen in Parteiensystemen. Arbeitspapiere. Institut für Wirschaftsforschung, Eidgenössische Technische Hochschule Zürich
66. Fano U (1957) Description of states in quantum mechanics by density matrix and operator techniques. Rev Modern Phys 29(1):74–93
67. Fechner GT (1860) Elemente der Psychophysik. Druck und Verlag von Breitkopf und Hartel, Leipzig
68. Feger H (ed) (1977) Studien zur intraindividuellen Konfliktforschung. Huber, Bern
69. Feger H (1978) Konflikterleben und Konfliktverhalten. Huber, Bern
70. Feigenbaum MJ (1978) Quantitative universality for a class of nonlinear transformations. J Stat Phys 19(1):25–52
71. Feigenbaum MJ (1979) The universal metric properties of nonlinear transformations. J Stat Phys 21(6):669–706
72. Feistel R, Ebeling W (1989) Evolution of complex systems. Kluwer Academic, Dordrecht
73. Felsager B (1981) Geometry, particles and fields, Chap. 2.4: Path integrals—the Feynman propagator. Odense University Press, Odense
74. Festinger L (1957) A theory of cognitive dissonance. Row & Peterson, Evanston, IL
75. Feynman RP (1948) Space-time approach to non-relativistic quantum mechanics. Rev Mod Phys 20(2):367–387
76. Fick E, Sauermann G (1990) The quantum statistics of dynamic processes. Springer, Berlin, pp 287–289, 377
77. Fishbein M, Ajzen I (1975) Belief, attitude, intention and behavior: an introduction to theory and research. Addison-Wesley, Reading, MA
78. Fisher RA (1930) The genetical theory of natural selection. Oxford University Press, Oxford
79. Fokker AD (1914) Die mittlere Energie rotierender elektrischer Dipole im Strahlungsfeld. Annalen der Physik 348:810–820
80. Forsythe GE, Malcolm MA, Moler CB (1977) Computer methods for mathematical computations. Prentice-Hall, Englewood Cliffs, NJ

81. Freund JE (1988) Modern elementary statistics, 7th edn. Prentice-Hall, Englewood Cliffs, NJ
82. Freund JE (1992) Mathematical statistics, 5th edn. Prentice-Hall, Englewood Cliffs, NJ
83. Gang H (1989) Solvable model of the Fokker-Planck equation without detailed balance. Phys Rev A 39(3):1286–1292
84. Gardiner CW (1985) Handbook of stochastic methods, 2nd edn. Springer, Berlin
85. Gardiner CW, McNeil KJ, Walls DF, Matheson IS (1976) Correlations in stochastic theories of chemical reactions. J Stat Phys 14(4):307–331
86. Garrido L, Masoliver J (1982) On a class of exact solutions to the Fokker-Planck equations. J Math Phys 23(6):1155–1158
87. Geest T, Olsen LF, Steinmetz CG, Larter R, Schaffer WM (1993) Nonlinear analyses of periodic and chaotic time series from the peroxidase-oxidase reaction. J Phys Chem 97: 8431–8441
88. Gelb A (ed) (1974) Applied optimal estimation, Chap. 6.3: Nonlinear least-squares estimation. MIT Press, Cambridge, MA
89. Glance N, Hogg T, Huberman BA (1991) Computational ecosystems in a changing environment. Int J Mod Phys C 2(3):735–753
90. Glance NS, Huberman BA (1992) Dynamics with expectations. Phys Lett A 165:432–440
91. Goel NS, Maitra SC, Montroll EW (1971) On the Volterra and other nonlinear models of interacting populations. Rev Mod Phys 43(2):231–276
92. Goel NS, Richter-Dyn N (1974) Stochastic models in biology. Academic Press, New York, NY
93. Grabert H, Weidlich W (1974) Masterequation, H-theorem and stationary solution for coupled quantum systems. Zeitschrift für Physik 268:139–143
94. Grad H (1958) Principles of the kinetic theory of gases, vol 12, Handbuch der Physik. Springer, Berlin
95. Graham R (1980) Forward Hopf bifurcation with multiplicative Gaussian white noise: exact Fokker-Planck-solution. Phys Lett A 80(5, 6):351–353
96. Graham R (1980) Solution of Fokker Planck equations with and without manifest detailed balance. Zeitschrift für Physik B 40:149–155
97. Granger CWJ, Newbold P (1977) Forecasting economic time series. Academic Press, New York, NY
98. Granovetter M, Soong R (1983) Threshold models of diffusion and collective behavior. J Math Sociol 9:165–179
99. Grassberger P (1981) On the Hausdorff dimension of fractal attrators. J Stat Phys 19:25
100. Greenspan HP, Benney DJ, Turner JE (1986) Calculus. An introduction to applied mathematics, 2nd edn. McGraw-Hill, Toronto, ON
101. Greenwood DT (1988) Principles of dynamics, 2nd edn. Prentice-Hall, Englewood Cliffs, NJ
102. Griffiths HB, Oldknow A (1993) Mathematics of models. Ellis Horwood, New York, NY
103. Haag G (1989) Dynamic decision theory: applications to urban and regional topics. Kluwer Academic, Dordrecht
104. Haag G, Weidlich W, Alber P (1977) Approximation methods for stationary solutions of discrete master equations. Zeitschrift für Physik B 26:207–215
105. Haken H (1976) Generalized Onsager-Machlup function and classes of path integral solutions of the Fokker-Planck equation and the master equation. Zeitschrift für Physik B 24: 321–326
106. Haken H (1983) Advanced synergetics. Springer, Berlin
107. Haken H (1983) Synergetics, 3rd edn. Springer, Berlin
108. Haken H (1988) Information and self-organization: a macroscopic approach to complex systems. Springer, Berlin
109. Hallam TG (1986) Community dynamics in a homogeneous environment. In: Hallam TG, Levin SA (eds) Mathematical ecology. Springer, Berlin
110. Hamblin RL, Jacobsen RB, Miller JLL (1973) A mathematical theory of social change. Wiley, New York, NY

111. Hänggi P, Haag G (1980) Continued fraction solutions of discrete master equations not obeying detailed balance. Zeitschrift für Physik B 39:269–279
112. Hao B-L, Wang G-R, Zhang S-Y (1983) U-sequences in the periodically forced Brusselator. Commun Theor Phys 2(3):1075–1080
113. Hao B-L, Zhang S-Y (1982) Hierarchy of chaotic bands. J Stat Phys 28(4):769–792
114. Hartree DR (1928) Proc Cambridge Philos Soc 24:111
115. Harvey JH (1976) Attribution of freedom. In: Harvey JH, Ickes WJ, Kidd RF (eds) New directions in attribution research, Chap. 4. Wiley, New York, NY
116. Hauk M (1994) Evolutorische Ökonomik und private Transaktionsmedien. Lang, Frankfurt/Main
117. Haus JW, Kehr KW (1987) Diffusion in regular and disordered lattices. Phys Rep 150(5, 6): 263–406
118. Hausdorff F (1919) Dimension und äußeres Maß. Mathematische Annalen 79:157–179
119. Heider F (1946) Attitudes and cognitive organization. J Psychol 21:107–112
120. Heider F (1958) The psychology of interpersonal relations. Wiley, New York, NY
121. Heisenberg W (1927) Über den anschaulichen Inhalt der quantummechanischen Kinematik und Mechanik. Z Phys 43:172–198
122. Heisenberg W (1930) The physical principles of quantum theory. University of Chicago Press, Chicago, IL
123. Helbing D (1990) Physikalische Modellierung des dynamischen Verhaltens von Fußgängern. Master's thesis, Georg-August University Göttingen
124. Helbing D (1991) A mathematical model for the behavior of pedestrians. Behav Sci 36: 298–310
125. Helbing D (1992) A fluid-dynamic model for the movement of pedestrians. Complex Systems 6:391–415
126. Helbing D (1992) Interrelations between stochastic equations for systems with pair interactions. Physica A 181:29–52
127. Helbing D (1992) A mathematical model for attitude formation by pair interactions. Behav Sci 37:190–214
128. Helbing D (1992) A mathematical model for behavioral changes by pair interactions. In: Haag G, Mueller U, Troitzsch KG (eds) Economic evolution and demographic change. Formal models in social sciences. Springer, Berlin, pp 330–348
129. Helbing D (1992) Models for pedestrian behavior. In: Natural structures. Principles, strategies, and models in architecture and nature. Sonderforschungsbereich 230, Stuttgart, Part II
130. Helbing D (1993) Boltzmann-like and Boltzmann-Fokker-Planck equations as a foundation of behavioral models. Physica A 196:546–573
131. Helbing D (1993) Stochastic and Boltzmann-like models for behavioral changes, and their relation to game theory. Physica A 193:241–258
132. Helbing D (1993) Stochastische Methoden, nichtlineare Dynamik und quantitative Modelle sozialer Prozesse. Shaker, Aachen. PhD thesis, University of Stuttgart, 1992
133. Helbing D (1994) A mathematical model for behavioral changes by pair interactions and its relation to game theory. Angewandte Sozialforschung 18(3):117–132
134. Helbing D, Molnár P (1995) Social force model for pedestrian dynamics. Phys Rev E 51:4282–4286
135. Hilborn RC (1994) Chaos and nonlinear dynamics. Oxford University Press, New York, NY
136. Hilfer R (1986) Renormierungsansätze in der Theorie ungeordneter Systeme. Harri Deutsch, Frankfurt/Main
137. Hofbauer J, Schuster P, Sigmund K (1979) A note on evolutionarily stable strategies and game dynamics. J Theor Biol 81:609–612
138. Hofbauer J, Sigmund K (1984) Evolutionstheorie und dynamische Systeme. Paul Parey, Berlin
139. Hofbauer J, Sigmund K (1988) The theory of evolution and dynamical systems. Cambridge University Press, Cambridge

140. Hoffmann LD, Bradley GL (1989) Calculus for business, economics, and the social and life sciences, 4th edn. McGraw-Hill, New York, NY
141. Hofstätter PR (1956) Farbsymbolik und Ambivalenz. Psychol Beitr 2:526–540
142. Hofstätter PR (1972) Psychologie. Fischer, Frankfurt/Main
143. Hohenberg PC, Halperin BI (1977) Theory of dynamic critical phenomena. Rev Mod Phys 49(3):435–479
144. Hopf E (1942) Abzweigungen einer periodischen Lösung von einer stationären Lösung eines Differentialgleichungssystems. Math Naturwiss Klasse 94:1
145. Horsthemke W, Lefever R (1984) Noise-induced transitions. Springer, Berlin
146. Hu B, Rudnick J (1982) Exact solutions to the Feigenbaum renormalization-group equations for intermittency. Phys Rev Lett 48(24):1645–1648
147. Huberman BA, Hogg T (1988) The behavior of computational ecologies. In: Huberman BA (ed) The ecology of computation. Elsevier, Amsterdam
148. Ising E (1925) Beitrag zur Theorie des Ferromagnetismus Zeitschrift der Physik. 31:253ff
149. Johnson RA, Wichern DW (1988) Applied multivariate statistical analysis, 2nd edn. Prentice Hall, Englewood Cliffs, NJ
150. Kac M, Logan J (1979) In: Montroll EW, Lebowitz JL (eds) Studies in statistical mechanics. North-Holland, Amsterdam, pp 46–53
151. van Kampen NG (1981) Stochastic processes in physics and chemistry. North-Holland, Amsterdam
152. Kaplan J, Yorke J (1979) In: Peitgen H-O, Walther H-O (eds) Functional differential equations and approximation of fixed points. Springer, New York, NY
153. Keener JP, Tyson JJ (1986) Spiral waves in the Belousov-Zhabotinskii reaction. Physica D 21:307–324
154. Keizer J (1987) Statistical thermodynamics of nonequilibrium processes, Chaps. 2.7 to 2.9. Springer, New York, NY
155. Kendall SM (1976) Time-series, 2nd edn. Griffin, London
156. Kenkre VM (1977) The generalized master equation and its applications. In: Landman U (ed) Statistical mechanics and statistical methods in theory and application. Plenum Press, New York, NY
157. Kennedy AM (1983) The adoption and diffusion of new industrial products: a literature review. Eur J Mark 17(3):31–88
158. Khintchine A (1934) Korrelationstheorie der stationären stochastischen Prozesse. Mathematische Annalen 109:604–615
159. Kolmogorov AN (1931) Über die analytischen Methoden in der Wahrscheinlichkeitsrechnung. Math Ann 104:415–458
160. Kramers HA (1940) Brownian motion in a field of force and the diffusion model of chemical reactions. Physica 7:284–304
161. Kruskal JB, Wish M (1978) Multidimensional scaling. Sage, Beverly Hills, CA
162. Kubo R, Toda M, Hashitsume N (1985) Statistical physics II. Nonequilibrium statistical mechanics, Chap. 2.8: Boltzmann equation. Springer, Berlin
163. Landau LD, Lifshitz EM (1981) Course of theoretical physics, vol 10, Physical kinetics. Pergamon Press, Oxford
164. Lang S (1985) Complex analysis, 2nd edn. Springer, New York, NY
165. Lang S (1986) Introduction to linear algebra, 2nd edn. Springer, New York, NY
166. Langevin P (1908) On the theory of Brownian motion. Comptes Rendues Acad Sci Paris 146:530–533
167. Levine IN (1988) Physical chemistry, Chap. 17: Reaction kinetics, 3rd edn. McGraw-Hill, New York, NY
168. Lewenstein M, Nowak A, Latané B (1992) Statistical mechanics of social impact. Phys Rev A 45(2):763–776
169. Lewin K (1951) Field theory in social science. Harper & Brothers, New York, NY

170. Liboff RL (1990) Kinetic theory. Classical, quantum, and relativistic descriptions. Prentice-Hall, Englewood Cliffs, NJ
171. Liouville J (1838) Note sur la théorie de la variation des constantes arbitraries. J de Math 3:342ff
172. Lotka AJ (1920) Analytical note on certain rhythmic relations in organic systems. Proc Nat Acad Sci USA 6:410–415
173. Lotka AJ (1956) Elements of mathematical biology. Dover, New York, NY
174. Luce RD (1959) Individual choice behavior, Chap. 2.A: Fechner's problem. Wiley, New York, NY
175. Luce RD, Raiffa H (1957) Games and decisions. Wiley, New York, NY
176. Luhmann N (1991) Soziale Systeme, 4th edn. Suhrkamp, Frankfurt/Main
177. Ma S-K (1976) Modern theory of critical phenomena. Benjamin, Reading, MA
178. Mahajan V, Peterson RA (1985) Models for innovation diffusion. Sage, London
179. Malchow H (1988) Spatial patterning of interacting and dispersing populations. Mem Fac Sci Kyoto Univ (Ser Biol) 13:83–100
180. Mandelbrot BB (1983) The fractal geometry of nature. Freeman, New York, NY
181. Matheson I, Walls DF, Gardiner CW (1975) Stochastic models of first-order nonequilibrium phase transitions in chemical reactions. J Stat Phys 12(1):21–34
182. May RM (1973) Stability and complexity in model ecosystems. Princeton University Press, Princeton, NJ, pp 50–53, 193–194
183. McCauley JL (1993) Chaos dynamics and fractals. Cambridge University Press, Cambridge
184. Montgomery D (1971) Brownian motion from Boltzmann's equation. Phys Fluids 14(10): 2088–2090
185. Montroll EW (1965) Model making in biological and behavioral sciences. In: Eringen AC (ed) Recent advances in engineering science, vol. I. Gordon and Breach, New York, NY
186. Montroll EW (1978) On some mathematical models of social phenomena. In: Lakshmikantham V (ed) Nonlinear equations in abstract spaces. Academic Press, New York, NY
187. Montroll EW, Badger WW (1974) Introduction to quantitative aspects of social phenomena. Gordon and Breach, New York, NY
188. Montroll EW, Shlesinger MF (1984) On the wonderful world of random walks. In: Lebowitz JL, Montroll EW (eds) Nonequilibrium phenomena II: from stochastics to hydrodynamics. North-Holland, Amsterdam
189. Montroll EW, West BJ (1979) On an enriched collection of stochastic processes. In: Montroll EW, Lebowitz JL (eds) Fluctuation phenomena. North-Holland, Amsterdam
190. Moore DS, McCabe GP (1989) Introduction to the practice of statistics. Freeman, New York, NY
191. Mosekilde E, Larsen E, Sterman JD (1991) Coping with complexity: deterministic chaos in human decisionmaking behavior. In: Casti JL, Karlqvist A (eds) Beyond belief: randomness, prediction and explanation in science. CRC Press, Boca Raton, FL, pp 199–229
192. Mosekilde E, Thomsen JS, Larsen ER, Sterman J (1992) Nonlinear interactions in the economy. In: Haag G, Mueller U, Troitzsch KG (eds) Economic evolution and demographic change. Formal models in social sciences. Springer, Berlin
193. Moyal JE (1949) Stochastic processes and statistical physics. J Roy Stat Soc B 11:150–210
194. Munz M, Weidlich W (1990) Settlement formation. Part II: Numerical simulation. Ann Reg Sci 24:177–196
195. von Neumann J (1927) Wahrscheinlichkeitstheoretischer Aufbau der Quantenmechanik. Göttinger Nachrichten 1(10):245–272
196. von Neumann J, Morgenstern O (1944) Theory of games and economic behavior. Princeton University Press, Princeton, NJ
197. Newhouse S, Ruelle D, Takens F (1978) Occurrence of strange axiom A attractors near quasiperiodic flow on Tm, m $\geq$ 3. Commun Math Phys 64:35
198. Newton I (1687) Philosophiae Naturalis Principia Mathematica. Available at: http://www.ntnu.no/ub/spesialsamlingene/ebok/02a019654.html

199. Nicholson W (1992) Microeconomic theory, 5th edn. Dryden Press, Fort Worth, pp 90ff, 111ff
200. Nicolis G, Malek-Mansour M, van Nypelseer A, Kitahara K (1976) The onset of instabilities in nonequilibrium systems. J Stat Phys 14(5):417–432
201. Nicolis G, Nicolis C (1988) Master-equation approach to deterministic chaos. Phys Rev A 38:427–433
202. Nicolis G, Prigogine I (1977) Self-organization in nonequilibrium systems. From dissipative structures to order through fluctuations. Wiley, New York, NY
203. Nitecki ZH, Guterman MM (1986) Differential equations with linear algebra, Chap. 3.10: Homogeneous systems with constant coefficients: multiple roots. Saunders, Philadelphia, PA
204. Oeding D (1963) Verkehrsbelastung und Dimensionierung von Gehwegen und anderen Anlagen des Fußgängerverkehrs, vol 22, Straßenbau und Straßenverkehrstechnik, Bonn. p 8, Fig. 10.
205. Older SJ (1968) Movement of pedestrians on footways in shopping streets. Traffic Eng Control 10:160–163
206. Olinick M (1978) An introduction to mathematical models in the social and life sciences. Addison-Wesley, Reading, MA, pp 59–65
207. Olmsted JMH (1961) Advanced calculus. Appleton-Century-Crofts, New York, NY
208. Oppenheim I, Schuler KE, Weiss GH (eds) (1977) Stochastic processes in chemical physics: the master equation. MIT Press, Cambridge, MA
209. Ortúzar JdeD, Willumsen LG (1990) Modelling transport, Chap. 7: Discrete-choice models. Wiley, Chichester
210. Oseledec VI (1968) A multiplicative ergodic theorem: Lyapunov characteristic numbers for dynamical systems. Trans Moscow Math Soc 19:197
211. Osgood CE (1953) Method and theory in experimental psychology. Oxford University Press, New York, NY, pp 708–714
212. Osgood CE, Suci GJ, Tannenbaum PH (1957) The measurement of meaning. University of Illinois Press, Urbana, IL
213. Osgood CE, Tannenbaum PH (1955) The principle of congruity in the prediction of attitude change. Psychol Rev 62:42–55
214. Parsons T (1967) Sociological theory and modern society. Free Press, New York, NY
215. Pauli H (1928) In: Debye P (ed) Probleme der Modernen Physik. Hirzel, Leipzig
216. Pawula RF (1967) Approximation of the linear Boltzmann equation by the Fokker-Planck equation. Phys Rev 162:186–188
217. Pearl R (1924) Studies in human biology. Williams & Wilkins, Baltimore, MD
218. Pearson CE (ed) (1974) Handbook of applied mathematics. Van Nostrand Reinhold, New York, NY, pp 585–596
219. Peschel M, Mende W (1986) The predator-prey model. Springer, Wien
220. Petty RE, Cacioppo JT (1986) Communication and persuasion. Central and peripheral routes to attitude change. Springer, New York, NY
221. Planck M (1917) In: Über einen Satz der statistischen Dynamik und seine Erweiterung in der Quantentheorie, Sitzungsber Preuss Akad Wiss, pp 324ff
222. Poincaré H (1892) Les Méthodes Nuovelles de la Méchanique Celeste. Gauthier-Villars, Paris
223. Pomeau Y, Manneville P (1980) Intermittent transition to turbulence in dissipative dynamical systems. Commun Math Phys 74:189–197
224. Prigogine I (1963) Non equilibrium statistical mechanics. Wiley, New York, NY
225. Prigogine I (1976) Order through fluctuation: self-organization and social system. In: Jantsch E, Waddington CH (eds) Evolution and consciousness. Human systems in transition. Addison-Wesley, Reading, MA
226. Prigogine I, Herman R (1971) Kinetic theory of vehicular traffic. Elsevier, New York, NY
227. Puu T (1989) Nonlinear economic dynamics, Chap. 3: Population dynamics. Springer, Berlin
228. Rachlin HC (1976) Behavior and learning. Freeman, San Francisco, CA

229. Rachlin HC, Baum WM (1972) Effects of alternative reinforcement: does the source matter? J Exp Anal Behav 18:231–241
230. Rapoport A (ed) (1974) Game theory as a theory of conflict resolution. Reidel, Dordrecht
231. Rapoport A (1983) Mathematical models in the social and behavioral sciences. Wiley, New York, NY
232. Rapoport A (1986) General system theory. Essential concepts and applications. Abacus Press, Tunbridge Wells, Kent
233. Rapoport A (1990) Experimental studies of interactive decisions. Kluwer Academic, Dordrecht
234. Rapoport A, Chammah MA (1965) Prisoner's dilemma. A study in conflict and cooperation. University of Michigan Press, Ann Arbor, MI
235. Ravenstein E (1876) The birthplaces of the people and the laws of migration. Geogr Mag III:173–177
236. Reif F (1965) Fundamentals of statistical and thermal physics. McGraw-Hill, Singapore, pp 202ff, 229ff
237. Reiner R (1988) Interregionale migration. PhD thesis, University of Stuttgart
238. Reiner R, Munz M (1990) Ranking regression analysis of spatiotemporal variables. Environ Plan A 22:507–526
239. Reiner R, Weidlich W (1993) Der Beitrag der Synergetik zum Naturverständnis. In: Bien G, Gil Th, Wilke J (eds) Natur im Umbruch. Frommann-Holzboog, Stuttgart
240. Rieckers A, Stumpf H (1977) Thermodynamik, Band 2, Chap. H: Ableitung der Boltzmanngleichung aus der Pauli-Master-Gleichung. Vieweg, Braunschweig
241. Risken H (1989) The Fokker-Planck equation, 2nd edn. Springer, Berlin
242. Risken H, Vollmer HD (1989) Methods for solving Fokker-Planck equations with applications to bistable and periodic potentials. In: Moss F, McClintock PVE (eds) Noise in nonlinear dynamical systems, Vol. 1: Theory of continuous Fokker-Planck systems. Cambridge University Press, Cambridge
243. Rosenberg MJ, Abelson RP (1960) An analysis of cognitive balancing. In: Rosenberg MJ, Hovland CI, McGuire WJ, Abelson RB, Brehm JW (eds) Attitude organization and change. Yale University Press, New Haven, CT, pp 112–163
244. Rosenblatt M (1971) Markov processes. Structure and asymptotic behavior. Springer, Berlin
245. Rosser WGV (1982) An introduction to statistical physics. Horwood, Chichester, pp 108f
246. Salas SL, Hille E, Anderson JT (1986) Calculus, 5th edn. Wiley, New York, NY
247. Schaffer WM, Olsen LF, Truty GL, Fulmer SL, Graser DJ (1988) Periodic and chaotic dynamics in childhood infections. In: Markus M, Müller StC, Nicolis G (eds) From chemical to biological organization. Springer, Berlin
248. Schnabl W, Stadler PF, Forst C, Schuster P (1991) Full characterization of a strange attractor. Chaotic dynamics in low-dimensional replicator systems. Physica D 48:65–90
249. Schnakenberg J (1976) Network theory of microscopic and macroscopic behavior of master equation systems. Rev Mod Phys 48(4):571–585
250. Schrödinger E (1926) Quantisierung als Eigenwertproblem. Annalen der Physik 384(6): 489–527
251. Schroeder SR, Holland JG (1969) Reinforcement of eye movement with concurrent schedules. J Exp Anal Behav 12:897–903
252. Schüßler R (1989) Exit threats and cooperation under anonymity. J Conflict Resolut 33(4):728–749
253. Schüßler R (1990) Threshold effects and the decline of cooperation. J Conflict Resolut 34(3):476–494
254. Schuster HG (1988) Deterministic chaos, 2nd edn. VCH, Weinheim
255. Schuster P (1983) Replicator dynamics. J Theor Biol 100:533–538
256. Schuster P, Sigmund K, Hofbauer J, Wolff R (1981) Selfregulation of behavior in animal societies. Biol Cybern 40:1–25

257. Schweitzer F, Bartels J, Pohlmann L (1991) Simulation of opinion structures in social systems. In: Ebeling W, Peschel M, Weidlich W (eds) Models of selforganization in complex systems (MOSES). Akademie, Berlin
258. Scott SK (1991) Chemical chaos. Clarendon Press, Oxford
259. Sgouridis S (1985) Zur Lösung der nichtlinearen Fokker-Planck-Gleichung mit Benutzung orthogonaler Polynome. Atomenergie·Kerntechnik 46(1):29–35
260. Shimada I, Nagashima T (1979) A numerical approach to ergodic problem of dissipative dynamical systems. Prog Theor Phys 61(6):1605–1616
261. Spiegel B (1961) Die Struktur der Meinungsverteilung im sozialen Feld. Huber, Bern
262. Stadler PF (1991) Dynamics of autocatalytic reaction networks IV: Inhomogeneous replicator networks. BioSystems 26:1–19
263. Stanley HE (1971) Introduction to phase transitions and critical phenomena. Oxford University Press, New York
264. Steeb W-H (1991) A handbook of terms used in chaos and quantum chaos. B.I.-Wissenschaftsverlag, Mannheim
265. Steiner ID (1970) Perceived freedom. In: Berkowitz L (ed) Advances in experimental social psychology. Academic Press, New York
266. Stewart I, Tall D (1983) Complex analysis (The Hitchhiker's guide to the plane). Cambridge University Press, Cambridge
267. Stratonovich RL (1963, 1967) Topics in the theory of random noise, vols 1, 2. Gordon and Breach, New York
268. Stübel G (1975) Sensitivity model generator as a tool for automatic credibility test of continuous system simulation. In: Proceedings of the 1975 summer simulation conference, San Francisco, pp 1189–1197
269. Szép J, Forgó F (1985) Introduction to the theory of games. Akadémiai Kiadó, Budapest
270. Taylor B (1717) Methodus incrementorum directa et inversa. Available at: http://books.google.ch/books?hl=de&lr=&id=c5o_AAAAcAAJ&oi=fnd&pg=PA8&dq=B+Taylor+Methodus+incrementorum&ots=R-NcLbLUvC&sig=fKuURhXye3JU0nQB1rWN4pPrg1M
271. Taylor P, Jonker L (1978) Evolutionarily stable strategies and game dynamics. Math Biosci 40:145–156
272. Theil H (1958) Economic forecasts and policy. North-Holland, Amsterdam, pp 31ff
273. Thom R (1975) Structural stability and morphogenesis. Benjamin, Reading, MA
274. Troitzsch KG (1989) Mikromodellierung der Bildung politischer Einstellungen in interagierenden Populationen. In: Falter JW, Rattinger H, Troitzsch KG (eds) Wahlen und politische Einstellungen. Lang, Frankfurt/Main
275. Tuma NB, Hannan MT (1984) Social dynamics. Models and methods. Academic Press, Orlando, FL
276. Uhlenbeck GE (1973) The validity and the limitations of the Boltzmann-equation. In: Cohen EGD, Thirring W (eds) The Boltzmann equation. Springer, Wien
277. Uhlenbeck GE, Ford GW (1963) Lectures in statistical mechanics. American Mathematical Society, Providence, RI
278. Valentin L (1981) Subatomic physics: nuclei and particles. North-Holland, Amsterdam, pp 127–130, 342–349
279. Vance RR (1978) Predation and resource partitioning in one predator-two prey model communities. Am Nat 112:797–813
280. Varian HR (1992) Microeconomic analysis, 3rd edn. Norton, New York, NY
281. Verhulst PF (1845) Nuov Mem Acad Roy Bruxelles 18:1ff
282. Volterra V (1931) Leçons sur la Théorie Mathématique de la Lutte pour la Vie. Gauthier-Villars, Paris
283. Vorob'ev NN (1977) Game theory. Lectures for economists and system scientists. Springer, New York, NY
284. Walls DF (1976) Non-equilibrium phase transitions in sociology. Collective Phenomena 2:125–130

285. Weidlich W (1971) The statistical description of polarization phenomena in society. Br J Math Stat Psychol 24:51
286. Weidlich W (1972) The use of statistical models in sociology. Collective Phenomena 1:51–59
287. Weidlich W (1978) On the structure of exact solutions of discrete masterequations. Zeitschrift für Physik B 30:345–353
288. Weidlich W (1987) Quantitative social science. Physica Scripta 35:380–387
289. Weidlich W (1991) Physics and social science—the approach of synergetics. Phys Rep 204:1–163
290. Weidlich W (1994) Synergetic modelling concepts for sociodynamics with application to collective political opinion formation. J Math Sociol 18:267–291
291. Weidlich W, Braun M (1992) The master equation approach to nonlinear economics. J Evol Econ 2:233–265
292. Weidlich W, Haag G (1983) Concepts and models of a quantitative sociology. The dynamics of interacting populations. Springer, Berlin
293. Weidlich W, Haag G (1987) A dynamic phase transition model for spatial agglomeration processes. J Reg Sci 27(4):529–569
294. Weidlich W, Haag G (eds) (1988) Interregional migration. Springer, Berlin
295. Weidlich W, Munz M (1990) Settlement formation. Part I: A dynamic theory. Ann Reg Sci 24:83–106
296. Wiener N (1930) Generalized harmonic analysis. Acta Math 55(1):117–258
297. Williams HCWL (1977) On the formation of travel demand models and economic evaluation measures of user benefit. Environ Plan A 9(3):285–344
298. Williams WSC (1991) Nuclear and particle physics. Clarendon Press, Oxford, pp 56–61
299. Wortman CB, Brehm JW (1975) Responses to uncontrollable outcomes: an integration of reactance theory and the learned helplessness model. In: Berkowitz L (ed) Advances in experimental social psychology. Academic Press, New York, NY, pp 277–336
300. Wunderlin A, Haken H (1984) Some applications of basic ideas and models of synergetics to sociology. In: Frehland E (ed) Synergetics—from microscopic to macroscopic order. Springer, Berlin
301. Young FW, Hamer RM (1987) Multidimensional scaling: history, theory, and applications. Lawrence Erlbaum Associates, Hillsdale, NJ
302. Zeeman EC (ed) (1977) Catastrophe theory. Addison-Wesley, London
303. Zeeman EC (1980) Population dynamics from game theory. In: Nitecki A, Robinson C (eds) Global theory of dynamical systems, Springer, Berlin
304. Ziman JM (1972) Principles of the theory of solids. Cambridge University Press, London
305. Zipf GK (1946) The P1P2/D hypothesis on the intercity movement of persons. Am Sociol Rev 11:677–686
306. Zwanzig R (1964) On the identity of three generalized master equations. Physica 30: 1109–1123
307. Zwicker E (1981) Simulation und Analyse dynamischer Systeme in den Wirtschafts- und Sozialwissenschaften. De Gruyter, Berlin, pp 444ff

Index

Locators in bold refer to sections and locators in italics refer to figures in the text.

MIX
Papier aus verantwortungsvollen Quellen
Paper from responsible sources
FSC® C105338

If you have any concerns about our products,
you can contact us on
ProductSafety@springernature.com

In case Publisher is established outside the EU,
the EU authorized representative is:
Springer Nature Customer Service Center GmbH
Europaplatz 3, 69115 Heidelberg, Germany

Printed by Libri Plureos GmbH
in Hamburg, Germany